Transportation Research, Economics and Policy

Volume 15

Series Editors
David Gillen
University of British Columbia Sauder School of Business
Vancouver, British Columbia, Canada

Werner Rothengatter
Karlsruhe Institute of Technology (KIT)
Karlsruhe, Germany

The objective of the series is to provide academics, students, policy analysts and practitioners with authoritative and accessible information to develop practical solutions, assist in policy formulation and inform debate on transportation issues and questions.
Topics in the series may include:

- transportation economics
- management of transportation by firms
- government bodies and regulators and the integration of transportation into cities, regions and countries

The unifying feature of each volume is the development or application of quantitative economic models and methods to examine current and emerging issues relating to transportation. Coverage includes the analysis of data, models of market structure, the evolution of institutions and emerging governance structures all in the context of transportation economics, policy and management. Both research monographs and edited volumes will be considered for publication in the series.

More information about this series at http://www.springer.com/series/6647

Yoshitsugu Hayashi • Shigeru Morichi • Tae Hoon Oum • Werner Rothengatter
Editors

Intercity Transport and Climate Change

Strategies for Reducing the Carbon Footprint

Springer

Editors
Yoshitsugu Hayashi
Graduate School of Environmental Studies
Nagoya University
Nagoya
Japan

Shigeru Morichi
Policy Research Center
National Graduate Institute for Policy Studies (GRIPS)
Tokyo
Japan

Tae Hoon Oum
Sauder School of Business
University of British Columbia
Vancouver
British Columbia
Canada

Werner Rothengatter
Karlsruhe Institute of Technology (KIT)
Karlsruhe
Germany

ISSN 1572-4387
ISBN 978-3-319-35258-9 ISBN 978-3-319-06523-6 (eBook)
DOI 10.1007/978-3-319-06523-6
Springer Cham Heidelberg New York Dordrecht London

Printed on acid-free paper

Springer is part of Springer Science+Business Media (www.springer.com)

Preface

Climate change caused by greenhouse gas is considered to be one of the most serious problems facing humankind. The transport sector contributes 20–30 % of the carbon footprint of developed countries. Since a large part of transport emissions comes from urban transport, emission reduction policies have generally focused on this sub-sector of transport. However, globalisation of the economic system and growing inter-regional exchanges have recently been increasing intercity travel. As a result, climate change policies related to intercity transport now give serious consideration to controlling emissions from air transport, promoting investment in high-speed rail, introducing road pricing schemes, etc. Nevertheless, there has been only limited research on the theme of intercity transport and climate change.

Railways are generally considered to be a low-emission mode. In Japan, the railway share in intercity transport is as high as that in urban transport and is much higher than in the US and the EU countries, mainly because of the "Shinkansen" high-speed rail (HSR) network. As the next generation of HSR, the plan of the Superconducting Maglev (magnetic levitation system) has already been approved for the Tokyo-Osaka route, for which the travel time will be only 67 min for 438 km. GHG emissions per passenger-km from the Maglev will be much less than those from air transport, but a little more than the latest model of Shinkansen vehicle, the emission coefficient of which is almost half of the first generation Shinkansen vehicle.

Against this background, the Institute for Transport Policy Studies (ITPS), a Tokyo-based policy think-tank, decided to initiate an international research project in 2007 on the theme of intercity transport and climate change (chair: Shigeru Morichi). The research project was intended to examine broad patterns in transport emissions and policy responses in the US, the EU and Japan. Relevant theoretical concepts have been discussed and quantitative analyses have been conducted in order to gain useful insights for practical polices. In particular, the project evaluated the Maglev system from the viewpoint of the environment, and the findings of the research are expected to contribute to future decisions on HSR investment in the US and other countries, including developing countries.

We would like to gratefully acknowledge the valuable contributions of various organisations, experts and academics to the process of this research project.

The kick-off symposium for this research was organized in December 2007 in Tokyo, and was supported by the Government of Japan, the Embassy of Germany, the Embassy of the UK and Nanzan University. In the symposium, Lord Nicholas Stern, Professor of the London School of Economics, and former Chief Economist and Senior Vice President of the World Bank, delivered the keynote speech. The contribution of Lord Stern, an eminent scholar and global leader in the field of climate change policy, served as an important driving force for this research. Among other prominent scholars and experts who made important presentations in the symposium were Dr. Ottmar Edenhofer, Chief Economist, the Potsdam Institute for Climate Impact Research; Mr. Jack Short, Secretary General, the International Transport Forum; Mr. Norman Fujisaki, President, Metron Aviation, Inc.; Mr. Yoshiyuki Kasai, Chairman, Central Japan Railway Company; and Professor Takehiko Sugiyama, Hitotsubashi University (and current President of ITPS). All of these distinguished speakers presented on the current situation of GHG emissions and related transport policies in the EU, the US or Japan. Professor Kazuhiro Ueta from Kyoto University and Professor Takafumi Matsui from the University of Tokyo commented on the presentations, and floor discussions followed (all titles are in 2007).

After the symposium, an international research team was formed and a different theme was allocated to each team member. Since then, we have planned to publish the research output as an edited volume, and an editorial team was formed comprising us: Yoshitsugu Hayashi, Shigeru Morichi, Tae Hoon Oum and Werner Rothengatter.

In the course of this research, we needed detailed background information on the related institutions and policies in the US and the EU. This was provided by Dr. Dawn L. Rhoades and Dr. Michael J. Williams of Embry-Riddle Aeronautical University for the US, and by Professor Rothengatter for the EU.

After the completion of the first draft of the book, an international symposium entitled "Climate Change and Intercity Transport Strategy" was organized in 2010 by ITPS in Tokyo with support from the Ministry of Land, Infrastructure, Transport and Tourism (MLIT) and the Ministry of Environment (MOE). The presentations in the symposium were based on the contents of each chapter of this book.

The drafts of the book chapters were reviewed by Professor Anthony May from the University of Leeds, Prof. David Banister from the University of Oxford, and Professor Lori Tayasszy from the Delft University of Technology. The reviewers' comments and feedback contributed much to improving the contents of the draft chapters.

We sincerely appreciate the important contributions made by presenters and supporters of the two symposiums. We also thank all of the researchers who have been involved in the research work and preparation of the book manuscript for their hard work.

We hope that this book, which contains the outcome of our project, will assist policy makers and experts to take better policy decisions in relation to the transport and environment agenda and thereby contribute to lowering the carbon footprint.

Finally, we are deeply indebted to the Nippon Foundation for their support throughout this research project and ITPS activities.

Introduction: How this book is organised

Most studies on transport and the environment treat the transport sector comprehensively as a whole or focus on the problems of urban transport. Intercity passenger transport, which is in the focus of this book, is a less explored transport segment because it lacks a common statistical definition and plays a minor role in travel surveys because it includes only a small share of overall trip generation. But as the share of overall passenger transport performance is high—because of the longer distance per trip—and the climate relevance is even higher because of the dominance of car and air modes, there is enough motivation to place intercity transport into the focus of analysis. Chapter 1 is, therefore, to document the importance of intercity passenger transport in the overall transport segment and to underline its role for climate change. This analysis is placed into the framework of international climate policy and enriched by examples of Japanese transport policy to meet the future challenges.

In Chap. 2, international comparison is made concerning the situation and development of intercity passenger transport according to the national transport plans. Problems and tipping points of climate change generated by intercity transport are highlighted by referring to intercity transport policies and planning systems in Japan, Europe and the US, and at the same time by discussing the necessity of giving incentives to the private firms to foster technological innovation in this market. This is supplemented by a report on the rapid development of intercity transport systems in China.

In Chap. 3, we extend the dimension of analysis to the overall external costs stemming from intercity transport. In Sect. 3.1 the external costs of climate changes are positioned within the overall external costs of transport sector, which provides a basis for discussing the general options for a harmonised internalisation of externalities including climate change in a consistent way. In Sect. 3.2, life-cycle CO_2 emissions from HSR and air transport are calculated on a trial bases and the results are compared. This gives an impression on the magnitude of the reduction potential for external costs through fostering a modal shift towards the railways. Section 3.3 focuses on selected instruments for internalisation, in particular the impact of imposing a carbon tax on regional transport demand is calculated taking Japan as a case study.

Chapter 4 discusses approaches to assess intercity passenger transport policies on the base of spatial economic analysis. The first part of this chapter sets out the framework of analysis for measuring the regional distribution of economic impacts of high speed rail (HSR) investments by constructing spatial computable general equilibrium (SCGE) models. The second part of this chapter predicts the spatial distribution of the economic effects and the CO2 emissions to be caused by the Japan's super-high speed Maglev rail project. The last part of this chapter computes the spatial economic impacts of the South Korea's KTX HSR investments and the Taiwan's HSR investments, and compares the results with those of Japan's Maglev investments.

Chapter 5 attempts to measure the 'social' economic efficiency of two major transport modes, aviation and railways, in Japan's domestic intercity passenger travel market, by taking into account of the life cycle CO_2 emissions caused by carrier operations, construction and operation of infrastructure, and construction of aircraft/rolling stocks as well as the time spent by passengers to use these two intercity transport modes. We draw conclusions for the future development of intercity transport systems. Japan is chosen as an example to show that also in a country with a well-developed intercity and HSR system there are still options for major improvements.

Chapter 6 will summarise the results and give policy proposals (see Fig. 1).

Yoshitsugu Hayashi
Shigeru Morichi
Tae Hoon Oum
Werner Rothengatter

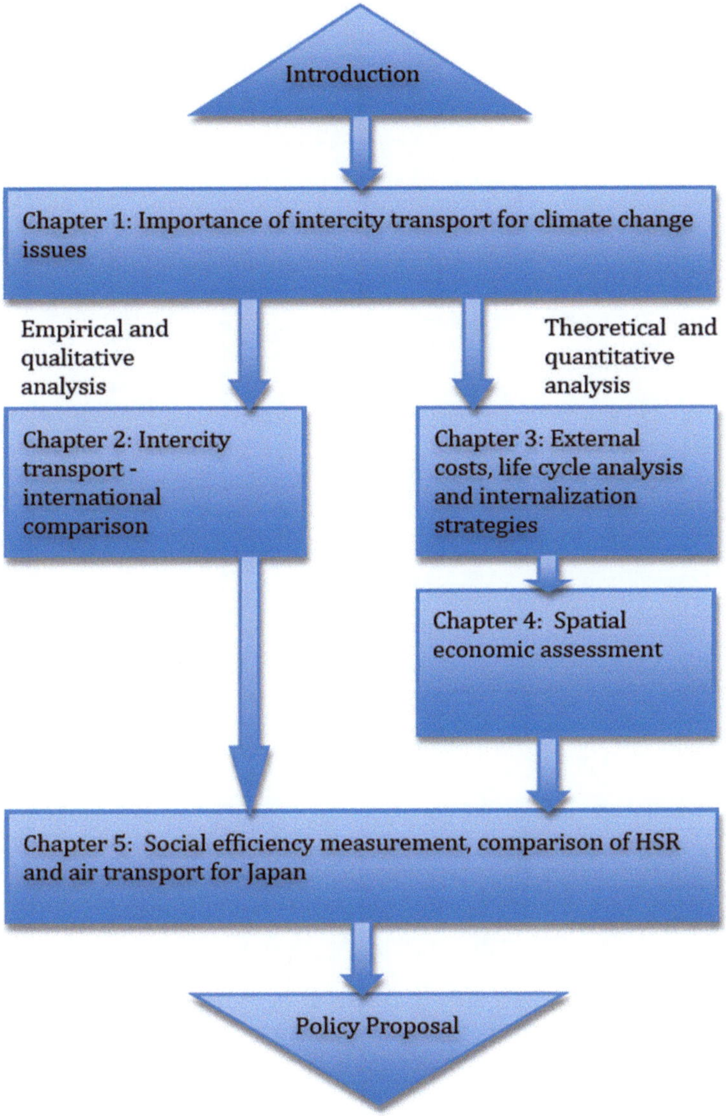

Fig. 1 Sequence of analysis in the book

Contents

Contributors

Surya Raj Acharya Institution for Transport Policy Studies, Minato-ku, Japan

Koichi Fujisaki (Formerly) Institution for Transport Policy Studies (ITPS), Tokyo, Japan

Yoshitsugu Hayashi Graduate School of Environmental Studies, Nagoya University, Nagoya, Japan

Tomoki Ishikura Graduate School of Urban Environmental Sciences, Tokyo Metropolitan University, Tokyo, Japan

Ryo Itoh Graduate School of Economics, Nagoya City University, Nagoya, Japan

Keiji Kariyazaki Kanto District Transport Bureau, Ministry of Land, Infrastructure, Transport and Tourism, Tokyo, Japan

Hirokazu Kato Graduate School of Environmental Studies, Nagoya University, Nagoya, Japan

Atsushi Koike Graduate School of Engineering, Kobe University, Kobe, Japan

Hidetoshi Kume JICA Expert, Ministry of Public Works and Transport, Phnom Penh, Cambodia

Iwao Matsuoka Office of Regional Funds, The Sasakawa Peace Foundation, Tokyo, Japan

Mitsuhiro Miyashita Mitsubishi UFJ Research and Consulting Co., Ltd, Tokyo, Japan

Shigeru Morichi Policy Research Center, National Graduate Institute for Policy Studies, Minato-ku, Japan

Takaaki Okuda Graduate School in Business Administration, Nanzan University, Nagoya, Japan

Tae Hoon Oum Sauder School of Business, University of British Columbia, Vancouver, Canada

Haixiao Pan Department of Urban Planning, Tongji University, Shanghai, China

Somchai Pathomsiri Department of Civil and Environmental Engineering, Faculty of Engineering, Mahidol University, Nakhon Pathom, Thailand

Werner Rothengatter Karlsruher Institut für Technologie, Karlsruhe, Germany

Naoki Shibahara Japan Environmental Management Association for Industry, Tokyo, Japan

Atsushi Suzuki Department of Civil Engineering, Meijo University, Nagoya, Japan

Hiroyuki Takeshita Institution for Transport Policy Studies, Tokyo, Japan

Naoyuki Tomari Institution for Transport Policy Studies, Tokyo, Japan

Kazuyuki Tsuchiya Mitsubishi Research Institute, Inc., Tokyo, Japan

Tetsuo Yai Graduate School of Engineering, Tokyo Institute of Technology, Tokyo, Japan

Yuichiro Yoshida Graduate School for International Development and Cooperation, Hiroshima University, Hiroshima, Japan

Chapter 1
Importance of Intercity Passenger Transport for Climate Change Issues

Yoshitsugu Hayashi, Iwao Matsuoka, Koichi Fujisaki, Ryo Itoh, Hirokazu Kato, Werner Rothengatter and Hiroyuki Takeshita

1.1 The Characteristics of Intercity Transport

1.1.1 Emerging Demand of Intercity Transport

Long-term forecasts expect an increase of world population to 9 billion by 2050 while it was 3.5 billion in 1960 and 7 billion in 2010.[1] Population and economic growth will lead to an increase of mobility from 5500 billion (1960) to

[1] Rounded figures.

Y. Hayashi (✉)
Graduate School of Environmental Studies, Nagoya University, Nagoya, Japan
e-mail: yhayashi@genv.nagoya-u.ac.jp

H. Kato
Graduate School of Environmental Studies, Nagoya University, Nagoya, Japan
e-mail: kato@genv.nagoya-u.ac.jp

I. Matsuoka
Office of Regional Funds, The Sasakawa Peace Foundation, Tokyo, Japan
e-mail: matsuoka@spf.or.jp

K. Fujisaki
(Formerly) Institute for Transport Policy Studies (ITPS), Tokyo, Japan
e-mail: fujisakiyk@gmail.com

R. Itoh
Graduate School of Economics, Nagoya City University, Nagoya, Japan
e-mail: itoh@econ.nagoya-cu.ac.jp

W. Rothengatter
Karlsruher Institut für Technologie (KIT), Karlsruhe, Germany
e-mail: werner.rothengatter@kit.edu

H. Takeshita
Institution for Transport Policy Studies (ITPS), Tokyo, Japan
e-mail: takeshita@jterc.or.jp

Y. Hayashi et al. (eds.), *Intercity Transport and Climate Change*,
Transportation Research, Economics and Policy 15, DOI 10.1007/978-3-319-06523-6_1

105,000 billion pass-km (2050, see Schafer and Victor 2000). North America and Europe accounted for 64 % of passenger kilometers in 1960, but their shares are expected to drop to 52 % by 2050 while the Asian share might go up from 8 to 32 %. In the developing countries, the major part of this growth will occur in urban areas because of the rapid developments of megacities and related urban transport. In emerging economies, the structural development of passenger transport will depend on income and age structure of population. Looking at the example of China, one discovers that despite the continuing growth of urban transport, the intercity travel over longer distances shows the highest growth rate. Higher income for increasing segments of population and improving travel conditions—see the development of air traffic and high-speed rail (HSR)—will in particular stimulate intercity travel activities. In economies with aging population like Europe or Japan, urban and short-distance traffic is beginning to decline, while intercity traffic will continue to increase.[2]

The reasons for the changing structure of spatial transport demand can be discovered when analysing the travel purposes and their drivers behind. While urban transport is dominated by the journeys to work, education, shopping and short-time leisure, the dominant intercity travel purposes are holidays and long-time leisure, business and long-time visits to friends and relatives. Rising income levels and aging of population are the main drivers for the intercity travel activities, supported by the improving quality of transport networks and the decreasing costs of long-distance travelling. Also, in a prosperous economic environment, people of senior age show increasing mobility dynamics and travel more and longer distances compared with previous generations. This leads to an interesting observation: While the number of trips—including non-motorised trips—is almost a constant in most countries of the world (around three trips per day), the trip lengths and the total distance travelled per year differ substantially depending on the magnitude and structure of the drivers mentioned above. Schafer and Victor (2000) estimate around 40,000 km per year per inhabitant for the USA, 30,000 km for Europe and 11,000 km for Asia in 2020. The percent differences between US/Europe and Asia will shrink substantially until 2050 because the drivers of intercity transport are developing much more dynamically in Asia.

1.1.2 Statistics on Intercity Transport

While it seems easy to depict the importance of intercity transport in qualitative terms, it turns out quite difficult to show the quantities in an international and intertemporal context. The main reasons are:

- An internationally common statistical definition of intercity transport does not exist.

[2] This is dependent on the rate of decrease of the population.

Table 1.1 Different definitions of intercity transport

Country/union	Definition of intercity transport
Japan	Inter-prefectural
USA	≥50 miles
Germany	≥50 km (planning); ^{3}100 km (survey)
EU ETIS base	Inter-zonal (NUTS3 zones)

- In travel surveys, intercity transport plays a minor role because of its low share of overall trip generation of households.

Only for air travel, there are no problems to allocate all travel activities to the intercity segment. For public transport, the classification can be made in a pragmatic way by looking at the public units responsible for its organisation. Therefore, all public transport supply which is organised between cities, by provinces, states or the federal government (or enterprises in charge of these units) can be classified as intercity public transport. However, subdividing bus and individual car travel into urban, rural and intercity transport leads to major difficulties. Lacking a common statistical definition, there exist a number of different country-related approaches to estimate the volume and performance of public and individual intercity transport. Table 1.1 gives some examples.

Some countries apply distance-related definitions: Travel activities with distances longer than 50, 75 and 100 km are the most frequently used classifications; occasionally, this is dependent on the purpose of analysis (e.g. forecasting long-distance travel to prepare federal transportation plans). Other countries use a regional base for definition, i.e. travel activities, which pass the boundaries of regions (prefectures in Japan, counties in the USA or Nomenclature of Units for Territorial Statistics 3, NUTS3, regions in Europe), are called inter-prefectural, interregional or intercity transport. Japanese statistics, for instance, exhibit inter-prefecture transport.

In Europe, there exists only a small number of surveys on intercity transport. In Germany, a supplementary survey (INVERMO 2004) on intercity travel (distances at least 100 km) has been conducted in addition to the periodical general traffic survey (MID 2008). According to these surveys, the share of long-distance transport (≥100 km) of total passenger transport volume is only about 2 %. A similar survey has been performed on the European level with the project DATELINE (2003), again for travel of at least 100 km distance. The results of such surveys are not sufficient for deriving an aggregate figure for the passenger kilometers such that it is necessary to feed the survey results into transport models to generate a complete pattern of spatial activities and to derive aggregate figures on the kilometres travelled in intercity transport.

The ETIS plus project of the European Commission has followed the above approach: European and national survey data have been inserted into a Four-Step transportation model to construct a consistent passenger transport matrix on an NUTS3 scale. NUTS3 regions are comparable to counties in other countries—while prefectures or provinces are more comparable to the higher EU aggregation level of NUTS2. By aggregating the matrix data from ETIS plus, we receive the results for the 27 EU countries shown in Fig. 1.1.

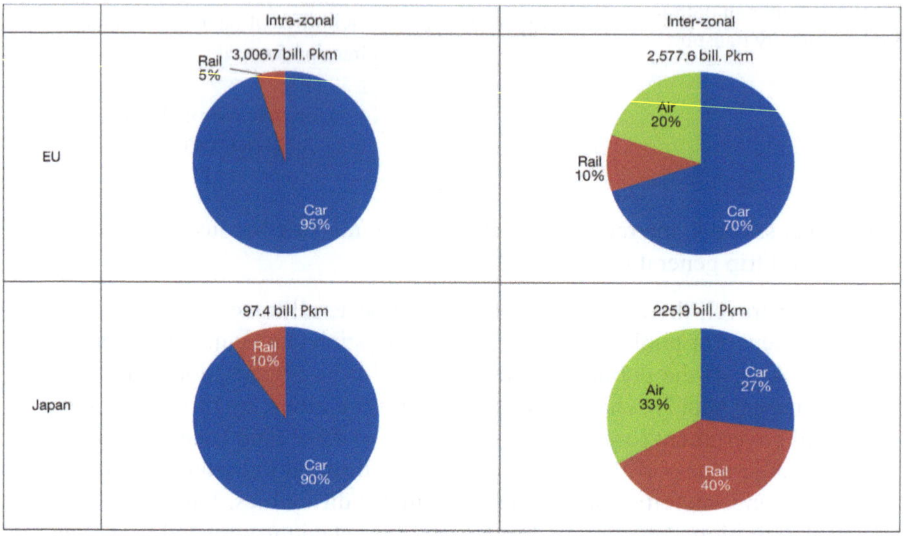

Fig. 1.1 Intra-and inter-zonal passenger transport in the EU (base: NUTS3) and Japan (base: prefectures). Bus transport is not included. (Japan) Excluding trips inside a prefecture or a metropolitan area for intra-zonal. (Japan) Zone is classified into eight divisions (Hokkaido, Tohoku, Kanto, Chubu, Kinki, Chugoku, Shikoku and Kyushu). (Source: ETIS plus 2012; MLIT 2010)

While the inter-zonal passenger transport measured in number of trips is only about 3 %, its share in passenger kilometers is 46 %, i.e. almost one half. It has to be considered that the inter-zonal transport includes all travel with distances of 50 km and more and also partly the travel with distances between 25 and 50 km. Therefore, the ETIS plus statistics can serve as an upper value estimate of the share of intercity transport in the total passenger transport activities.

The data of intra-zonal passenger transport for Japan in the Fig. 1.1 include only those in the three major metropolitan areas of Kanto (including Tokyo), Chubu (including Nagoya) and Kinki (including Osaka), the population of which accounts for 68.1 % of Japan's total population. The share of railways in Japanese metropolitan areas is 10 %, which is far higher than 5 % in the EU.

As the definition of intercity transport in Japan is restricted to inter-prefecture (comparable to inter-provincial or inter-zonal NUTS2 in Europe), it includes transport relationships of at least 100-km distance such that the share of intercity transport will naturally drop. Therefore, intercity transport in Japan accounts for only about 35 % of the overall passenger kilometers. This figure can be regarded as a lower bound for the share of intercity transport. Therefore, we can conclude that the share of intercity passenger transport in the total passenger transport can be estimated between one third and one half, on the average.

The Organisation for Economic Cooperation and Development (OECD) statistics differentiates between "urban" and "intercity + local" transport. This adds a further dimension to the statistical interpretation problems, and thus leads to an over-estimation of intercity transport. These statistics are used by International

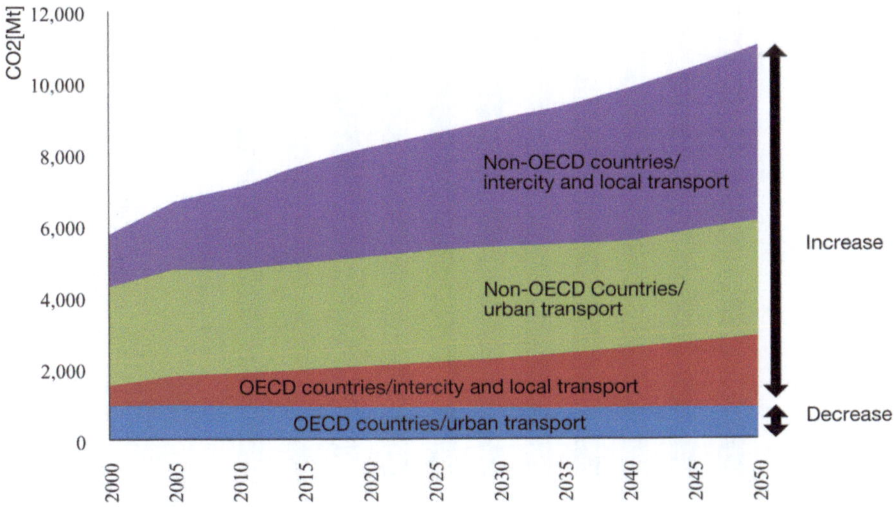

Fig. 1.2 Prediction of CO_2 emissions of transport by region in the world. (Data source: International Energy Agency; ETP 2008)

Energy Agency (IEA) for their prediction of long-term CO_2 emissions of transport, differentiated by regions of the world and transport segments (see Sect. 1.1.3).

To conclude, the statistical treatment of intercity transport is very heterogeneous and does not allow for cross-country comparisons. In most countries, the travel surveys only capture a small sample of overall intercity trip making such that aggregation procedures and models have to be used to result in a rough estimate. According to the reasoning in this section, one can estimate that the share of intercity passenger transport of the total passenger transport is between one third and one half or in the order of magnitude of about 40 %.

1.1.3 Contribution of Intercity Transport to the Carbon Footprint

As in most countries, the car and air travel modes show the highest share of traffic and the highest growth rates (in particular air); under the conditions of business as usual (BAU), one can expect that the percentage impact of the intercity traffic in the transport global warming impacts is much higher than the share of total passenger transport which has been estimated at between one third and one half of the total in Sect. 1.1.2. This is reconfirmed by the IEA Energy Technology Perspectives (2008), although the definition of transport segments is not comparable to the EU and Japanese studies.

According to the IEA (2008), the CO_2 emissions of urban transport (blue and green areas in Fig. 1.2) will increase only slowly between 2010 and 2050, even declining in OECD countries, while intercity and local transport are growing most dynamically, in particular in the non-OECD countries (purple area in Fig. 1.2). But

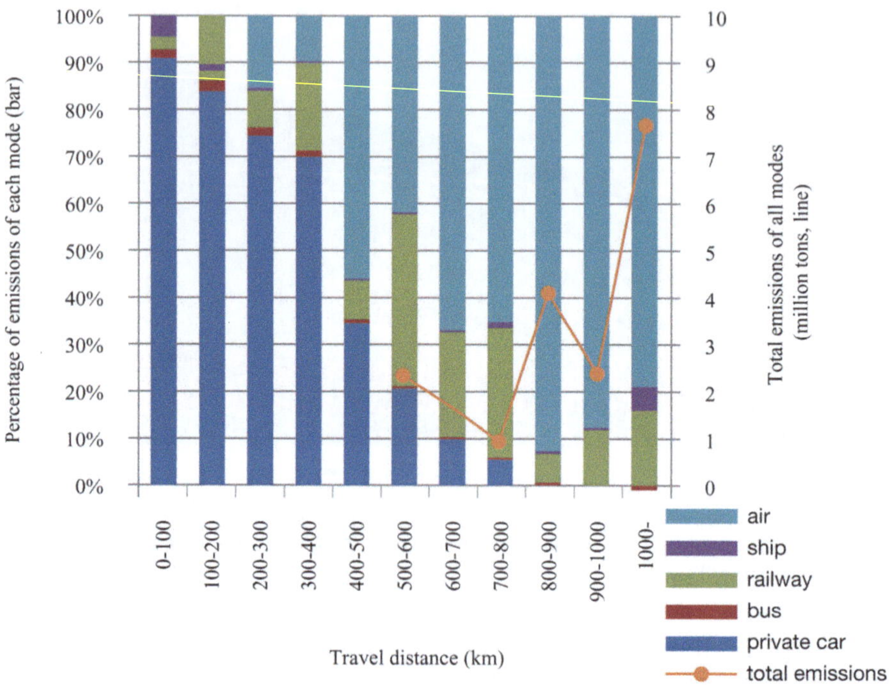

Fig. 1.3 Ratio of transport mode-specific CO_2 emissions by travel distance in Japan (2005)

also OECD countries show considerable growth (red area in Fig. 1.2). The share of intercity and local transport of total transport CO_2 emissions thus rises from less than 50 % in 2010 to about 70 % in 2050.

Although it is obvious that the OECD/IEA statistics lead to an over-estimation of intercity transport, the resulting figures underline the high and growing importance of this transport segment for CO_2 emissions of transport. The rapidly increasing travel distances in non-OECD countries, in particular in emerging economies, will boost car and air transport in all countries in which rail transport plays a minor role.

The case of Japan illustrates that a well-developed rail system for intercity transport can help to dampen the growth of CO_2 emissions. Nevertheless, also in Japan, the dynamic growth of air transport gives rise to a concern. Figure 1.3 shows some results of the project "Low-Carbon Society Scenarios Towards 2050" in 2007.[3] CO_2 emissions were estimated for the year 2005 by travel distance using a model of the Civil Aviation Bureau of the Ministry of Land, Infrastructure, Transport and Tourism (MLIT). This underlines that the share of emissions from aircraft rises overall depending on the increase in travel distance.

[3] Carried out by researchers from throughout Japan with the National Institute for Environment Studies.

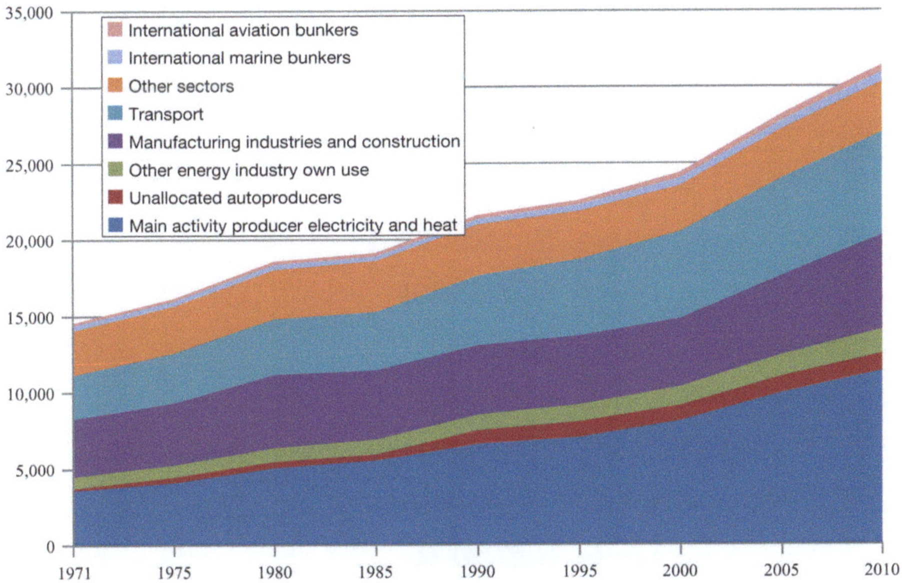

Fig. 1.4 World CO_2 emissions (million tonnes of CO_2). (Data source: IEA 2012)

The share of emissions from railroads becomes large at about 500–800 km, and the total amount of CO_2 emissions is kept low. On the other hand, railroads give way to cars in shorter distance and to aircraft in longer distance, and the total amount of CO_2 emissions are also larger. The effect of share of CO_2 emissions from intercity transport, particularly the effect of reducing CO_2 emissions at a distance where the share of railroads is relative large, can be seen.

We can conclude from these figures that a better organisation of intercity transport with a higher share of railways on distances between 100 and 1000 km is a necessary element of a CO_2 reduction strategy, and that even in countries with high shares of railway transport like Japan, the further potential of fostering the modal shift towards the railways has to be explored.

1.2 Intercity Transport in the Global Framework of Climate Change Issues

1.2.1 World Greenhouse Gas Emissions

According to IEA statistics, during the period from the Kyoto Protocol's base year of 1990–2004, world greenhouse gas (GHG) emissions increased about 70 % from 28.7 billion to 49 billion t. Carbon dioxide (CO_2) had the highest rate of increase, at 80 %. Figure 1.4 shows CO_2 emissions over time. Although from 1971 to the

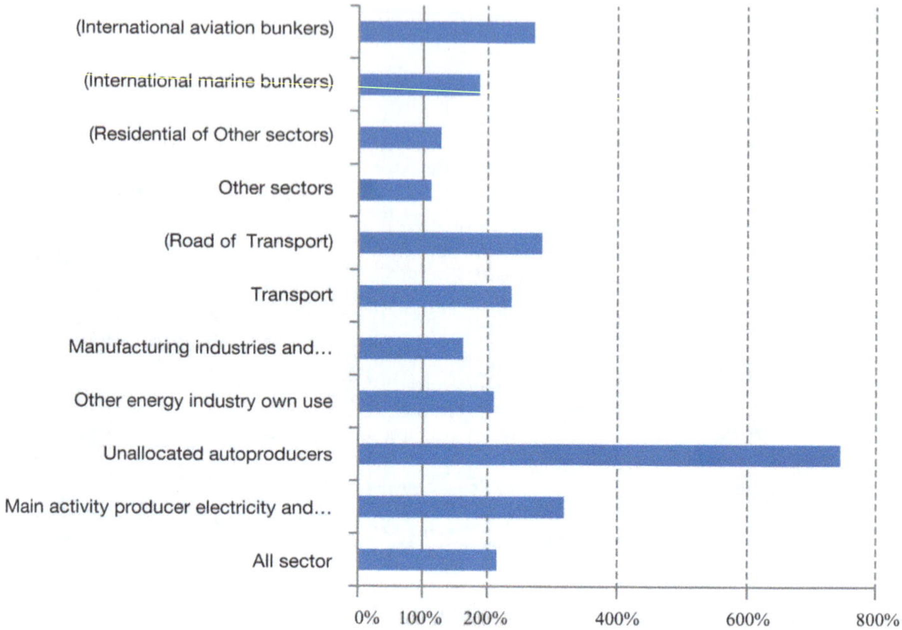

Fig. 1.5 Percentage change (1971–2010) in CO_2 emissions by sector (1971 = 100 %). (Data source: IEA 2012)

present the energy sector (power production and heat supply operations, self-produced power, other energy industries) contributed the highest share of emissions, in the mid-1990s, the CO_2 emissions from transport sector surpassed that of the manufacturing and construction industries and assumed second place, with a 23.4 % share as of 2005. In particular, emissions from road transport accounted for as much as 17 % of total CO_2 emissions as of 2005, approaching the percentage of the manufacturing and construction industries.

Figure 1.5 shows the percentage of increase in CO_2 emissions by sector from 1971 to 2005. During these 35 years, traffic tended to increase steadily, and it can be seen that in particular the increase in CO_2 emissions from surface transport comes after increases in the energy industry and in international aviation. Of note are the high increases in international transport (aviation and shipping). Especially, the emissions from international aviation are growing steadily, with an increase of over 240 % in the same period. However, note also that this sector still accounts for only about 2 % of total CO_2 emissions.

1.2.2 Per capita GHG Emissions by Region and by Country

In the transport sectors, GHG emissions from OECD countries accounted for 74 % of the worldwide CO_2 emissions in 2004. However, in scenario projections of IEA

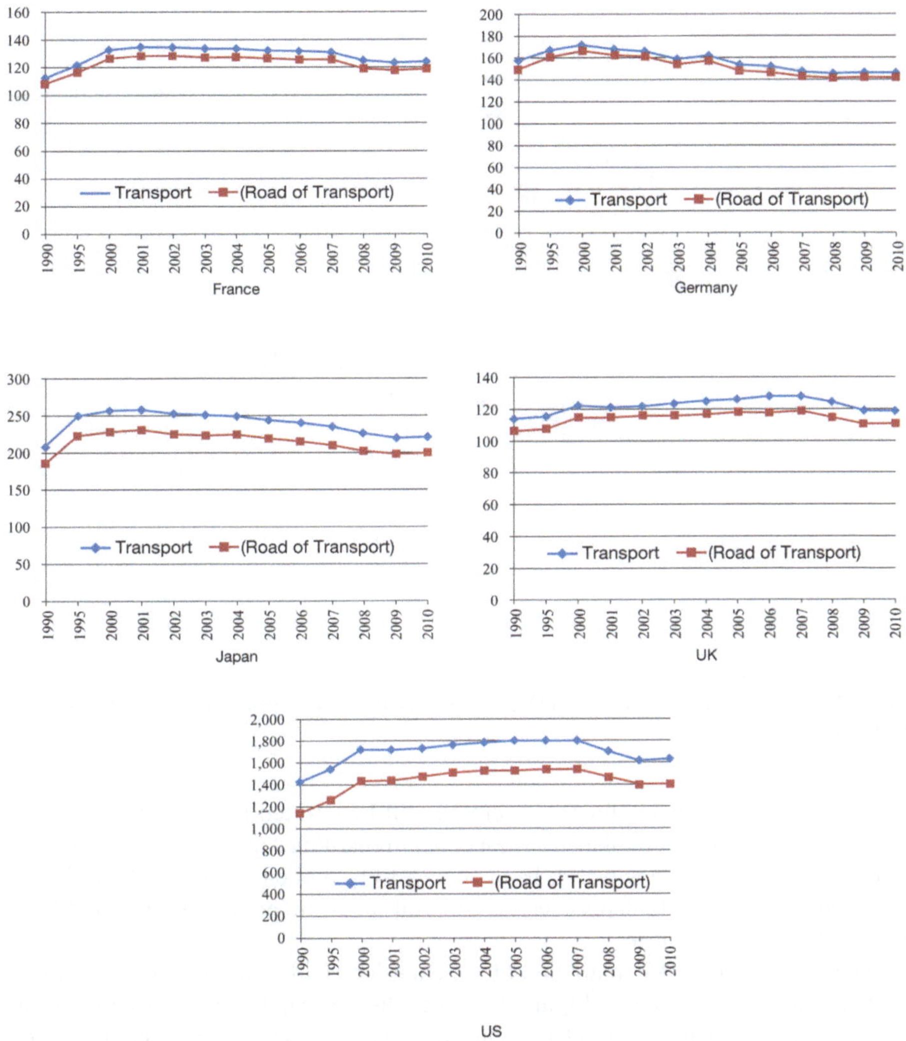

Fig. 1.6 Transport-sector CO_2 emissions of France, Germany, Japan, UK and USA (million tonnes of CO2). (Data source: IEA 2012)

(2013a; New Policies Scenario), it is estimated that OECD countries would account for 40 % of the whole energy demand in the world in 2030 in the transport sector (OECD: 1085 million t, Non-OECD: 1611 million t in 2030).

There are some countries whose CO_2 emissions from transport sector appear to be trending downward. Figure 1.6 shows the transport-sector CO_2 emissions of Germany, France and Japan. The CO_2 emissions from transport sector of Germany in particular have been falling rapidly since adoption of the Kyoto Protocol in 1997. It is perhaps also evident that decreasing CO_2 emissions from automobiles is a reason for the decrease of the CO_2 emissions from transport sector in these countries.

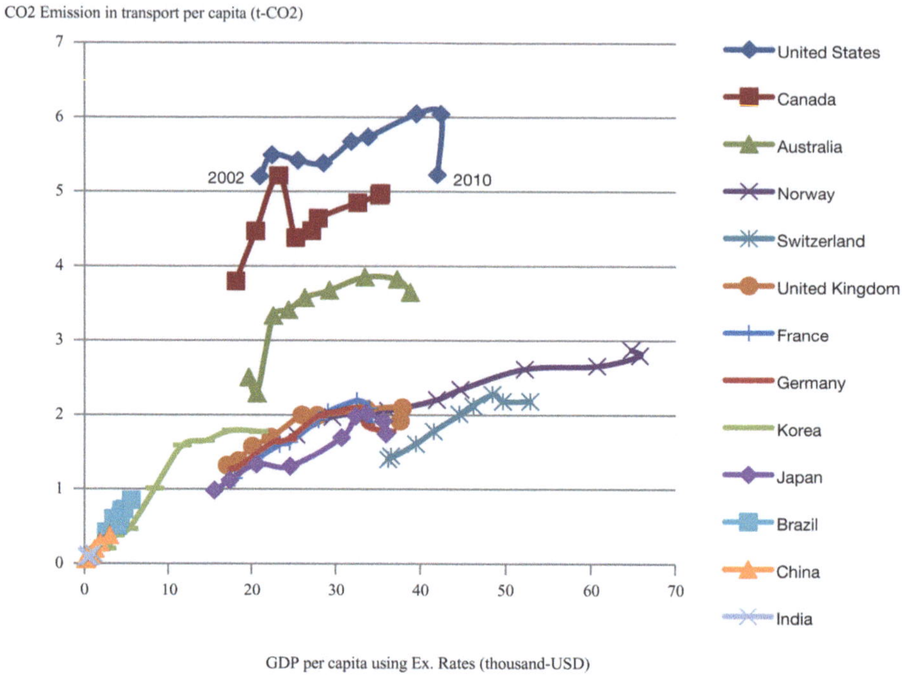

Fig. 1.7 Per capita transport-sector CO2 emissions. GDP gross domestic product. (Data source: IEA 2012)

Discussion of the reasons for this decrease will be omitted here because the later chapter will provide an explanation, especially regarding Japan. What is important is that there are countries which have successfully broken the rising trend in the GHG emissions from transport sector, and that the same possibility exists at least for the developed countries.

The final part of this explanation of the emissions will examine per capita transport-related CO_2 emissions. Figure 1.7 plots the per capita CO_2 emissions from transport sector of many countries. One can discern several characteristics from this graph. First, geographically large countries like the United States tend to have large per capita emissions. In the future, this will likely apply also to geographically large developing countries such as China and India. Next, if we exclude those countries, per capita CO_2 emissions tend to converge in the range of about 2–3 t of CO_2. Considering that the trend in emissions by region discussed above was more or less rising, population growth is likely a major influence on the past trends towards increasing emissions. It is thought that population will decline from now on not only in developed countries but also in China, and despite factors such as the time lag, at the least, the population effect of the overall rising trend will likely diminish.

As the concluding part of the CO_2 emissions explanation, let us examine the IEA's predictions by way of representative data. Figure 1.8 presents the CO_2 emissions from transport sector (ETP 2008 baseline case) by mode predicted using the

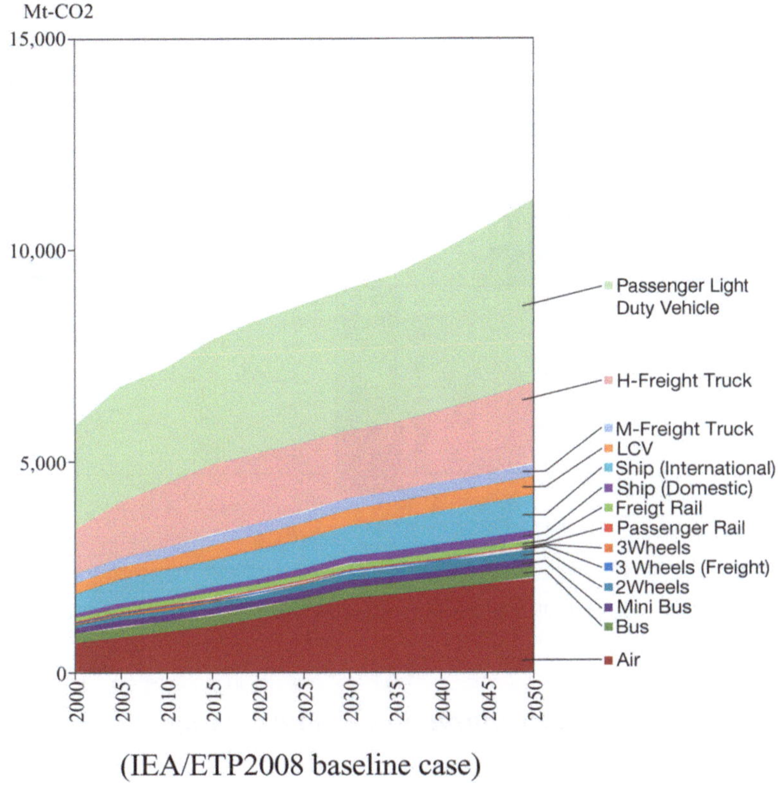

Fig. 1.8 Predicted world transport-sector CO_2 emissions. IEA/ETP 2008 baseline case. *LCV* light commercial vehicle

IEA mobility model. Compared with the emissions in 2000, the CO_2 emissions in this case are predicted to nearly double in 2050. The figure clearly shows that emissions from passenger cars and aircraft will increase dynamically, and that emissions from large freight trucks will also account for a very large percentage.

Figure 1.9 shows the percentage of change from 1990 to 2005 in transport-sector CO_2 emissions by region. Except Former Soviet Union, all regions have higher CO_2 emissions than the emissions in 1990, which is the base year of the Kyoto Protocol. Yet, even in the region where emissions declined, the decrease may be attributed to the effects of factors such as economic stagnation, certainly not to the results of any effective measures.

Anthropogenic GHG emissions tend to consistently increase not only overall but also in just the transport sector, and it is believed that this trend will continue. What actions have the world taken amid this emissions increase? Below, this chapter will focus on the international initiatives that have significantly influenced governments' current limitation actions and commitments (targets) for GHG emissions, and consider their effects on curbing future transport-sector emissions.

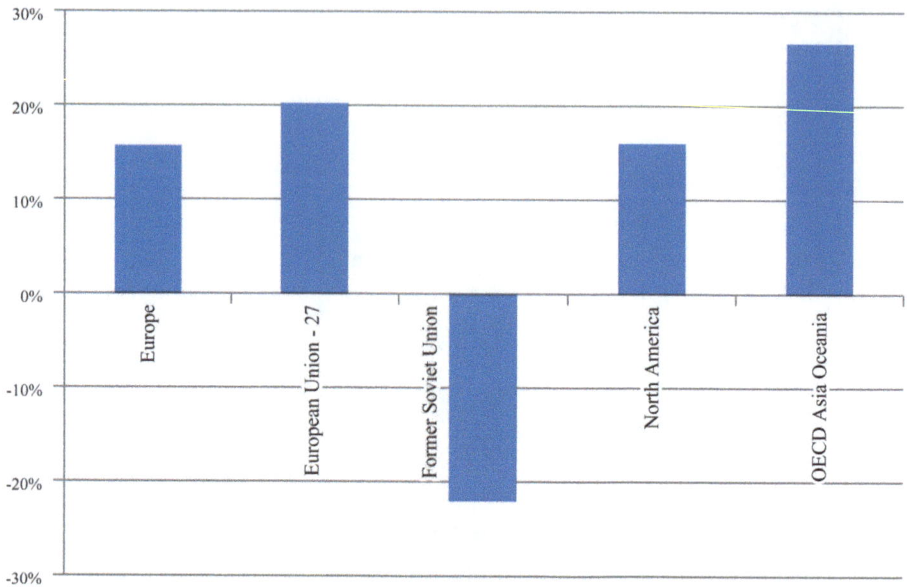

Fig. 1.9 Percentage change (1990–2010) in transport-sector CO_2 emissions, by region. (Data source: IEA 2012)

1.2.3 Framework Convention on Climate Change

Currently, the biggest international framework on global warming is the United Nations Framework Convention on Climate Change (UNFCCC), which was adopted at the 1992 UN Conference on Environment and Development (commonly known as the "Earth Summit") held in Rio de Janeiro, and which entered into force 2 years later in March 1994. To achieve the convention's objective of stabilising atmospheric GHG concentrations at a level that poses no dangerous interference with the climate system (Article 2), the convention sets forth a number of principles and methodologies that greatly influenced subsequent international activities. In fact, although the convention does specify international arrangements for combating global warming, one finds no specific mention of transport-sector emissions, which are part of the whole. However, there are several provisions which indirectly affect subsequent transport-sector emissions. The major ones are the principles of "common but differentiated responsibilities" (CBDR, Article 3) and "sustainable economic growth" (Article 3). The former came to be used as justification for sanctioning differences between the remedial measures and emissions of developing and developed countries. When considering sector-specific international initiatives or other actions in, for example, the transport sector, and especially when attempting to create uniform world indicators or the like, CBDR is often cited as grounds for arguing against them. With respect to the latter, countries are supposed to cooperate in trying to establish an international economic system that will make it possible to

Table 1.2 Greenhouse gas emission reduction targets in 2008–2012 versus 1990 in the Kyoto protocol

Country and region	Reduction target	
	Kyoto protocol (%)	EU region (%)
EU	−8	
Germany		−21
France		±0
UK		−13
Ireland		13
Spain		15
Portugal		27
Greece		25
Sweden		4
Denmark		−21
Norway	1	
Russia	±0	
Ukraine	±0	
Japan	−6	
Australia	8	
New Zealand	±0	
Canada	−6	

better address problems by means of sustainable economic growth, and this is one major reason for the Kyoto Protocol's clean development mechanism (CDM), and for recent assistance to developing countries.

1.2.4 Kyoto Protocol

The Kyoto Protocol, which sets forth the detailed rules for the FCCC, was adopted at the Third Conference of the Parties to the UNFCCC (COP3) held in Kyoto in 1997, 5 years after the FCCC was adopted, and finally entered into force in 2007, 10 years after adoption. The protocol's major characteristics are that it sets specific reduction targets for some countries, and that it incorporates flexibility mechanisms for attainment of those targets. A brief discussion follows:

1. GHG Reduction Targets

The Kyoto Protocol specifies six GHGs (carbon dioxide, nitrous oxide, methane, hydrofluorocarbons, perfluorocarbons and sulphur hexafluoride) that are considered to cause global warming and should be reduced. The Kyoto Protocol sets 1990 as the emissions base year, and states a specific reduction target of 5 % for the emissions from the parties listed in Annex I (Annex I parties), that is, developed countries. See Table 1.2 for the current reduction targets of each country. Although the United States was participating at the time of adoption (its initial reduction target was 7 %), it withdrew from the protocol in 2001 and is therefore omitted from this table. Criticism has been levelled at these reduction targets, one being that they have no scientific basis. In fact, the COP, which is the largest venue for discussing global

Table 1.3 Country-specific targets for greenhouse gas emissions reduction

Country	Long-term reduction target	
	Target gas	Reduction target (%)
UK	CO_2	− 60
France	GHG	− 75
Germany	GHG	− 80

warming issues, is now a venue for diplomacy, and that trend will probably continue. Regarding the overall GHG emission reductions, however, it has been observed that current reduction targets are woefully inadequate to attain a CO_2 concentration that would stabilise the climate, and this will heavily influence post-Kyoto targets.

2. Flexibility Mechanisms

A second characteristic of the Kyoto Protocol is that three economic instruments generally called the "Kyoto mechanisms" are recognized. The three are: Emissions trading, which enables trading in the right to emit GHGs (tradable permits); joint implementation, which makes it possible for Annex I parties to provide other Annex I parties with assistance in funding and technology and to allocate the jointly reduced emissions and the clean development mechanism, which is similar to joint implementation but enables Annex I countries to implement projects with developing countries and obtain part of the emission reductions. The reason these mechanisms are allowed is that the impacts of GHGs, that is, climate change, are global in nature and not confined to regions, and are not dependent on the emitting country or region.

1.2.5 Moves Surrounding Post-Kyoto Protocol

From 2009 through 2012, it was a period to decide the direction of GHG emissions reduction target of the so-called Post-Kyoto from 2013 onward. Major international moves surrounding "Post-Kyoto" in recent years are as follows:

1. Heiligendamm G8 Summit

At the 33rd Summit meeting held in Heiligendamm, Germany in June 2007, major countries agreed to consider reducing GHG emissions by 2050 to half of the present level. This value is set in consideration of the CO_2 concentration to stabilise climate conditions. In response to this agreement, some European countries started moves to consider stricter targets on their own. For reference, such examples are shown in Table 1.3. Base year is 1990.

2. COP13

COP13, which was held in Bali, Indonesia in December 2007, made discussions about an international framework to reduce GHG emissions after the commitment period of the Kyoto Protocol, and reached an agreement on the "Bali Action Plan" that is a timeline for such framework. Though COP13 could not decide any specific reduction targets, agreements on some important directions were reached. Particularly, Nationally Appropriate Mitigation Action (NAMA) is considered as an important agreement in the transport sector. NAMA represents the decision to take Measurable, Reportable and Verifiable (MRV) actions not only in advanced countries but also in developing countries. Following this Bali action plan, COP15 took over the moves, as one possibility, to create an agreement that is not a framework based on the past top-down reduction targets.

3. COP15

At COP15/MOP5 held in Copenhagen, Denmark in 2009, foundations of the Post-Kyoto Protocol were planned to be established based on the Bali action plan. After all, the COP15 meeting ended in the form that resulted in the so-called Copenhagen Accord only "to take note of" agreements of certain countries. It can be hardly said that this meeting was a great success. However, this accord has a possibility to have a great impact on the international trend of measures to be taken in the future against global warming. This accord has some important matters. Three matters that are considered being particularly related to emissions from the transport sector are briefly mentioned as follows. One is the point that the scientific knowledge to control the increase in temperature by less than 2 °C is used as a premise of this accord. With Intergovernmental Panel on Climate Change (IPCC's) Fourth Assessment Report as a guide, for example, to keep the increase in the global average temperature since the beginning of industrialisation by 2 °C or less, it is necessary to reduce CO_2 emissions from all sectors in 2050 by 50 % to about 15 % from the 2000 level. In other words, in the case where a superlong period, until 2050, is assumed, it is considered, as a premise, necessary to reduce CO_2 emissions by at least half from the 2000 level. Moreover, this accord also clearly took over the NAMA, which was declared in the Bali action plan. Particularly, as the specific amount of funds proposed to be contributed from advanced countries to developing countries was declared as the third characteristic, this NAMA has suddenly become real rather than concept. In particular, because it is difficult to set baselines and evaluate linkages and also because it is not easy to recover costs even in consideration of carbon prices, as far as the measures for reduction in CO_2 emissions in the transport sector, where assistance based on CDM as an existing assistance measure is said to be difficult, are concerned, it would be no doubt that this NAMA is one of the few international assistance plans that are now broadly proposed. In fact, the share of transport-related projects drastically increased to 29 % in the action plans to reduce the CO_2 emissions collected from developing countries by the end of January 2010, whereas the share thereof based on CDM was only 0.3 %. It exceeded the 27 % share of energy supply projects based on CDM that had the highest share of 71.8 %. For reference, examples are shown in Table 1.4.

Although specific measures are not stated, Costa Rica, Cote d'Ivoire, Egypt, Mongolia and Papua New Guinea indicate the transport sector as a target sector for NAMA.

Table 1.4 List of GHG reduction actions in developing countries (2011)

Country name	Content of GHG reduction action (transport sector)
Armenia	Conversion to electricity and gas in transport sector
Botswana	Energy saving in transport sector
Central African Republic	Control of automobile exhaust gas in metropolitan area
Chad	Development of transportation means with lower emissions of contaminated substances and development and wider use of biofuels
Colombia	Utilization of CDM in energy, forestry, industry, transport and waste disposal sectors
Ethiopia	Planning of railway
Eritrea	Investigation, development, application and proliferation of technology, practice and procedure which manage, reduce and prevent greenhouse gas (transport sector is also one of target sectors)
Ghana	Development of public transportation systems and introduction of Euro 3
Indonesia	Conversion to low-emission-type transport mode
Jordan	Railway project, light railway project, etc.
Macedonia	Improvement of efficiency, improvement of public urban transportation, etc.
Madagascar	Utilization of biofuels and introduction and development of low-pollution transportation means (public transportation, etc.)
Mauritania	Promotion of public transportation
Morocco	Development of streetcars, electrification of railways, etc.
San Marino	Reduction of final energy consumption and implementation of enlightenment campaign in each sector (including transport sector)
Togo	Public transportation system, fuel conversion, etc.

4. COP16

The Cancun Agreements are the set of agreements adopted at the COP 16 meeting held in Cancun, Mexico, in December 2010.

First of all, the long-term objective, which is to limit the increase in global temperature to less than 2 °C by reducing GHGs, became official agreement. As shown in the above section, this means it is necessary to reduce CO_2 emissions in 2050 by 50–15 % of the 2000 level. Therefore, developed countries would be required to enforce hikes in their reduction targets. The documents which summarize the submitted reduction target of developed countries and the NAMAs of developing countries to the UNFCCC secretariat that were in agreement with the Copenhagen Accord were taken note of in the official document of COP.

Regarding NAMAs, a registry was decided to match the plan to limit GHG emissions in developing countries, as well as the degree of support needed to achieve them. When receiving international support, this would be in conjunction with MRV (Measurement, Reporting and Verification). Even when not receiving international support, it was agreed to provide updates through international consultations and analysis biennially. In order to support developing countries, the establishment of a Green Climate Fund was agreed upon. This was based on the funds of developed countries which agreed upon the Copenhagen Accord and were placed officially under the Cancun agreements.

It was also agreed to establish a Cancun Adaptation Framework, which is the international support mechanism for adaptation policy for varied climate change issues and technological transfer schemes. In addition, it was also decided to consider developing a new market mechanism in COP17. Furthermore, the issue of the Kyoto protocol expansion was postponed after COP17.

5. COP17

COP17 held in Durban, South Africa, successfully accomplished four achievements, i.e. a roadmap to a future framework, an agreement about the second commitment period of the Kyoto Protocol, the Green Climate Fund and a series of decisions to implement the Cancun Agreements. COP17 made strides towards the creation of a fair and effective new international framework in which all of major countries participated.

Particularly, a roadmap was agreed to set up the Ad Hoc Working Group on the Durban Platform for Enhanced Action (ADP) as a new process to prepare legal instruments for the future framework, and to complete relevant works as early as possible or by 2015 at the latest and enter into force the protocol, legal instruments or legally binding results of agreement and start implementing them in 2020.

In respect of implementation of the Cancun Agreements, moreover, COP17 successfully made achievements such as the basic design of the Green Climate Fund and development of the guidelines for MRV of emissions reduction measures of respective countries.

As for the Kyoto Protocol, though the agreement to set the second commitment period was adopted, some countries including Japan expressed their intentions not to participate in the second commitment period. The Ad Hoc Working Group on Kyoto Protocol (AWG-KP) set the emissions reduction targets of the advanced countries that participate in the second commitment period at COP18 and completed its roles.

6. COP18

After meetings of "ADP", "Ad Hoc Working Group on Further Commitment for Annex I Parties (Note) under the Kyoto Protocol (AWG-KP)", "Ad Hoc Working Group on Long-Term Cooperative Action under the Convention (AWG-LCA)" and working-level negotiations at meetings of two subsidiary organs, further discussions were repeatedly made between ministers in a high-level segment on December 5–7. As a result, the following series of decisions by COP and CMP were finally adopted as the "Doha Climate Gateway", and decisions by COP and CMP were also adopted concerning other topics.

Note: Annex I Parties include those countries that are obliged to reduce and report GHG emissions, including OECD countries, the former Soviet Union and East European countries.

As a result of adoption of the Doha Climate Gateway, it was decided to extend the Kyoto Protocol into the second commitment period. The second commitment period is an 8-year period from 2013 through 2020. With this extension, a blank period could be avoided for a while. According to the agreement made under AWG-KP, participating advanced countries as a whole aim to reduce their emissions from 2013 through 2020 by 18 % from the 1990 level. With this agreement, AWG-KP completed its work. However, Japan expressed its stance not to participate in the second commitment period. This is reflected in the revised Annex B. Furthermore, the Japanese government declared not to purchase Assigned Amount Unit (AAU) that is repeatedly carried over in the first commitment period, together with the EU, Australia, Switzerland and others.

As for the CDM, it was confirmed that those countries that do not participate in the second commitment period would be allowed to participate in CDM projects and primitively acquire CDM credits (CER; transfer CER to their own countries) in and after 2013. However, it was decided to allow only those countries that participate in the second commitment period to internationally acquire and transfer CER by participating in Joint Implementation (Article 6 of the protocol) and international emissions trading (Article 17 of the protocol) in the second commitment period (for reference, Japan can continuously participate in international emissions trading during the adjustment period—from 2013 to the latter half of 2015 onward—of the first commitment period).

It was decided to leave the study on the new framework that has binding force upon all countries to ADP. It was decided that a basis for the new framework would be designed by the time when COP20 meeting is held. Additionally, it was also decided to declare measures by COP19 to fill the gap between the pledged amount of emissions reductions and the amount of reductions required to keep the temperature increase within 2 °C from the level before the Industrial Revolution. With such measures, advanced countries are expected to further reduce their emissions.

Putting focus on assistance to developing countries, it was decided that advanced countries would be required to "pledge to give assistance to developing countries… when the financial conditions of respective countries permit" and to "present information on the strategy and methodology to increase the amount of funds for measures against climate change to US$ 100 billion by 2020" (the amount presented by advanced countries at COP15 meeting) at COP19 meeting.

For comparison: The IEA (IEA 2013b) estimated that US$ 54,500 billion would be required in order to achieve the 2 °C target for non-OECD countries in transport sector. Even if focusing on railway (include HSR), introduction cost, US$ 7,300 billion, would be required. From this estimation, given this information, it is obvious that the $ 100 billion target amount for developing countries would be insufficient.

7. COP19

COP18 presented a roadmap to create the new framework. This roadmap finally made a step forward while moves remained slow in recent years. However, it is suggested that the new framework may possibly change drastically in the process. It is highly conceivable that the new framework can be a bottom-up type that set reduction targets set by respective countries on their own and appropriate reduction actions are set, rather than a top-down type under the Kyoto Protocol.

As for the target value, the common vision to "keep the temperature increase within two degrees in Celsius from the level before the Industrial Revolution" was

Table 1.5 Reduction case of CO_2 emissions from transport sector (50% reduction by 2050 from 2000 level)

Regions	CO_2 emissions (base year;2000) [million tonnes CO_2]	Reduction target of CO_2 emissions (target year: 2050) [million tonnes CO_2]
North America OECD	1996	159 (−92%)
Europe OECD	1146	135 (−88%)
Asia-Pacific OECD	489	56 (−89%)
FSU	177	72 (−59%)
East Europe	84	25 (−71%)
China	252	427 (+70%)
India	361	467 (+30%)
Other Asia	138	456 (+231%)
Middle East	215	125 (−42%)
Latin America	348	191 (−45%)
Africa	159	580 (+264%)

adopted in the Cancun Agreements. As already mentioned, this shows that drastic reduction in GHG emissions in all sectors from 50 to 15% in 2050 has become the world's target in the Fourth Assessment Report of IPCC. However, it is clear that there are big gaps between the required amounts of reductions to achieve the target values.

Transport has not been successful in CDM framework. By the end of 2012, there have been only 18 transport projects approved, which shares only 0.3% among whole projects in CDM. Instead, many efforts have been made to prepare NAMAs for transport projects to reach 19% of whole NAMA projects until COP19.

1.2.6 Implication to the Challenge in Transport Sector

As COP limits the increase in global temperature to less than 2 °C, the transport sector will probably have to make further efforts to reduce CO_2 emissions. At the end of this section, let us think about how much GHG emissions are required to be reduced in the transport sector over a long period of time. For example, in the case where 2050 is set as the target year, if IEA's prediction (this is used as a baseline case) is used, CO_2 emissions from all sectors in the world are considered increasing by about twice the 2000 level, and CO_2 emissions from the transport sector are also considered to increase by the same extent. However, to keep the temperature increase within 2 °C, emissions must be reduced to at least 50% from the 2000 level. Table 1.5 shows the emissions reduction targets in respective regions in the case where CO_2 emissions from the transport sector are halved by 2050 from the 2000 level. The targets are calculated on the assumption that the amount of CO_2 emissions from the transport sector per capita in 2050 is made equal all over the world. As a result, in the USA, for example, CO_2 emissions will have to be decreased by 90% or more in total. To meet such a dramatic challenge, it will be necessary to review not only the transport system but also the economic and social system itself.

As for developing countries, it is obvious that the concept of NAMA becomes one of the cores of future discussions. The proposed NAMAs include plans for transportation. As "Green Climate Fund" was established, assistance to developing countries is also expected in the transport sector where almost no progress was seen in respect of the Kyoto Mechanism. However, many of the proposed plans for transportation are related to the urban transport sector in reality. In consideration of the increase in auto transport and air transport demands in developing countries, it is absolutely necessary to establish means of intercity transport. Therefore, developing countries are required to pay attention to a broad range of plans, including intercity transport, and the direction to evaluate them is required. In addition, it is also an important challenge to develop MRV methodologies by implementing transport NAMA in respective countries and by not repeating the same mistake as the CDM in transport sector.

1.3 Principles of GHG Reduction Policy in the Transport Sector

1.3.1 Basic Requirements for a Success of GHG Reduction Policy

Many scenarios have been constructed worldwide to explore whether a change of trends is possible in the transport sector, and if so, under which conditions. It is widely agreed that the following three conditions have to come together to achieve a major shift to reducing the carbon footprint of intercity transport:

- A clear political will and strategy to reduce the carbon footprint in all transport segments manifested in taxation, charging, regulation, infrastructure and education policy.
- A private willingness to change behaviour and accept state's policy actions to foster this in a balanced way.
- A technology development which offers a host of choices to perform intercity transport activities at low carbon emissions.

All three conditions are not easily fulfilled. Although most countries have declared their agreement with the long-term reduction goals suggested by the IPCC (2007), major GHG-producing countries like China and the USA are still not accepting obligatory reduction targets. The World society is far from agreeing on a post-Kyoto declaration which would set clear guidelines for all countries to achieve the overall reduction target of −50 % (−80 % for industrialised countries) until 2050 compared with 1990. Even the strong protagonists of CO_2 savings, Japan and the EU, who some years ago saw themselves in the role of top runners, show a declining effort to continue initiatives, which they had started some years ago. Examples are the suspension of the EU Directive to include aviation into the European trading scheme (ETS) for CO_2 emissions or the last minute interventions of a few member countries (Germany among them) to delay a Directive for reducing the passenger car CO_2 emissions from presently 130 to 95 g/km after 2020 with the aim to achieve more exceptions for their car manufacturing industries. There are also strong indications that the theoretical test cycles applied by the automotive industry to measure fuel consumption deviate from practical test cycles by around 25 % (ICCT 2013). This underlines that the declared goals and targets of policy as well as the declared implementation by the industry are not necessarily corresponding with revealed facts, so a strict monitoring of targets and their achievements by independent institutions is required.

A latent private willingness to change behaviour is a necessary precondition of successful state policy actions towards reducing carbon footprint because in democracies the voters decide on the success of political programs. 10 years ago, there was a broad social consensus in many countries that substantial changes of transport behaviour are necessary, following the embarrassing publications of IPCC and strengthening the green movements in these countries. But now, the stimulating effects of climate change warnings seem to have weakened. In important countries like the USA, Canada or Australia, people seem to be increasingly convinced of the effectiveness of adaptation measures rather than of mitigation measures, and this also holds for parts of their scientific societies. This means that the propensity of societies has increased to exploit the potential of available resources now and leave the problems arising to the future generations.

Fortunately, the increasing climate awareness of people in the last decade, supported by appropriate political legacy and regulation, has achieved a remarkable progress in transport technology. Within one decade, the average fuel consumption of newly licensed cars dropped from 170 to 130 g/km in the EU, and a further reduction has been prepared by legislation (in the EU to 95 g/km from 2020 onwards). In the USA, the fuel economy standard (CAFÉ) has been tightened from 27.5 mpg (miles per gallon; by 2010) to 35.5 mpg (by 2016) and a further reduction is foreseen to a level of 54.5 mpg by the year 2025, which would imply a doubling of fuel efficiency compared with 2010.

In Japan, the target value based on the fuel consumption of top runner car, taking into account the improvement by technological developments until FY2020, has been determined for each vehicle weight category. If the target values are achieved, it is estimated that the fuel consumption would be improved from the actual value 16.3 km/l in the FY2009—20.3 km/l as average value of the fuel consumption by the FY2020.

In aviation, new aircraft types have been developed, saving up to 25 % of energy compared with old technology (Boeing 787, Airbus A 350). In HSR, the new-generation trains are able to recycle energy in braking phases, show improved aerodynamics and are controlled by more efficient train control systems. Some of the developments go more straight towards energy savings rather than higher speeds (e.g. the new Siemens ICx generation). These examples reveal the high potential of technological development, but the challenge remains to implement the technical progress with high penetration rates throughout the transport markets. With respect to this challenge, one can observe a declining effort in the recent years: The car manufacturing industry shows an increasing resistance against tighter emission limits; the airline industry is fighting against its integration into the EU ETS and continues to use old technology in developing and emerging economies. Also, the progress through modern HSR train technology is hampering because of the missing infrastructure development.

In the industrialised countries, the investment in the rehabilitation and maintenance of existing networks is lagging behind and the planning and construction of new links takes decades of time because of lack of funds. For Germany, the financial gap for the existing transport networks was estimated 7.2 billion € per year, about 2 billion € of this figure applies for the railways. As this country was evaluated very positively in an infrastructure comparison report of Ernst and Young (2010), one can follow that the situation in other countries like the USA or the UK is even worse. From this follows that the new transport technologies can only exploit their reduction potential if the infrastructure allows for moving the vehicles at energy-efficient speed cycles. Furthermore, technologies, which are able to shift traffic from high to low energy consumption modes, need new or upgraded infrastructures. This holds in particular for HSR which is able to attract traffic from the car mode on distances below 500 km and from aviation on distances between 500 and 1000 km.

1.3.2 Principles of Climate Policy in the Japanese Transport Sector

1.3.2.1 Trends of Domestic Transport-Sector CO_2 Emissions in Japan

In order to achieve the 6 % reduction target committed to under the Kyoto Protocol, Japan has set the individual reduction target for each sector according to its domestic Kyoto Protocol Target Achievement Plan (Ministry of the Environment (MOE) 2008), and currently set the target indicator of 240 million t for the domestic transport sector in 2010. Japan's transport sector emits approximately one fifth of the national CO_2 emissions. As Fig. 1.10 shows, Japan has successfully reversed an upward trend in transport emissions. These have been declining after peaking at 267 million t in 2001 and had already fallen to 232 million t in 2010 under the target indicator for 2010, despite the fact that the growth of the Japanese economy has basically not stopped throughout this period (Fig. 1.11). CO_2 emissions from freight vehicles have decreased since 1996 and those from private passenger cars since 2001. Regarding the phenomena, the MLIT has explained that this is "the result of Japan's systematic and well-organised climate change policies" (MLIT 2009a, p. 1), including "improvements in the fuel efficiency of vehicles according to the top-runner standards, promotion of dissemination of environment-friendly vehicles through green taxes for vehicles, streamlining of physical distribution by enlargement of lorries, etc." (my translation) (MLIT 2009b, p. 6).

Based on examples in Japan, CO_2 emissions are compared in breakdown by competing transport mode. Figure 1.12 shows data announced by the MLIT. CO_2 emissions from passenger trains are one ninth of CO_2 emissions from private

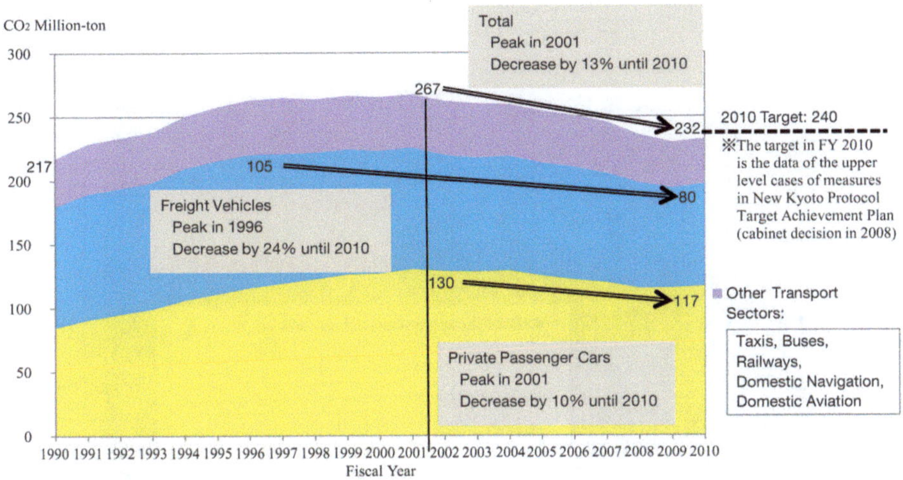

Fig. 1.10 Changes in CO_2 emissions in domestic transport sector in Japan. (Data source: Prepared based on the data from Greenhouse Gas Inventory Office of Japan, National Institute for Environmental Studies 2013; adapted from the graphic of MLIT)

passenger cars and one sixth of CO_2 emissions from aircraft. This shows that trains are environment-friendly transport mode.

Transport Research Laboratory in the UK, which actively promotes research activities at COP and ad hoc working groups of COP, has classified the strategy for transport-sector CO_2 emissions reduction into three categories, i.e. "Technology Improvement (Improve)", "Modal Shift (Shift)" and "Traffic Avoidance (Avoid)". The measures to shift a large share of traffic volume to those modes that have relatively low CO_2 emissions are classified into "Modal Shift (Shift)". According to Fig. 1.12, shift from private cars and aircraft to trains can obviously contribute to CO_2 emissions reduction. In this figure, the comparison among 2004, 2007 and 2010 shows that CO_2 emissions from private cars and aircraft decreased in the medium term. This seems to be largely attributable to "Technology Improvement (Improve)" such as improvement of fuel efficiency. As for the remaining "Traffic Avoidance (Avoid)", its index can be the decrease in traffic volume itself as shown in Fig. 1.11.

It should be noted here that the relative degrees of superiority between transport modes can be changed by technological improvement such as improvement of fuel efficiency and usage conditions as represented by vehicle occupancy. For example, it is also highly likely that the degree of superiority of trains may lower because of significant improvement of CO_2 emissions intensity of private cars with the advent of electric vehicles and hybrid vehicles or early replacement of aircraft with higher priority given to weight reduction and fuel efficiency. Moreover, though direct comparison cannot be made because there is no unified equation among countries to calculate CO_2 emissions, comparison with Japan, which boasts high vehicle occupancy of commuter lines in metropolitan areas and that of the Shinkansen line (particularly in the section between Tokyo and Osaka), shows that CO_2 emissions per unit of transport of trains in the EU tend to be larger.

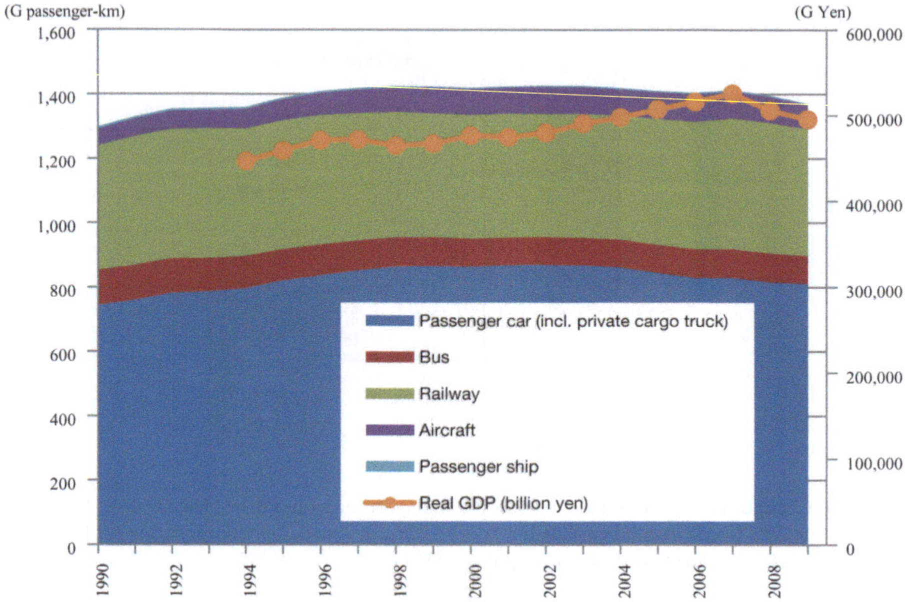

Fig. 1.11 Changes in traffic volume by transport system and real gross domestic product (*GDP*) in Japan. (Data source: (traffic volume) MLIT, Policy Bureau, 2011; (real GDP) Cabinet Office (n.d.a))

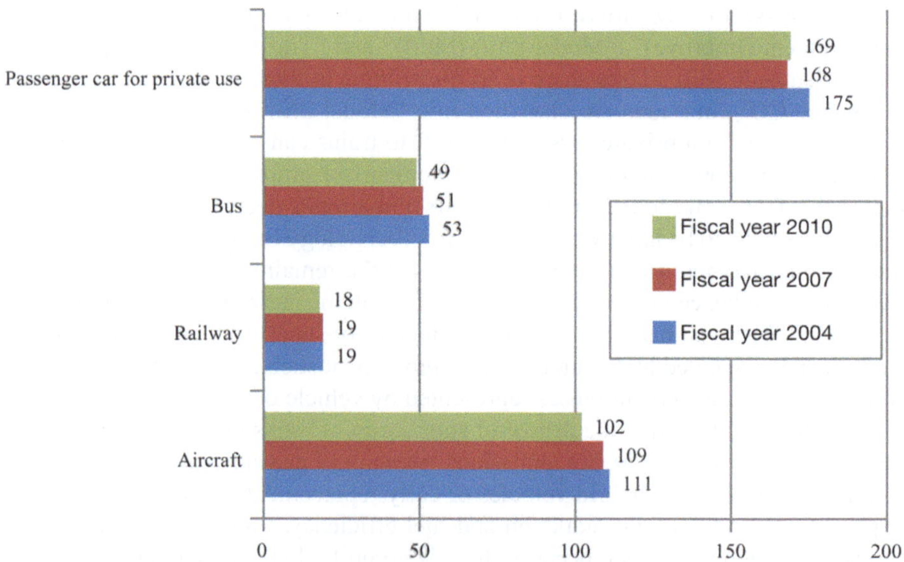

Fig. 1.12 CO_2 emissions per unit of passenger transport quantity in Japan. (Data source: MLIT 2006; MLIT, Shikoku Transport and Tourism Bureau 2009; Japan Railway Construction, Transport and Technology Agency (n.d.))

In response to the impact of the Great East Japan Earthquake and accidents at Fukushima Nuclear Power Station, the Japanese government is now undertaking radical review of energy and environment policies. As a result of this move, it is possible for the shares of power generation sources to drastically change. It is also assumed that the relative degrees of superiority between transport modes will change.

1.3.2.2 Classification of Measures to Reduce Greenhouse Gas Emissions in Japan's Transport Sector

The MLIT explained Japan's approaches in the transport sector as follows:

> The MLIT is implementing measures aimed at individual motor vehicles, including improving fuel consumption, developing systems to higher fuel performance, a green tax plan for motor vehicles and developing the next generation of low emission vehicles. At the same time, it is making efforts to increase the awareness of driving methods through the promotion and dissemination of economical driving. The ministry is also proceeding to promote environmental measures in the transport sector by achieving smooth road traffic flows and enhanced coordination between logistics business operators and corporate shippers through the holding of Green Logistics Partnership Conferences and similar initiatives. The ministry is promoting the use of public transport by making it more convenient through the introduction of integrated circuit (IC) cards and the improvement of public transport connections. (MLIT n.d. a, part.2, Chap. 7, p. 65)

Those measures are promoted by a mixture of instruments such as regulations, subsidies and taxes, and by developing databases and statistics so that the government, industry and individuals might devote all their resources towards the goal while keeping harmony with economic and social growth.

Here, taking account of the measures classified by Transport Research Laboratory as mentioned earlier, the CUTE matrix (methodology of classification by crossing three strategies and policy and technical means), which was developed in Nakamura, Hayashi and May (2004) is applied to Table 1.6. Then, A-1 corresponds to "Technology Improvement (Improve)", A-2 to "Modal Shift (Shift)" and A-3 to "Technology Improvement (Improve)" and "Traffic Avoidance (Avoid)", and B to "Traffic Avoidance (Avoid)". If the CUTE matrix is applied to measures in respective countries, their characteristics can be seen well (Hayashi and Nakamura 2013). In Nakamura, Hayashi and May (2004), measures for intracity transport are classified and examples of classification of intercity transport are also introduced (Hayashi and Nakamura 2013).

In this book, the CUTE matrix in Hayashi and Nakamura, K. (2013) is retouched and revised, and then the measures announced by the Japanese government for intercity transport are allocated (Table 1.6). In that case, almost all of the measures can be categorized as "Technology Improvement (Improve)" and it can be found that measures clearly categorized as "Reduction (Avoid)" and "Shift (Shift)" are few. Also, of the measures categorized as "Shift (Shift)", many of the measures for "construction of trunk lines such as railroads" and "wider utilization of existing railroads and buses" are related to intracity transport measures. The measures for

Table 1.6 Measures in transport sector in Japan against global warming (CUTE matrix)

Strategy means	Avoid	Shift	Improve
Technology	TV conference technology	Creating an environment for bicycling (jg)	Improving fuel efficiency of vehicles (top-runner standard) (jg)
		Improving efficiency of cargo transshipment facility and facility planning	Biomass fuels, etc. (jg)
		Improving velocity of train cars	Promoting ITS (jg)
		Harmonizing travel IC cards for public transportation systems	Measures for bottleneck railroad crossings, etc. (jg)
			Improving the efficiency of truck transport (jg)
			Improving energy consumption efficiency of railways, ships and aircraft (jg)
			Flight navigation system
Information	Promoting teleworking (jg)	Green logistics partnership conference	Promoting eco-driving (jg)
Regulatory		Strengthening crackdown on illegal parking	Defining a new fuel efficiency standard (jg)
			Control with using flight navigation system
Economy	Reducing the distance of land transportation for marine container cargo (jg)	Modal shift to railway and maritime transportation (physical distribution between cities) (jg)	Promoting low-emission vehicles (jg)
	Realizing compact cities (jg)	Modal shift to railway transportation (passenger transport between cities)	
		Developing newrailway lines, etc.(passenger transport in urban areas) (jg)	
		Promoting utilization of existing railways and buses (jg)	

jg measures indicated in White Paper on MLIT, Japan

"modal shift to railway and marine transportation" are related to freight transportation. The construction of Shinkansen lines has been promoted by public entities as intercity transport measures in Japan, while the operations of Shinkansen lines are at present undertaken by private JR companies, although modal shift measures for intercity passenger transportation are not clearly positioned.

The following are examples of some distinctive measures:

1. "Technology Improvement (Improve)"

a. Measures for improving vehicle performance (individual vehicles)

Since "automobiles account for the majority of energy consumption in the transport sector" (MOE 2008, p. 48), performance improvement of automobiles is highly effective to reduce CO_2 emissions. In Japan, major automakers compete for development of hybrid vehicles and fuel cell vehicles. In consideration of such conditions, "the Government promotes automobile measures such as further improvements in fuel efficiency through the world's highest technology and dissemination of highly fuel-efficient or clean energy vehicles"(*ibid.*, p.48). Those measures include the top-runner approach for fuel efficiency regulations, automobile green tax scheme and subsidies for low-pollution and energy-efficient vehicles. Referring to MLIT (2009a, p. 2),

"A top-runner approach requires setting fuel efficiency standards at an ambitious level higher than the performance of the best vehicles currently available in the market, taking future technological improvements in account. In 1999, Japan introduced the top-runner approach" for light duty vehicles to "motor vehicle regulations to further encourage improvement of fuel efficiency. Thanks to the top-runner standards, the average fuel efficiency of new gasoline" passenger cars "in Japan was improved by 26 % from 1995 through 2006, and the 2007 new standards" for light duty vehicles "are expected to bring about a 50 % improvement in fuel efficiency" of new gasoline passenger cars by "2015, compared with that in 1995".

Further new standards towards 2020 for light-duty vehicles were promulgated in 2013. "In addition, in 2006, Japan introduced the world's first standards for heavy-duty vehicles [such as buses and trucks], and would like to encourage other countries to introduce similar standards" (*ibid.*, p. 2).

b. Promotion of environment-friendly driving behaviour

Complementary to the measures for improving vehicle performance, the government has promoted environmentally friendly usage of vehicles. For example, eco-driving has been disseminated and promoted, which includes "idling stop" while stopping or parked, and driving at safe and constant speeds appropriate for the traffic conditions.

c. Promotion of introduction of energy-efficient railway vehicles

As for railways, the government has applied tax incentives such as halving the rate of fixed property tax for 5 years to such measures as improvement of speed control method by means of introduction of new railway vehicles with a Variable Voltage Variable Frequency (VVVF) device, reduction of weight by 1/8 or more or introduction of electricity regeneration braking. Subsidy has been applicable to the introduction of energy-efficient railway vehicles and locomotives as well as regenerated electricity batteries.

For example, in the case of Japan Railway (JR) companies that actually exert such efforts, the power consumption of the N700 series Shinkansen train has declined to 51 % compared with that of the 0 series Shinkansen train as a result of weight reduction, reduction in running resistance and improved efficiency of power regeneration. As for train cars for narrow-gauge lines, the power consumption of the 103 series has decreased to 47 % compared with that of the 209 series and E231 series.

d. Support for introduction of energy-efficient equipment, etc. into domestic shipping

Dissemination of Super-Eco-Ship, in which a generation unit can supply power not only for the propulsion unit but also for the distribution unit so that increased freedom in hull forms and configuration enables optimization of the forms to increase fuel efficiency, has been promoted.

2. "Modal Shift (Shift)"

a. Improvement of logistics efficiency

In Japan, CO_2 emissions per unit of railway freight transport and that of maritime freight transport are one sixth and one third of that of trucks, respectively, in fiscal 2009, according to calculation by MLIT (2012a, p. 15) based on the data from Greenhouse Gas Inventory Office of Japan and the transport statistics from MLIT, so that modal shift is recommended also in freight transport.

With regard to the Act on Promotion of Comprehensive and Efficient Logistics Operations, the administration shall examine whether a plan spontaneously submitted by a freight carrier contributes to such shifts, or promotion of joint order placement, before the certification is provided. Against the backdrop of the Green Logistics Partnership Conference to promote projects through alliances and joint efforts between freight transport business operators and consigners, the Act on Promotion of Comprehensive and Efficient Logistics Operations, and the Act on the Rational Use of Energy, improvement of truck transport, modal shift and other relevant measures are promoted. In addition, a subsidy scheme has been started in 2011, in which projects undertaken by consultative committee consisting of logistics-relevant entities such as freight transport business operator and consigner according to their plan to promote modal-shift etc. is supported regarding the implementation costs.

b. Promotion of usage of public transport, etc.

As is the case with physical distribution, we can understand the effectiveness of modal shift in passenger transportation because of a large difference in CO_2 emissions per (capita and km) between modes as shown in Fig. 1.12. Though details are explained in Sect. 2.2.1.3, as for intercity transport, the Shinkansen lines operated by JR companies that are now private companies have been extended by public entities. The government is making efforts not only to extend the Shinkansen lines but also to promote upgrading the conventional trunk lines by means of subsidies.

In addition, regarding intracity transport, it is essential to further extend the accessibility, attractiveness and convenience of public transport, based on the sound development of railway networks in the midst of the competition with other modes such as private passenger cars and air transport. The government recognizes that Japan still must construct new railway lines and maintains a system to subsidise construction of such railway lines.

Against the convenience of private passenger cars, which can bring us from door to door, it is important to increase interconnectivity and seamlessness among various public transport operators by constructing station or stops where passengers can change between different modes of public transport directly and smoothly. Park&Ride is also promoted. The government has promoted introduction of IC cards commonly used among multiple public transport operators and across multiple modes of public transport. Also, bus and LRT location system enables passengers to be informed of updated detailed operation schedules.

In general, railways in Japan are characterised as punctual, densely networked and reliable, with which we are able to associate the high modal share of surface passenger public transport in Japan in the era of tough competition with private passenger cars, which are getting more and more comfortable through technological development and road construction, as well as with aviation. The high modal share of railways has contributed to reduction in CO_2 emissions. This point is indicated as follows, including some background information.

In metropolitan areas, intracity tramways and other lines extended to suburban areas were built or a lot of simple narrow-gauge railways were constructed by the private sector in and after the latter half of the Meiji era (in the early twentieth century). They were changed to subways in urban areas and improved to HSRs in suburban areas after World War II, and played the roles to help expansion of metropolitan areas along their lines. Particularly, major private railroad companies that operate suburban railroad services absorbed development profits of construction of railroad lines by making use of the business model to simultaneously enhance suburban development along railroad lines and operate terminal buildings in urban areas. Against the background of such circumstances, because of the railroad networks that densely expanded, Tokyo's 23 wards enjoy a remarkably high share of railway passenger transport, at 77 % of passengers transported in fiscal 2007, when calculated on the data from Institution for Transport Policy Studies (ITPS) and MLIT in ITPS's "Urban Transport Annual Report 2009" (2010, p. 132–133). Indeed, that district is not a small "compact city" but rather an agglomeration, which demonstrates one possibility of future megacities in developing economies. The urban railways have been developed and operated not only by JR (or their predecessor Japan National Railways (JNR)) and public entities but also by various other private railway companies, which simultaneously manage the development of real estate, commercial shops and leisure facilities around the stations along the railway lines. Even within the railway networks, operation of direct trains going through different operators has been developed by means of collaboration, supported by the central and local government.

We see similar phenomena in other large metropolitan areas such as Osaka. In addition, those urban networks of public transport have been well and conveniently connected through intercity railway networks with a focus on safe and earthquake-proof Shinkansen, as partly described in Sect. 2.3.1.3. That is an element which may not have been sufficiently realised in some emerging countries.

On the other hand, there were many private tramways in local cities with a population of 200,000 or more. However, most of them were done away with because of the progress of motorisation. Keeping pace with motorisation, urban areas increasingly tend to expand in a low-density manner. At present, mobility in most of cities relies on cars.

In hilly or mountainous areas and suburban areas facing rapid aging society, private bus operators have pulled out of their business due to decreasing population. The Chugoku District Transport Bureau, one of the regional bureaus under the MLIT responsible for such areas, has supported local governments lacking in sufficient know-how to maintain local bus operation service by providing for free standardised software for preparation of planning and management of local bus operation. Similar efforts have been started for sea routes for isolated islands.

3. "Traffic Avoidance (Avoid)"

Compact city

According to the classification of the MLIT, though it is not classified into the transport sector, it can also be regarded as a transport measure in a sense to shorten the transport trip required for urban life and urban activity. Moreover, it is possible to create an environment where public transport systems can be easily established by establishing space-compact densely populated areas.

Chapter 2
Intercity Transport Policy and Planning System: International Comparison Between the EU, USA, China and Japan

Tetsuo Yai, Koichi Fujisaki, Ryo Itoh, Keiji Kariyazaki, Hidetoshi Kume, Haixiao Pan, Werner Rothengatter, Atsushi Suzuki and Naoyuki Tomari

2.1 Introduction

It has been pointed out in Sect. 1.1.2 that the share of intercity transport of total passenger-kilometres is estimated to be more than 40 % and will probably increase substantially in the future. In freight transport, the share of intercity transport is much higher and can account for more than 80 % of the total tonne-kilometres in industrialised countries. This underlines—in spite of the statistical difficulties for international comparisons—that intercity transport is a highly relevant segment of transport when it comes to analysing the responsibility of the transport sector to global warming and the potential for mitigation.

This chapter gives an overview of the current status and the intercity transport policies concerning greenhouse gas (GHG) reduction. Our focal points here are how the impact of global warming is treated in the existing transport planning processes and how to integrate the avoid/shift/improve strategies described in the previous chapter into the long-term planning context. International comparison will be made, including Japan, the USA and Europe, partly extended by incorporating Chinese examples. As a consistent intercountry comparison is possible only for a

T. Yai (✉)
Graduate School of Engineering, Tokyo Institute of Technology, Tokyo, Japan
e-mail: tyai@enveng.titech.ac.jp

K. Fujisaki
(Formerly) Institution for Transport Policy Studies (ITPS), Tokyo, Japan
e-mail: fujisakiyk@gmail.com

R. Itoh
Graduate School of Economics, Nagoya City University, Nagoya, Japan
e-mail: itoh@econ.nagoya-cu.ac.jp

K. Kariyazaki
Kanto District Transport Bureau, Ministry of Land, Infrastructure, Transport and Tourism, Tokyo, Japan
e-mail: kariyazaki-k55qn@ktt.mlit.go.jp

Y. Hayashi et al. (eds.), *Intercity Transport and Climate Change*,
Transportation Research, Economics and Policy 15, DOI 10.1007/978-3-319-06523-6_2

few characteristics, we also chose to give more detailed information and data on a specific country only when an international comparison was not possible.

2.2 Current Situation and Issues of Intercity Transport

2.2.1 Current Situation of Intercity Transport

2.2.1.1 Key Trends of Intercity Transport

In the past 40 years, people's mobility has significantly increased globally as shown by the aggregate national trends for selected countries in Fig. 2.1. The share of road transport has increased all over the world due to the progress of motorisation while the share of railways in total passenger transport has continuously decreased in most countries. Currently, the railway's share in passenger transport is small, except for Japan.

Regarding intercity transport, in general the share of car use decreases and those of railways and air increase with the travel distance, but the situation is different across some of the countries. For example, railways are dominant in Japan for medium distance travel (between 300 and 750 km), while air transport is dominant for longer distances (see upper left figure of Fig. 2.1).

In contrast to Japan, in the USA, the rail share is almost zero pretty much for all distances. Only on the Northeast Corridor (NEC) between Washington and Boston, intercity rail transport has a relevant modal share. Private car is the most dominant mode even for longer trips up to 1000 km while most people use air transport for longer trips. Also in the UK, road transport is high on average, but about 10% of passengers of all distances use railways. Therefore, the characteristics of the modal split of the UK are somewhere between Japan and the USA. Germany and, in

H. Kume
JICA Expert, Ministry of Public Works and Transport, Phnom Penh, Cambodia
e-mail: kumehidetoshi@gmail.com

H. Pan
Department of Urban Planning, Tongji University, Shanghai, China
e-mail: hxpank@gmail.com

W. Rothengatter
Karlsruher Institut für Technologie (KIT), Karlsruhe, Germany
e-mail: werner.rothengatter@kit.edu

A. Suzuki
Department of Civil Engineering, Meijo University, Nagoya, Japan
e-mail: atsuzuki@meijo-u.ac.jp

N. Tomari
Institution for Transport Policy Studies (ITPS), Tokyo, Japan
e-mail: tomari@jterc.or.jp

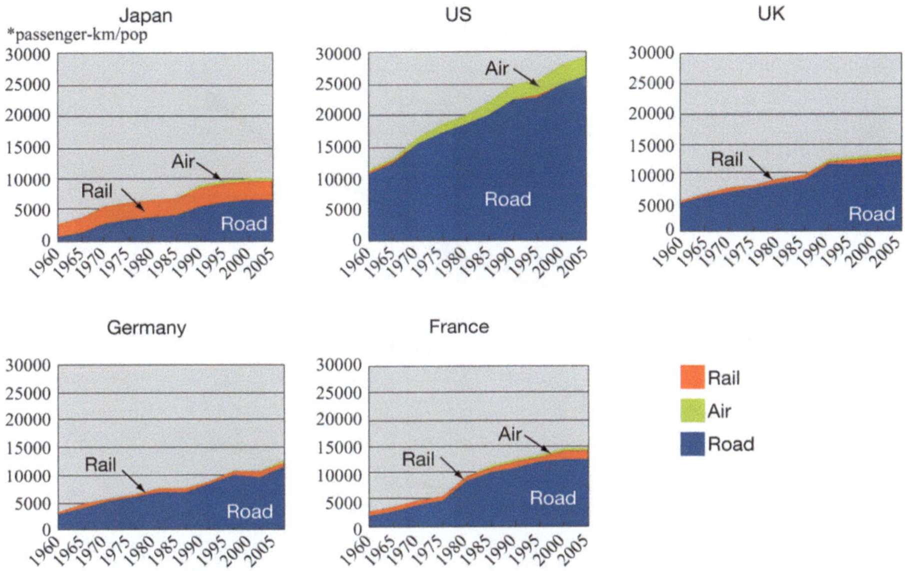

Fig. 2.1 Passenger mobility trends for selected countries. (Source: MLIT 2012b)

particular, France have partly modernised their railway systems and show higher railway patronage for intercity passenger transport. Nevertheless, the average modal share over all distance ranges is only 8 % for Germany and 9.2 % for France (2011).

However, we should note that the railways in European countries and the USA play an important role in freight transport, which is not treated in this book. The shares of railways in freight transport in the USA and Germany were over 40 and 17 % in 2009, respectively, while in Japan, it was just 6 %. On average, in the EU, the share of freight rail is about 11 %, while it is higher in countries with high foreign trade and efficient railway freight service.

In the process of rapid economic progress, China has experienced a rapid growth of transport, as shown in Fig. 2.2, for the years 2000–2010. Total transport activities have increased in the decade by 146 %; in particular, 316 % for civil aviation and 126 % for public roads. Also the railway transport increased strongly by 93 % due to massive investments in high-speed rail (HSR) which now offers competitive rail services between major agglomerations.

The extension of the HSR networks and services has in particular influenced the railway transport development. Figure 2.3 shows the extension of the HSR networks over time in different countries.

According to Fig. 2.3, the HSR operation started in 1964 in Japan which has kept the largest HSR network of the world for more than 40 years. But it was recently taken over by China, which has increased its HSR network since 2007, exceeding 8000 km in 2011 and approaching 10,000 km in 2013. Now Japan, France and Spain have almost the same HSR network length of 2000–2400 km, which is about

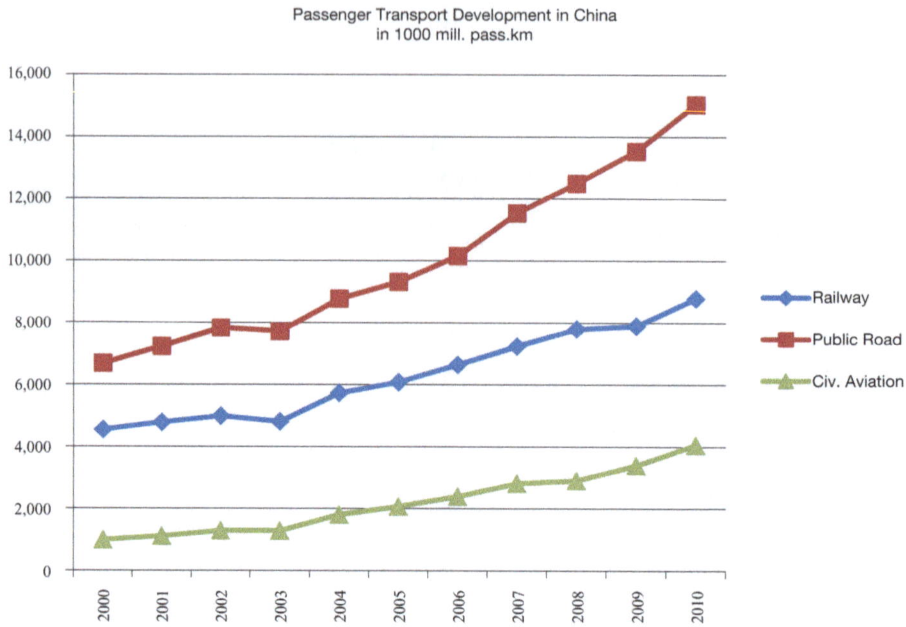

Fig. 2.2 Development of passenger transport in China from 2000 to 2010. (Source: National Bureau of Statistics of China 2002–2011)

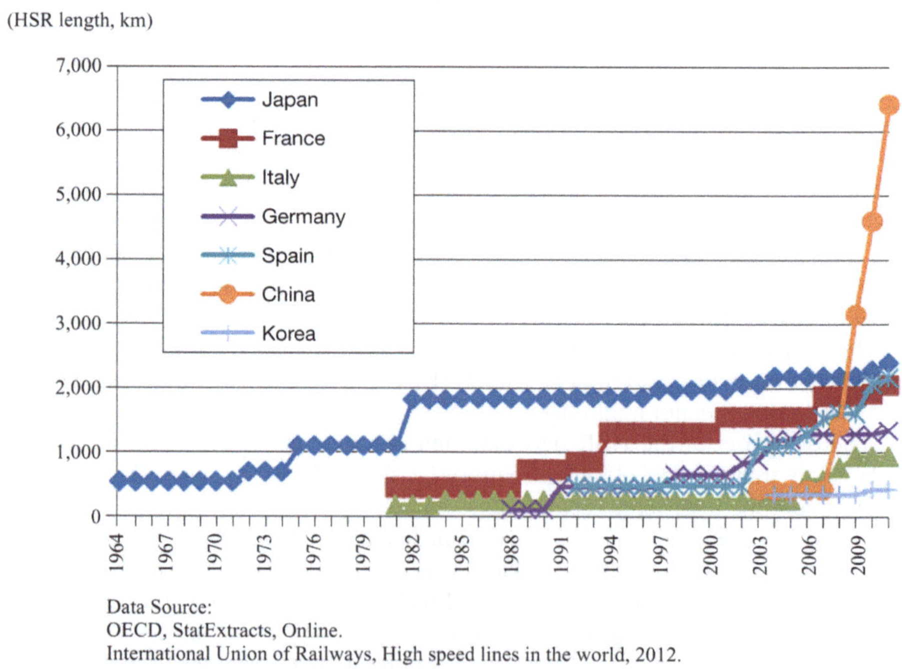

Fig. 2.3 Development of high-speed railways (HSR) in several countries. (Source: International Union of Railways 2013)

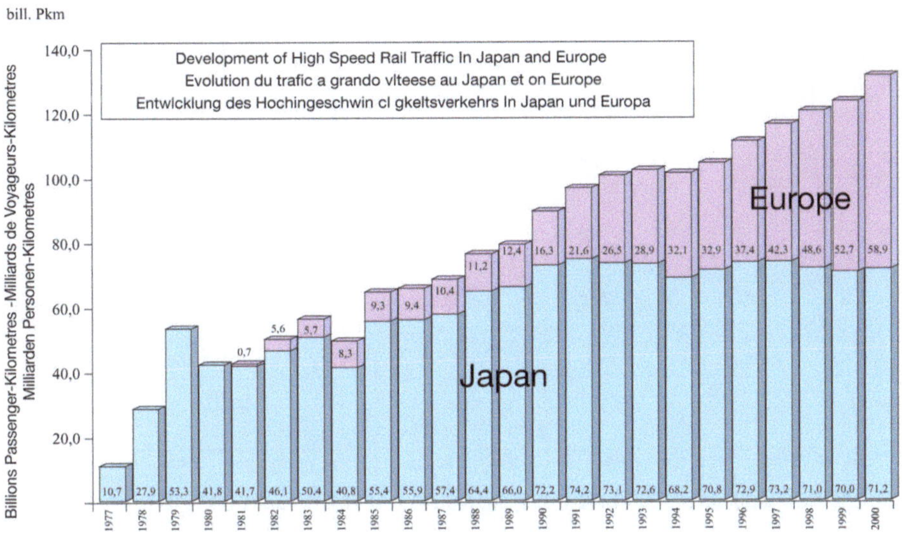

Fig. 2.4 Comparison of the development of HSR patronage in Japan and Europe (billion passenger-km). (Source: UIC 2011)

double the length of HSR lines of Germany and Italy in 2010. Spain has made the most significant progress in the past decade and has developed the longest HSR network in Europe (2144 km in 2012).

The development of HSR infrastructure and services had a significant influence on demand. Figure 2.4 compares the development of HSR in terms of passenger-kilometres in Europe and Japan. Japan started in 1964 with HSR on the most densely populated Honshu corridor between Tokyo and Osaka, followed by the Western and Eastern lines (Sanyo, Tohoku, Joetsu Shinkansen until 1990). After accomplishment of this important HSR network, the patronage rose to a high level, and remained almost stable until year 2000, from which point some complementary links were added (Nagano, Kyushu Shinkansen). In the years following the Asian HSR traffic, development was further driven by the investments in Korea and Taiwan (opened in 2005 and 2007, respectively; see Fig. 2.5).

The European HSR development started in 1981 in France with the first TGV line between Paris and Lyon (425 km) followed by Germany, where the first HSR lines (Hanover–Würzburg; Stuttgart–Mannheim) were opened 10 years later. Figure 2.6 adds complementary information on the HSR performance in Europe. It is obvious that the French TGV is carrying by far the most HSR passengers in Europe (around 52 billion pkm in 2009). The traffic volume of the German HSR service is less than one half (22.6 billion pkm) and that of Spain and Italy is less than one quarter (11.5 and 10.7 billion pkm, respectively) of the French figure. Spain shows the steepest traffic increase in the last few years, due to a rapid extension of their HSR network (see Fig. 2.3).

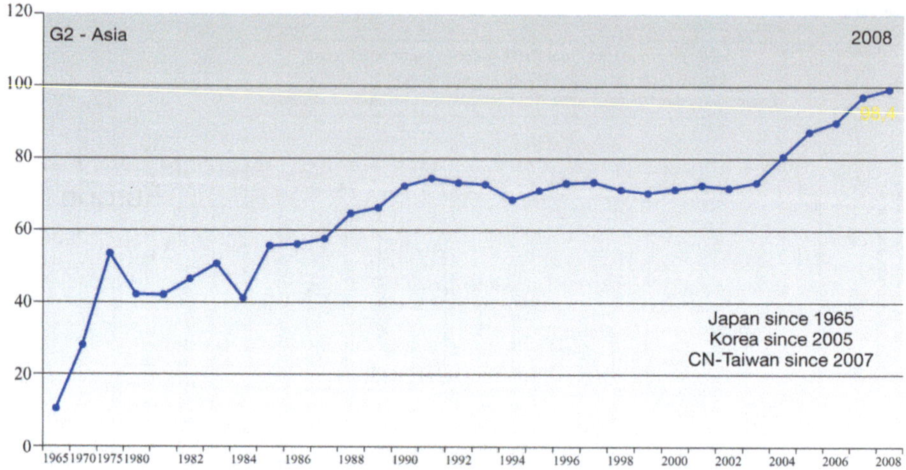

Fig. 2.5 Development of HSR in Asia (Japan, Korea, Taiwan). (Source: UIC 2011)

The upturn of rail passenger traffic in Spain in the past two decades is in particular due to the rapid HSR development. Figure 2.7 demonstrates that HSR in Spain makes already more than 50% of the overall railway patronage. As only 46 million people live in Spain—compared to 127 million in Japan—the overall HSR patronage is still modest, and it is obvious that HSR in Spain will—despite its very successful development in the past 10 years—never reach the financial viability of the Japanese Shinkansen system. (Source: UIC 2011)

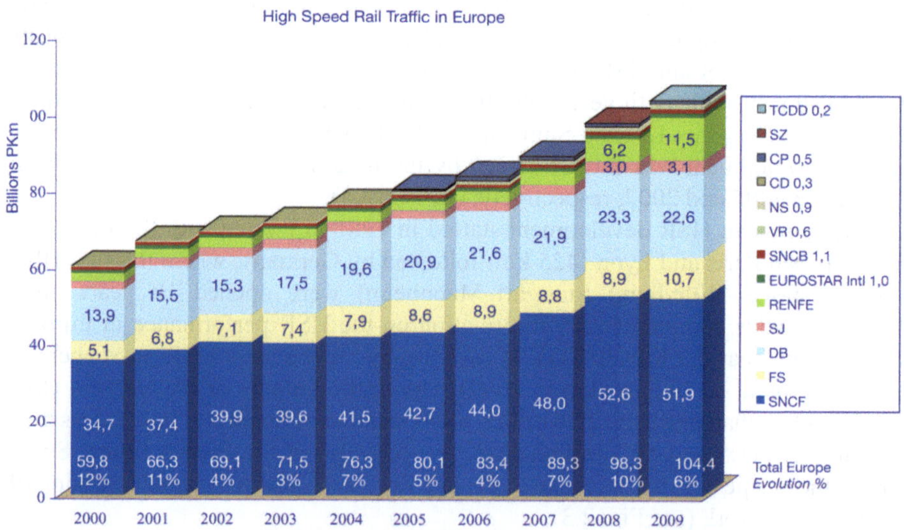

Fig. 2.6 Development of HSR performance in Europe by rail companies. Biggest companies: SNCF (France), DB (Germany), FS (Italy), RENFE (Spain). (Source: UIC 2011)

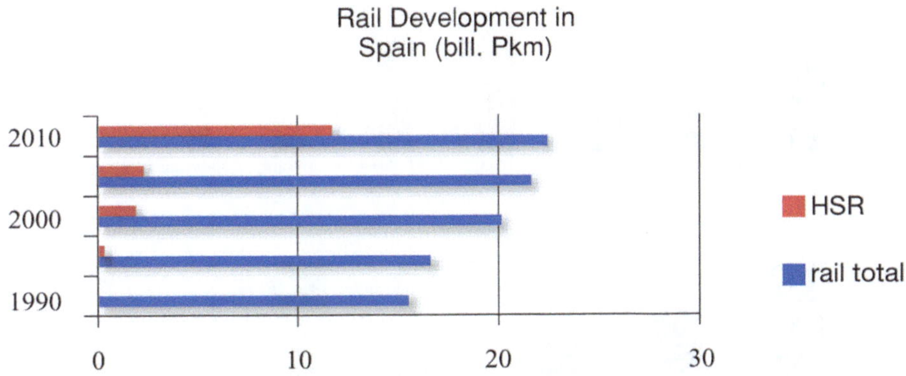

Fig. 2.7 Development of rail passenger transport and HSR in Spain (1990–2010). (Source: EU Commission 2012)

2.2.1.2 Modal Competition in Intercity Transport

The examples in the EU and other countries show the possibility of modal shift from airlines to railways in intercity transport or cooperation between airlines and railways. Figure 2.8 exhibits that the HSR link Madrid/Seville (471 km, opened in 1991) generated a remarkable modal shift from air to rail within 10 years from 33 % in 1991 to 83.6 % in 2000. This is higher than the HSR share, 72.9 %, between Tokyo and Osaka (515 km) in 2010.

Figure 2.9 gives a comparison of the passenger modal split by distance categories in Japan, the USA and the UK. HSR is dominating the intercity travel patterns in Japan up to a distance of 750 km, while the competition with low-cost carrier

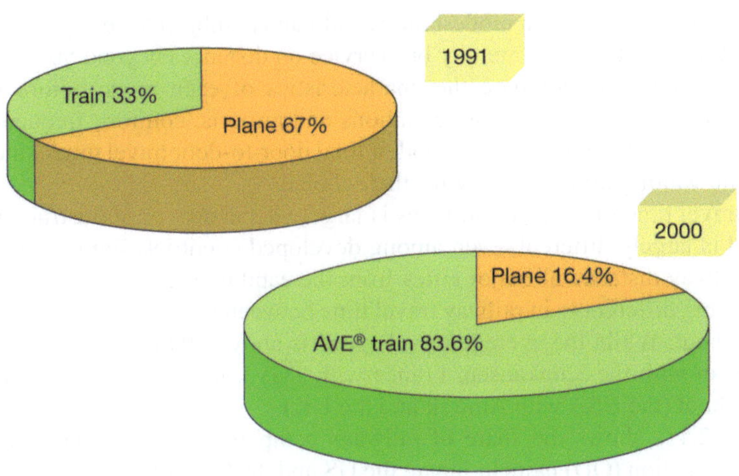

Fig. 2.8 Modal impact of HSR connection between Madrid and Seville within 10 years. (Source: UIC 2010)

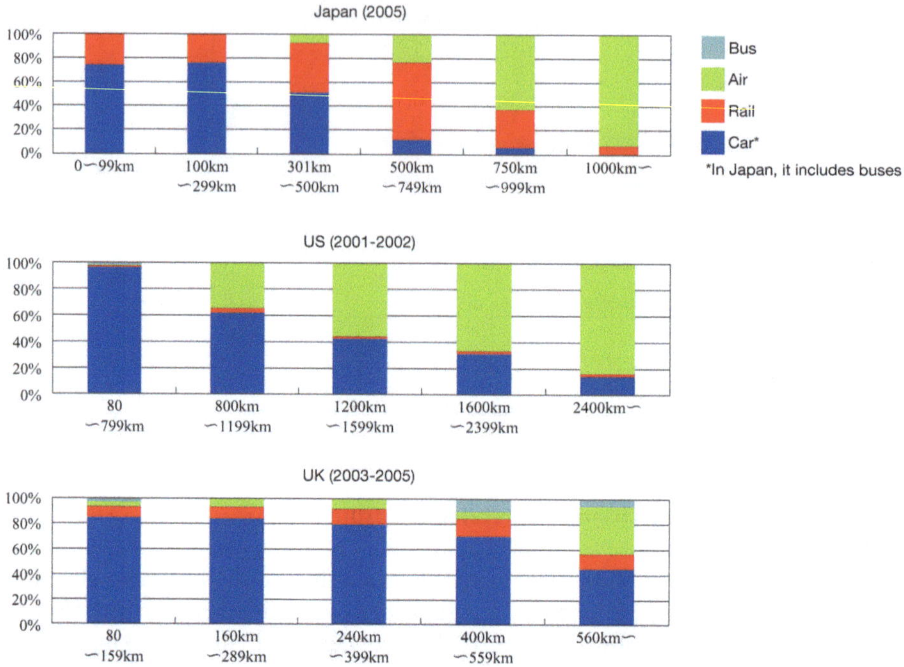

Fig. 2.9 Modal share by travel distance. (Source: Japan—MLIT 2010, USA—US Dept. of Transportation 2001, UK—Dept. of Transportation 2006)

(LCC) on distances between 500 and 1000 km has increased significantly. The US intercity travel is dominated by car and air transport; rail only plays a role in a few corridors like the NEC (Boston–New York–Philadelphia–Baltimore–Washington). Long-distance bus transport has been widely crowded out by LCC. In the UK, intercity rail transport shows a modest modal split and is subject to strong competition from LCC and partly from intercity bus service on distances beyond 500 km.

Intercity travellers compare the characteristics of each mode before making choices: fare, travel time, access to stations and airports, comfort, frequency, etc. Among these attributes of intercity modes, total door-to-door travel time—including access and waiting time—is a key for their choice.

The travel time of intercity railways is largely dependent on infrastructure policies, and is largely different, even among developed countries. Figure 2.10 exhibits the railway distance of major cities from the capital cities in various countries. There are big differences in railway travel time between the countries with HSR and those without. While the average travel speed is above 200 km/h in the countries with (Japan with the Shinkansen, France with TGV), it is just about 100 km/h in those without (the USA with Amtrak and the UK).

Figure 2.11a shows the share of railways compared to air on major domestic origin–destination (OD) pairs in Japan, the US and the UK. Railway shares are high in the regions along the Tokaido/Sanyo Shinkansen lines and low in the regions

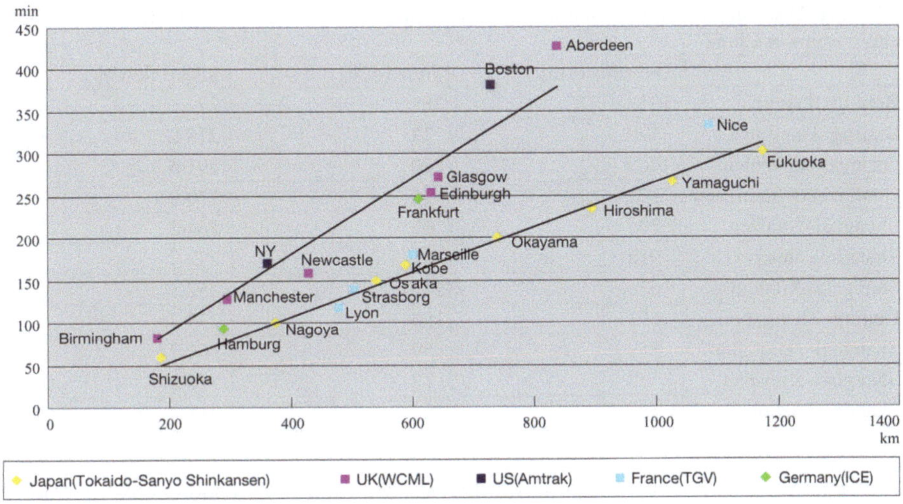

Fig. 2.10 Travel time versus distance from capital city by railway by country. (The figure was prepared according to the timetables of the respective railway companies in 2009)

without Shinkansen. Figure 2.11b shows the travel times by railway and its modal shares for selected OD pairs of Fig. 2.11a. It is possible to anticipate that railways are generally competitive with air when the travel time is no longer than 3–4 h. However, there are other factors such as fare, frequency and the level of service of rail companies and the competing airlines. In countries where LCCs are popular, such as the USA and the UK, the share of air transport is relatively higher than that in Japanese areas without Shinkansen service which indicates that in Japan LCC have not reached the low fare level of the USA and the UK.

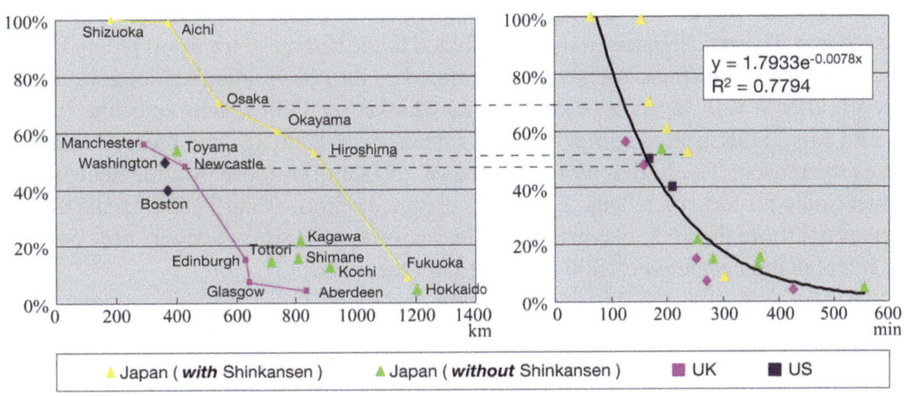

Fig. 2.11 **a** Distance and share of railway for selected OD pairs. **b** Travel time and share of railway (the origin of each journey is Tokyo for Japan, New York for the USA, London for the UK). (Source: Japan—Ministry of Land, Infrastructure, Transportation and Tourism 2009, USA—US Dept. of Transportation 2001, UK—Dept. for Transport 2007a.)

Table 2.1 Distance, travel time, and average speed of trains on selected origin–destination (OD) relationships in China

	Distance (km)	Time (min)	Speed (km/h)
Beijing–Shanghai	1318	288	274.6
Beijing–Tianjing	120	33	218.2
Guangzhou–Wuhan	1069	220	291.5
Guangzhou–Shenzhen	102	36	170.0
Shanghai–Nanjing	295	67	264.2
Shanghai–Jinan	912	201	272.2
Shanghai–Xuzhou	626	149	252.1
Beijing–Zhengzhou	693	150	277.2
Shanghai–Hangzhou	159	59	161.7
Shanghai–Kunshan	50	17	176.5
Shangai–Suzhou	81	23	211.3

The comparison of the Tokaido Shinkansen between Tokyo and Osaka (515 km) and the NEC between Boston and Washington DC (730 km) is a typical example to explain the railway conditions of Japan and the USA. While the fastest Shinkansen train takes 2 h 25 min from Tokyo to Osaka, the Boston–Washington one takes more than 6 h, even with the fastest Acela Express. Although the maximum speed of Acela Express is 240 km/h in some sections, the speed is often restricted due to commuter trains using the NEC. Moreover, the frequency of the Tokaido Shinkansen (between Tokyo and Osaka) is about 330 trains per day, while that of Acela Express is less than 20 trains per day, even on weekdays. The difference in level of service has obviously affected the demand. Accordingly, the number of passengers for the Tokaido Shinkansen is 400,000 per day, while that of the Amtrak NEC is only 32,000.

Despite the similarities in geography and population distribution along the two megacity corridors, these two major railways corridors show remarkable differences with respect to travel speed, frequency, service quality and the resulting number of passengers. These differences may be attributed to the history of transport policy of the two countries. While the USA have concentrated on the construction of motorways and airports after WWII, Japan has developed the Shinkansen HSR system since the 1960s.

In China, HSR is also competitive on distances of 1000 km and more. Table 2.1 gives the reason: The high average speeds of 250 km/h or more provide acceptable travel times for long distances. Even on a distance of more than 1300 km, the travel time is not more than 5 h. Nevertheless, the majority of HSR passengers are counted on travel distances between 300 and 700 km (see Fig. 2.12).

2.2.2 CO_2 Emissions from the Transport Sector

A comparison of the trends of CO_2 emissions from various countries' transport sector, shown in Fig. 2.13a, b, indicates that per capita emission from transport sector is increasing while emission per gross domestic product (GDP) is decreasing. Japan

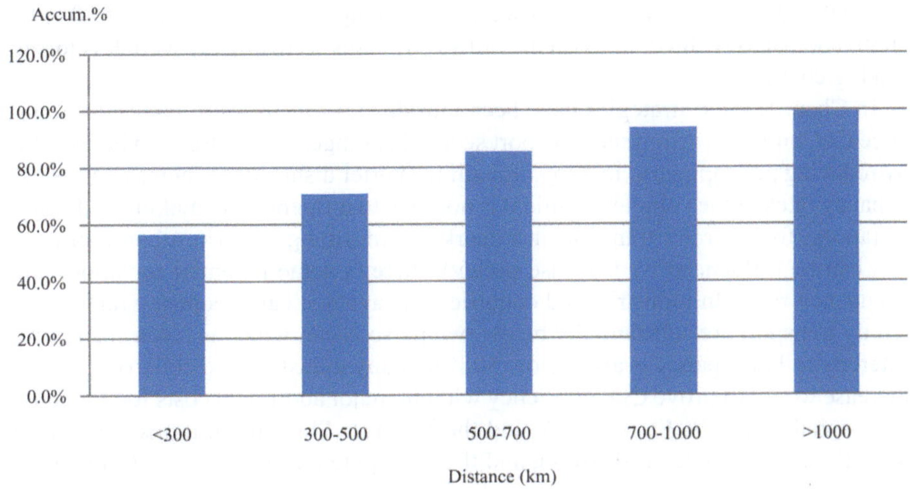

Fig. 2.12 Travel distance distribution from Shanghai Hongqiao HSR station

marks the lowest value in both indices. Improvement of the latter index becomes diminished because of Japan's continuous recession since the early 1990s. The USA and Germany especially improved the index between 1990 and 2006. However, in all the countries, the current trend indicates how difficult it is to realise the 2050 target of reducing GHG by 80% as compared to 1990.

Japan shows the lowest CO_2 emissions per GDP in the transport sector in 1990, but more than one decade of stagnation of the economy and of CO_2 efficiency improvement have reduced the gap between Japan and European countries until 2006. Nevertheless, emissions from the Japanese transport sector are still substantially lower than other countries, due to the highly developed railway systems in urban,

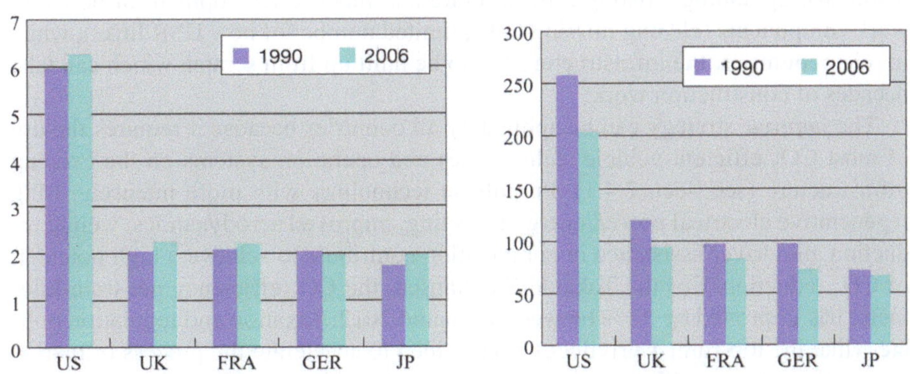

Fig. 2.13 **a** Per capita CO_2 emission from transport sector, including freight transport (tonnes-CO_2/population). **b** Transport CO_2 emission per unit GDP (thousand tonnes-CO_2/GDP in US$). (Data source: OECD/ITF 2011)

regional and intercity transport. However, achieving a reduction target of 80 % by 2050, compared with 1990, when the railway systems were already well developed, is a big challenge.

In Chap. 1, three strategies have been introduced—*avoid, shift, improve*—to reduce CO_2 emissions from the transport sector. Passenger-kilometres can be *avoided* by reducing per capita trip making, travelling shorter distances or increasing the occupancy rates of the vehicles. While it is not easy to influence trip making and travel distances[1] for intercity transport in a market-conforming way (in urban transport, this can be influenced by land-use policy), there is some potential for improving occupancy rates. In industrialised countries, the average car occupancy rates in intercity transport are substantially below two persons/car, which is lower than 50 %. Intercity rail occupancy is also below 50 % in many industrialised countries, and its increase would improve CO_2 efficiency without major additional costs for operation and capital. Only the French TGV and the Japanese Shinkansen trains have much higher occupancy rates (70–80 %), and thus, the potential for the *avoid* strategy in intercity transport in these countries is limited.

The *shift* strategy applies to all countries which have—or are on the way to develop—a well-developed railway system. Other countries like the USA, which have neglected passenger railway development for a long time, require huge infrastructure investments if they wish to build up a competitive railway system. Therefore, even under optimistic assumptions on economic development and funding prospects, only a few corridors might be developed for HSR service in the coming decades. In emerging economies with higher economic growth rates and a rapid development of the transport demand—for instance, India or Brazil—the chances seem better to install HSR systems on major transport corridors between mega-agglomerations. But it is not probable that such large emerging economies to follow the example of China and build up an HSR network of 10,000 km and more within a short period of time. Therefore, the *shift* strategy will need different speeds of implementation: In those countries with well-developed railway systems, it can start now by taking railway-friendly policy actions which includes operational adjustments, upgrading existing infrastructure and modest development of new network components (closing missing links, limited number of new HSR links), while in other countries, the infrastructure has to be built up from scratch which can take decades of construction work.

The *improve* strategy can be applied by all countries because it requires the use of most CO_2 efficient vehicle technologies and operation systems on the existing infrastructure (see Sect. 2.4.1). Propulsion technology with more intensely using regenerative electrical power, energy recycling, improved aerodynamics, weight reduction and driver assistance in combination contribute to achieve a high potential of CO_2 reduction. For the Tokaido Shinkansen, the CO_2 efficiency per train-kilometre has improved by 32 % between 1990 and 2012. Taxation and regulation policies (like the top-runner privilege) have helped to accelerate the process of market penetration of most efficient technologies.

[1] It has often been argued in the literature that telecommunication could substitute physical transport, but in reality, it has added to the communication options of people and not reduced transport activities.

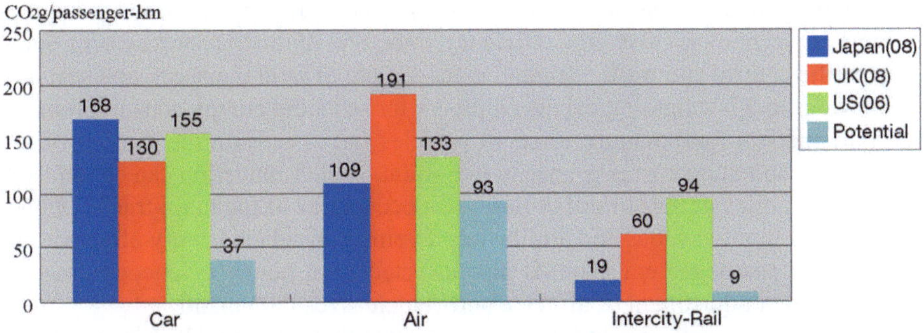

Fig. 2.14 CO_2 emissions by transport mode. We assume the potential cases as follows: driving at 15.4 km/l and four people riding in a car (which is the Japanese regulation of fuel efficiency by 2010); flying with a B777–200 between Tokyo–Osaka (600 km) for air; travelling in an N–700-type train between Tokyo–Osaka for intercity rail. For intercity rail, we use all JR trains for Japan, National Rail for the UK and Amtrak for the USA. The authors calculated CO_2 emission according to the data of "Energy Intensity of Transportation Mode" and a standard CO_2 emission/BTU of each energy source. (Source: Japan—Ministry of Land, Infrastructure, Transport and Tourism 2008a, USA—US Dept. of Transportation (n.d.), UK—Dept. for Transport 2008a)

The three strategies are exploited to differing degrees by different countries of the world and by different transport modes. In Fig. 2.14, we compare the CO_2 emission per passenger-kilometre by mode and country. The same figure shows the recent per-seat emissions of cars and airplanes in order to indicate the potential efficiency gain for each mode.

The current value for each mode and country is compared with the potential value achievable if the most efficient technologies and operation schemes, which are currently available, were to be employed. For passenger cars, Japan is the least-efficient country since just 1.4 people occupy a vehicle on average, while the UK and the USA are relatively more efficient, since their average number of passengers per vehicle is 1.6 and 1.7, respectively. The CO_2 intensity of air transport is higher in the UK and the USA, since many small airplanes are used in short distance routes in those countries. The difference from the potential value is smaller in this sector because the load factor of most main routes is more than 70 %, and because airlines have worked towards the renewal of airplanes in order to reduce fuel consumption and operating cost. Finally, we will find large differences in efficiency of railways between countries due to the different infrastructure, vehicle technologies and load factors. While there are a number of trunk lines in the UK and the USA that are still not electrified and use old diesel cars, Japanese railway companies are actively introducing new technologies such as regenerating brakes.

Furthermore, although the data are not strictly comparable between countries, the Japanese total average load factor of railways is about 50 %, while those of the UK and the USA are less than 40 %. On the other hand, the coaches (intercity buses) in the USA achieve a high CO_2 emission efficiency (50 g/pkm) because of their high load factor, but intercity buses have been massively crowded out by LCC.

From the above discussion, differences in efficiency of CO_2 emission are mainly attributed to technological energy efficiency, operation of the transport system, such as air traffic control and traffic demand management of road transport, and seat occupancy. Although technological development for reducing energy consumption by transport operators and manufacturers of vehicles and vessels for the purpose of reducing their operation cost is primarily important, further reduction can be induced by adequate policy actions. Furthermore, seat occupancy of the mass-transit public transport (PT), railway, bus and air, the improvement of which is very effective for reducing per-passenger emission, is also affected by a policy of encouraging an energy efficient adjustment of airplane and vehicle sizes to demand.

Bearing these points in mind, one can conclude that intercity HSR is an important component of a transport network, which contributes to CO_2 reduction in the passenger transport sector. HSR can be the backbone of transport between major agglomerations. But it will be necessary to integrate HSR in an intermodal transport system in which all modes are integrated according to their optimal functionality with respect to energy and CO_2 savings under viable economic conditions. Seamless interconnections with regional and urban PT, car traffic (park and ride) and air transport are as important as high-speed performances on route. The extension of HSR networks is a matter requiring careful long-term strategic planning, including all aspects of demand development, socioeconomic impacts, regional policy, financial issues and global warming.

2.3 Policies and Planning Systems to Reduce GHG in Intercity Transport

This section focuses on policies and planning systems of intercity transport to tackle climate change. Since the 1990s, several countries worldwide have started to adapt their policies and planning systems of intercity transport to address climate change issues.

This section covers consideration of GHG in transport policies (Sect. 2.3.1) and planning systems (Sect. 2.3.2) related to intercity transport, and appraisal of GHG impacts in the process of intercity transport planning (Sect. 2.3.2) in Europe, the USA, Japan and China.

2.3.1 Intercity Transport Policies to Reduce GHG

2.3.1.1 Intercity Transport Policies in Europe

1. The European Union

The European Commission (EC) published its first white paper (Commission of the European Communities 1992) on the future development of a common transport policy in December 1992. Furthermore, the Trans-European Networks (TEN) were introduced by the Maastricht Treaty in 1993—which established the legal framework for a single European market with free movement of people, goods and capital—and guidelines (European Parliament, Council of the European Union 1996) published in 1996. The TEN have included Trans-European Network for Transport (TEN-T) as one of its three network policies (the other two are for energy and communication), which has aimed at integrating the transport network and developing an efficient transport infrastructure for the sake of economic and social cohesion as well as enhancing competitiveness. In the 1990s, however, the transport policies of the EU has put lower emphasis on climate change than on fostering social cohesion and enhancing competitiveness by developing an efficient transport network.

In 2001, the EC published a new white paper, "European Transport Policy for 2010: Time to Decide" (EC 2001), which stressed the need for a balance between different modes of transport to protect the environment. However, although the white paper considered all sustainability issues—economic, social and environmental—it did not dedicatedly emphasise on the impacts on climate change. The recently published white paper of 2011 "Road Map to a Single European Transport Area—White Paper on Competitive and Sustainable Transport" (EC 2011) includes explicit CO_2 reduction targets for the transport sector (−60% by 2050 compared with 1990) and a number of policy actions (infrastructure, regulation, pricing) to guide the transport development to a sustainable path.

The EU transport infrastructure policy is based on the TEN-T concept, which serves as a guideline for the member countries. This means that the TEN-T is not obligatory, but the member states will only get co-funding from the EU if a project is part of the TEN-T and fulfils the EU requirements. In the case of railways, these requirements are stated in the so-called railway packages which aim at unifying the fragmented European railway technology ("interoperability issue"), introducing "intramodal" competition on the railway networks and making the organisations more competitive compared with road and air transport. A TEN-T impact assessment was conducted in 2004 (European Parliament, Council of the European Union 2004) for the revision of the TEN-T guideline published in 1996. This assessment presented an evaluation of CO_2 emissions from intercity transport in Europe and suggested a 17-million t reduction per year under the new transport policy. Meanwhile, the TEN-T has been revised for the third time by the end of 2011 and consists of a core network and a comprehensive network (see Fig. 2.15). The backbone of the core network is represented by nine "core network corridors".[2]

A further development of the HSR network is an integral part of TEN-T. In the "High-speed Europe" document of the EC (2010), most ambitious prospects for HSR development are announced which foresee up to 30,000 km of HSR in the EU using interoperable technology (ERTMS) and operated by commercial companies under intramodal competition (open network access). The overall savings of CO_2 are estimated to be 34 million t per annum after the core network will be completed in 2030.

[2] The EU has introduced a cap-and-trade system for CO_2 in 2005, which includes many industrial and power-generating sectors but not international aviation and maritime shipping because of international law.

Fig. 2.15 Nine core network corridors of TEN-T. (Source: European Commission 2013)

The 2009 green paper "TEN-T policy review: Towards a better-integrated trans-European transport network at the service of the common transport policy" refers to the importance of TEN-T's contribution to solving the problem of climate change. Thus, climate change targets have been explicitly mentioned in the EU transport policy documents since around 10 years. They prepared the basis for the white paper of 2011 (EC 2011) which sets explicit targets for CO_2 reduction by 2050 (−60 % for transport) and defines actions to be taken for achieving these targets (e.g. dedicated shift strategies from road and air to rail and waterways).

While the EC is favouring railways and inland waterways through their TEN-T policy, it tries to introduce fair conditions for competition between the transport modes. It is remarkable in this context that international air and maritime transport do not pay national fuel or excise taxes because of international conventions. Railways, on the other hand, have to pay energy and value-added taxes, either partly or full (like in Germany). Against this background, the EC has proposed in 2008 to include emissions from aviation in the EU Emission Trading System (EU-ETS),[3] and the EU Parliament as well as the EU Council adopted a corresponding directive. After heavy protests, in particular from the USA and China (the latter announcing to suspend Airbus orders), the EC has decided to postpone the implementation of the directive and to prepare a concerted action with the International Civil Aviation Organisation (ICAO). ICAO has meanwhile decided to agree on an international Emission Trading System (ETS) to be introduced not before 2020. The conditions of such a worldwide ETS are left open, while ICAO has definitively decided that isolated actions of countries or country unions before 2020 are not allowed.

While it seems to be very hard to implement direct CO_2 policies towards air carriers and maritime shippers, the countries have the right to impose taxes on air passengers and freight movement in ports. Germany, for instance, has introduced an air passenger tax in January 2011 to raise more revenues for the public budget in a difficult fiscal situation. Although the tax is not called a "climate tax", it has an appropriate effect. The possibilities of taxation and emission trading are discussed further in Chap. 3.

2. United Kingdom

The UK has been particularly advanced in dealing with the problem of climate change among the EU member countries. In the UK, the central government provides guiding policies for transport and planning through documents, such as white papers for regional and local governments, which in turn develop specific plans and projects in line with the national government's direction.

The white paper "New Deal for Transport: Better for Everyone" (Department for Transport 1998) was published in 1998. It sets out the fundamental direction of transport policy which has relevance throughout the UK, and promoted an integrated transport policy to contribute to the quality of life:

1. Integration within and between different types of transport
2. Integration with the environment
3. Integration with land-use planning
4. Integration with policies for education, health and wealth creation

In response to the white paper, the Transport Act 2000 was enacted in 2000, which introduced policy measures such as local transport plans (LTPs) and road user charging. Furthermore, the action programme "The Ten Year Transport Plan" (Department for Transport 2000) was published in the same year. This programme,

[3] Originally ten corridors had been defined in 2011 which were regrouped to nine corridors in 2013.

which would deliver the scale of resources required to put integrated transport into practice from 2001 to 2010, ranked climate change as one of the key challenges for the next 10 years. However, climate change was not treated as the main political issue but as one of the issues regarding the impacts of transport in the context of reduction of CO_2 at that point in time. Since the Stern Review (HM TREASURY 2006), developed under the leadership of Sir Nicholas Stern, was published in 2006, transport's impacts on climate change have been considered stronger than ever.

The Stern Review pointed out that the benefits of strong and early action on climate change far outweigh the economic costs of not acting, as well as that the costs of stabilising the climate are significant but manageable. In response to the Stern Review, the Department for Transport (DfT) published "Towards a Sustainable Transport System" (Department for Transport 2007b) in 2007 and "Delivering a Sustainable Transport System" (Department for Transport 2008b) in 2008. In these documents, reducing transport's emissions of CO_2 and other GHGs is set as one of the five main goals, and short- and long-term strategies are proposed for tackling climate change.

Moreover, the Climate Change Bill was introduced in the Parliament on November 14, 2007 and adopted on November 26, 2008 as the Climate Change Act. It sets out a target that the net UK carbon account for 2050 is at least 80 % lower than the 1990 baseline. It also introduces "carbon budgets" which is legally binding and limits the UK emissions over successive 5-year periods. The DfT published "Low Carbon Transport: A Greener Future" (HM Government 2009a) in July 2009 and set out the strategies required to meet the requirement of carbon budgets under the Climate Change Act.

The impacts on climate change have been considered even more strongly in the transport sector, and railway transport has been discussed favourably since it has been a transport mode with less impact on the environment. In the UK especially, the intercity railway network has been improved in recent years. The West Coast Main Line has been modernised such as with high-speed improvement. In January 2009, the government published a document that proposed a new HSR plan (Department for Transport 2010) and established High Speed Two (HS2) Ltd to develop the project. In March 2010, an HS2 report was published with the government policy for the next 20–30 years. The new HSR network is designed with a Y-shaped route from London to Birmingham and the two spurs, one to Manchester and the other to Leeds via the East Midlands shown in Fig. 2.16. The maximum speed of this route would be 250 mph (400 km/h), which would make it possible to move from Leeds to London in 75 min. The new Conservative–Liberal Democrat coalition government that came to power in May 2010 basically inherited the HSR plan, other than high-speed link to Heathrow.

Thus, the UK government under its broad agenda of promoting low-carbon world is increasingly considering climate change in the context of the transport system.

5. France

France is one of the leading nations to take policy initiatives for environmental issues. A conference, "Le Grenelle de l'environnement", was organised in France in

Fig. 2.16 British new high-speed railway network. (Source: Department for Transport, High Speed Rail 2010)

July 2007, inviting participants such as representatives of national and local government, industry, and non-governmental organisations. This conference organised six working groups devoted to the themes of climate change and energy, biodiversity and natural resources, and in these areas, set concrete action programmes decided by consensus between participants. In the conference, action plans were announced by the working groups in many areas such as housing and transport, for the purpose of a 75 % reduction of GHGs by 2050. The results of the conference were reflected in a bill, "Projet de loi de programme relative a la mise en oeuvre du Grenelle de l'environnement (Grenelle I)", adopted by the National Assembly in October 2008 and by the Senate in February 2009.

Transport measures specified in the law include investment of 16 billion € to construct 2000 km of HSR until 2020, extension of the tramway network from an existing 329 km to over 1500 km within 15 years, extension of the railway network in France and modal shift of freight.

Schéma national des infrastructures de transport (SNIT) was introduced as a new national transport plan by the law of Grenelle. The scheme sets the orientation of environmentally relevant public policies for the next 20–30 years to come, including

Fig. 2.17 Existing (*blue*) and planned French HSR network

modernisation and maintenance of infrastructure networks as well as reducing the impact of these networks on the environment. The draft version of the scheme was published in 2011, which stresses on the strategies that focus on alternative modes of transport to road and air transport, including HSR. The development plans for the French HSR network are shown in Fig. 2.17.

6. Germany

The development of German transport infrastructure systems after the WWII was dominated by the different regimes in the West (Federal Republic of Germany) and East (German Democratic Republic). After the political changes in 1989, the German unification was realised on October 3, 1990. Major decisions were taken to integrate the transport sector into the national strategy of fast economic recovery of East Germany:

- Seventeen transportation projects "German Unification" were decided on, nine for the railways, seven for the motorways and one for inland waterways. Reason was that the condition of transport infrastructure in East Germany was desolate because of missing rehabilitation and modernisation of the networks.
- A first transport master plan for the unified Germany was prepared following the principles of master plan development ("Bundesverkehrswegeplan, BVWP").
- Basic reforms were prepared such as the railway reform to unify the West German, East German and Berlin railway systems and privatise the—still state-owned—company under the name Deutsche Bahn AG.

The BVWP considers environmental aspects and CO_2 emissions, but does not include a strategic objective for CO_2 reduction (see the following section). A major revision of the BVWP is foreseen for 2015. The Ministry of Transport has published the basic principles for the new concept, which are:

1. Evaluation of network configurations in addition to the presently applied evaluations of single projects.
2. Comprehensive strategic environmental assessment (SEA; on a network base).
3. Consideration of alternatives, in particular for large investments.
4. Consideration of barriers to implementation in the priority setting.
5. Consideration of risk and uncertainty.
6. Realistic coordination with financial resources under consideration of increased share of user finance.

The integration of the BVWP into the political process will be explained in the following section. As the plan development and its implementation require the cooperation between all political levels, the process is complex and does not always lead to the most efficient outcomes, as for instance

- The planning processes take a long time, partly more than two decades for important projects like the HSR link Cologne–Frankfurt.
- High-cost overruns between first cost estimations and final cost realisations are not the exception but the rule.
- Many parties are involved and compromises have to be made not based on rationality from an economic point of view (e.g. additional stops for the HSR line Cologne–Frankfurt at small cities like Montabaur and Limburg).
- Possibilities of concerned stakeholders to influence the project design until the legal approval of the project.

Figure 2.18 shows the existing and planned HSR network in Germany, including all links which allow for speeds higher than 200 km/h. Most links are designed for a speed up to 250 km/h, while there are only two links allowing for a maximum speed of 300 km/h (Cologne–Frankfurt; Ingolstadt–Munich).

The railway projects, which are presently discussed most controversially, are

- Stuttgart 21 and HSR Stuttgart–Ulm
- Rhine valley Karlsruhe–Basel

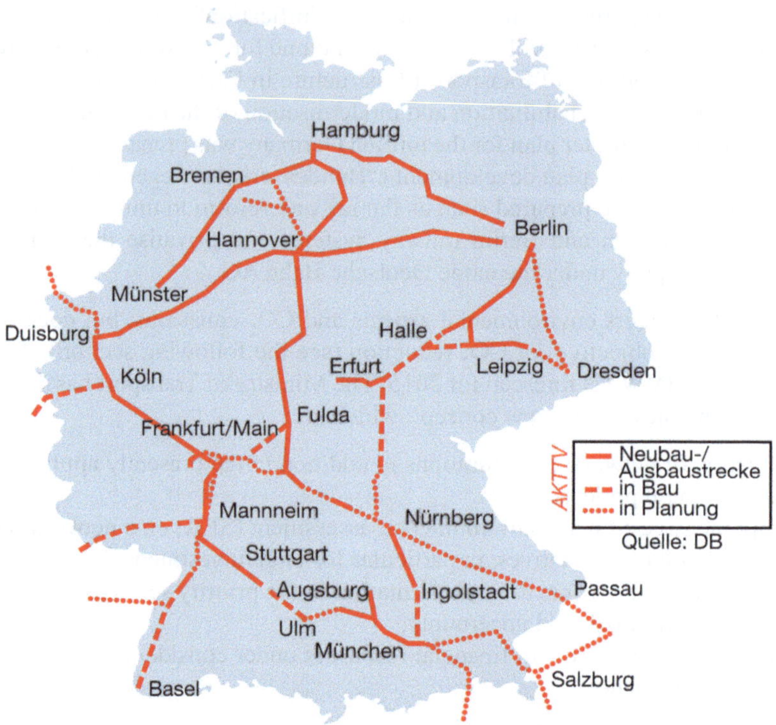

Fig. 2.18 HSR network in Germany (existing new/upgraded links, links under construction and planned projects)

The Stuttgart 21 project includes the abolition of the present dead-end station Stuttgart Central and its replacement by a through-bound underground station. As this includes a redesign of many links connecting the central station, partly HSR, partly regional, partly connection to the airport with a new airport station, the project is very complex and touches a very critical geological area. Therefore, the project was doubted already in the early phase of development by the mid-1990s, and led to open protest movements in 2010 with up to 60,000 participants and violent protest actions. After a phase of moderation, the decision was taken to continue with the project, which was confirmed by a referendum in 2011. Nevertheless, the project is still regarded very critical because of the increase of estimated costs, which jumped from 3 billion € in 2009 to 6.8 billion € in 2013.

The Rhine valley project Karlsruhe–Basel is planned to achieve two goals: First, a fast HSR connection on Frankfurt–Karlsruhe–Basel–Zurich/Italy, and second, to increase the capacity of freight train services alongside the most heavily used freight railway link Rotterdam–Basel–northern Italy (Geneva). The link would establish a connection to the Gotthard Tunnel in Switzerland and create a most important transalpine railway route. People in the area of Offenburg (between Karlsruhe and Freiburg) complain about the expected additional railway noise and interrupt the planning process by launching legal complaints. In this case, the alternatives for the railway company and the political planners were to wait for the judgements of courts or to redesign the guideway, e.g. to plan very expensive tunnels and noise protection walls in the area concerned.

These examples are to illustrate that people in Germany are in favour of environmental progress, fighting against climate change and fostering traffic diversion towards the railways. But as soon as they are concerned by the negative impacts of railway, primarily the railway noise, they establish powerful movements against more railway transport through their regions. The Rhine corridor between Cologne/Bonn and Basel is meanwhile the corridor with the most powerful anti-railway noise movement in Europe.

2.3.1.2 Intercity Transport Policies in the USA

In 1999, the Department of Transportation established the Centre for Climate Change and Environmental Forecasting to address issues associated with climate change. The Centre introduced the "Strategic Plan 2006–2010" which proposed its vision: "A transportation system that minimises greenhouse gas emissions and is prepared for the impacts of climate change" and three related goals. Although the plan has not given concrete quantitative indicators, it has proposed reducing GHG emissions. But it has not been correlated to intercity transportation policy. And in 2008, the Federal Highway Administration compiled a report titled "Integrating Climate Change into the Transportation Planning Process—Final Report", summarising advanced cases in which states or metropolitan planning organisations had treated climate change in the transport planning processes, especially in long-range plans. This seems to imply that the federal government has recognised the importance of such approaches.

Under the Obama administration since 2009, the US government has shown more concern for climate change. In June 2009, the legislation titled "American Clean Energy and Security Act of 2009" was passed by the House of Representatives. This legislation would require the USA to reduce GHG emissions by 17% from 2005 levels by 2020. President Obama declared the same target at the United Nations Climate Change Conference 2009 (COP15). However, the legislation has not referred to reduction targets in specific areas, including transport.

In April 2009, President Obama released his strategic plan for HSR in the USA, identifying ten potential corridors for federal funding. Those lines are California, the Pacific Northwest, South Central, Gulf Coast, Chicago Hub Network, Florida, Southeast, Keystone, Empire and northern New England (see Fig. 2.19). It may also include the NEC running from Washington to Boston. The budget for the railway service will be US$ 8 billion from the American Recovery and Reinvestment Act of 2009 enacted in February 2009 and US$ 1 billion per year for 5 years requested in the president's budget. However, the vision of the HSR aims mainly to expand employment or stimulate economies and is not linked clearly to consideration for climate change or reduction of GHG emissions.

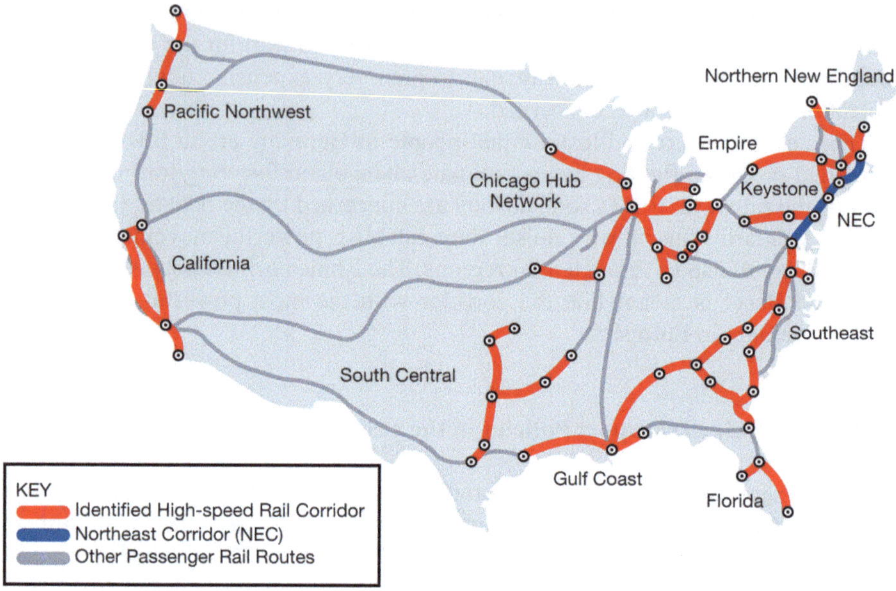

Fig. 2.19 Vision for high-speed rail in the USA. (Source: The White House 2009)*

Policies in the state of California

Contrasting the federal government, some states have been proactive on the problem of climate change. In particular, the states of California and New York have developed ambitious plans for reducing GHG emissions.

For example, California's intention of reducing GHG emissions was declared in Assembly Bill No. 32 (AB32), the "Global Warming Solution Act", passed in 2006. In concrete terms, it clearly has specified reduction of state-wide GHG emission levels to those of 1990 by 2020. Under such a concrete target, Senate Bill No. 375 (SB375) was enacted in 2008. It has been a state law, which aims at a new community design to reduce GHG emissions. It has tried to control GHG emissions from vehicles by reining in urban sprawl, and expects reduction of GHG emissions through land-use planning and its consistency with transport planning. SB375 has required each MPO to include a sustainable communities strategy (SCS) in its regional transportation plan. SCS is a strategy to clarify how each region achieves the reduction target of GHG emissions that the California Air Resource Board has established for the region. If the region judges that it is hard to achieve the target with SCS, the region is required to develop an alternative planning strategy, which specifies difficulties, feasible development patterns, means and policies. In this way, in the state of California, it is being institutionalised to treat reduction of GHG emissions in transport planning.

* The Figure is created based on the White House at http://www.whitehouse.gov/blog/09/04/16/A-Vision-for-High-Speed-Rail/ then some changes were made.

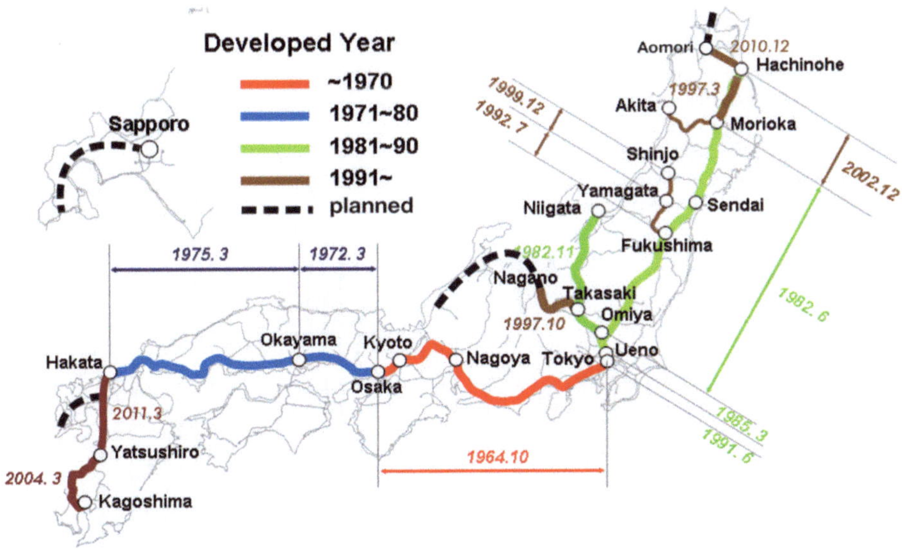

Fig. 2.20 Development process of Shinkansen

The USA is now developing provisions regarding global warming relevant to intercity transport policy. Some proactive states such as California address the problem as a state-wide one, but there are no references to concrete intercity transport policy. California HSR is, in this sense, only one of the recommended measures to reduce GHG emissions, but not an intercity transport policy.

2.3.1.3 Intercity Transport Policies in Japan

Despite the worldwide trend of motorisation and increasing demand for high-speed intercity transport, Japanese intercity railways have sustained competitiveness with road and air transport due to the development of the Shinkansen. We review and analyse the history and features of intercity transport of Japan from various angles.

1. History of Japanese intercity transport

The Tokaido Shinkansen, the first Shinkansen line in Japan linking Tokyo and Osaka, was completed in 1964. In those days, the frequency was two train services per hour and the travel time between Tokyo and Osaka was 4 h. In the subsequent 50 years, Shinkansen lines connecting major cities in the Pacific coastal area, such as the Sanyo (Osaka–Hakata) and Tohoku (Tokyo–Aomori) Shinkansen (see Fig. 2.20), were developed. At the same time, the speed of existing lines was significantly improved to the extent that the current fastest train (Nozomi) connecting Tokyo and Osaka takes just 2 h 25 min.

Although there is a gradual increase in the share of air transport, especially for long-distance trips where high speed is more important (Fig. 2.21), newly devel-

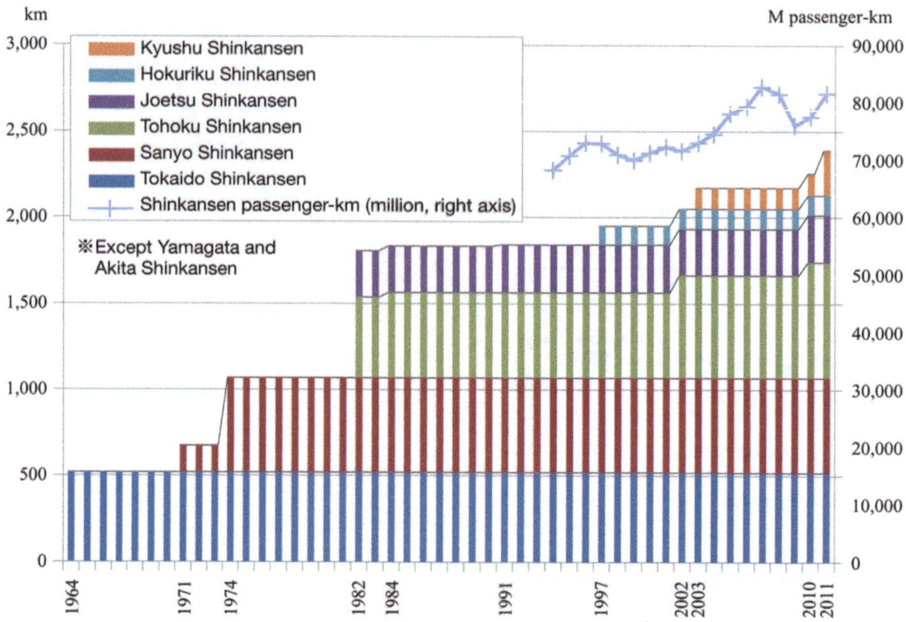

Fig. 2.21 Changes in operation-kilometres and passenger-kilometres of high-speed railways (Shinkansen) in Japan

oped Shinkansen lines have attracted a certain market share. Some domestic flights such as Tokyo–Nagoya or Tokyo–Sendai were actually abandoned in the past. Even today, the railway companies have continued the reform since the privatisation and separation of the Japanese National Railways in 1987 in order to maintain their competitiveness with the other modes. In 2003, the Tokaido–Sanyo Shinkansen reduced the special fare of the *Nozomi* (the fastest Shinkansen) super-express train, and increased the frequency in order to enhance its competitiveness.

Railway networks in Japan, excluding Shinkansen lines and those in the metropolitan areas, are almost completed by the early Showa era (the 1930s). Major regions all over Japan were covered by the Japan National Railways (JNR), and interregional networks were established. Moreover, as for intraregional transport, including feeders to JNR's stations, private railways were developed by private businesses. Together with JNR, such private railways became the mainstream of passenger and freight transport in Japan, largely contributing to development of local communities. This point is the unique characteristic that other countries do not have.

Under such circumstances, the rail passenger transport services are developed by railways because the benefits of higher efficiency and speed of railways became obvious in intercity transport, exceeding the cost of infrastructure construction by far in the densely populated corridors. Japan's Shinkansen lines that were established ahead of all others in the world lead motorisation and air passenger transportation. Though competition with airliners has been intensifying, the increase in speed of the Shinkansen trains has maintained its competitiveness against other transport modes. As a result, railways are the much noticeable mode for middle distances between 300 and 750 km and enjoy more than 60 % of passenger share in the distance range between 500 and 749 km (see Fig. 2.9). Construction of intraregional and interregional railways was promoted in Japan initially for economic reasons. Railway's contribution to environment by lowering CO_2 emissions per person transported as compared to car and air modes was not an important reason at that time why railways were developed in Japan.

Particularly, the extension of the Shinkansen lines has played a major role in intercity transport in Japan. Figure 2.21 shows the changes in passenger-kilometres that represent the network kilometres and passenger-kilometres of the Shinkansen lines. The increase in trains on Shinkansen line allowed for increased frequency of service, which also contributed to traffic growth. Although the overall CO_2 emissions on the Shinkansen lines have increased, this is partly due to the additional rail services substituting air transport such that the total CO_2 footprint could be reduced. For instance, the air route between Haneda and Sendai was abandoned after Shinkansen lines started commercial operation to Sendai. Assuming that all passengers on flights between Haneda and Sendai were shifted to the Shinkansen lines, CO_2 emissions are estimated to have dropped by about 19,000 t per year.

Although there are some examples for a traffic diversion from air to rail, one has to notice that a competitive aviation market has emerged due to globalisation and the deregulation of air transport which has been introduced since the 1980s. Double tracking of air services was permitted in 1986, and discount fares were allowed gradually since 1996. Also, Japan's air transport regulation was liberalised significantly in 2000, and subsequently, airfares were basically deregulated. Then, the discount airfares and LCCs have made air transport fairly competitive vis-à-vis railways. The impacts of increased competition led to lower rail passenger shares as shown in Fig. 2.22.

On the other hand, the abandonment of low-density local air routes has become a problem. In order to maintain reasonable level of air services on local air routes, the requirements for maintaining the minimum daily flights between Tokyo (Haneda Airport) and some local airports have been set up. Furthermore, in the 2000s, the national government allocated the number of departure and arrival slots at Tokyo Haneda Airport, which was significantly lower than the demand, to airlines taking into account of their fare discounts, safety and maintenance of service frequencies in local routes. While these policies have achieved a balance between efficiency and maintenance of minimum services, they do not explicitly consider parallel services provided by railways before the expansion of Haneda Airport runways since 2010. What is required is a multimodal review of the whole intercity transport network of Japan with a long-term and multimodal point of view.

2. Factors in choosing intercity transport modes

The most significant factor affecting a choice of intercity transport mode is time of travel. As we have seen above, railways are overwhelmingly dominant over other transport modes in an OD pair if its travel time is less than 3 h. However, various

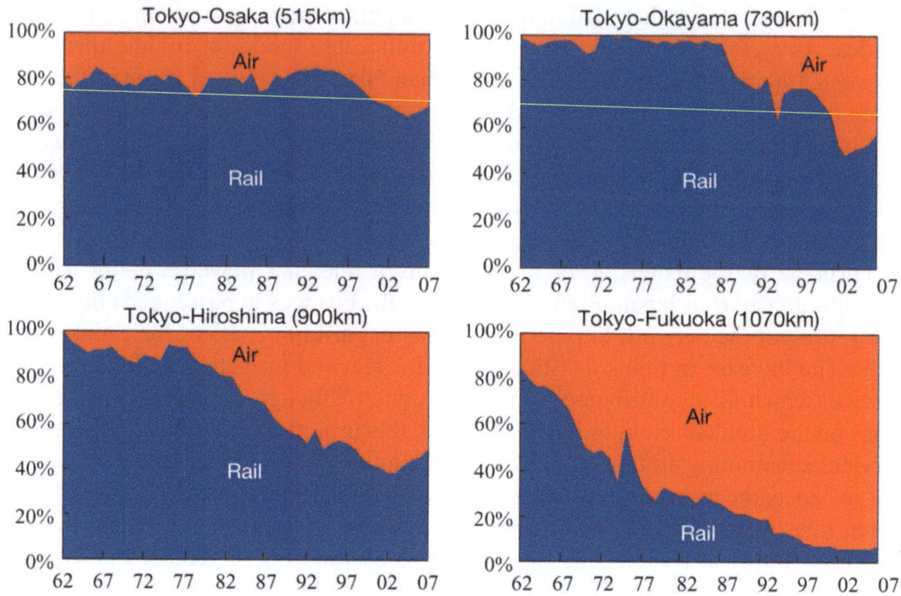

Fig. 2.22 Trends in number of passengers (Tokaido–Sanyo Shinkansen)

other factors such as the service level of competing airlines, time of intermodal exchange and access cost actually affect the modal choice.

Figure 2.23 shows the modal share of three OD pairs with a distance of 900 km by purpose of travel in 2005. Air significantly dominates for Aomori–Osaka, as the

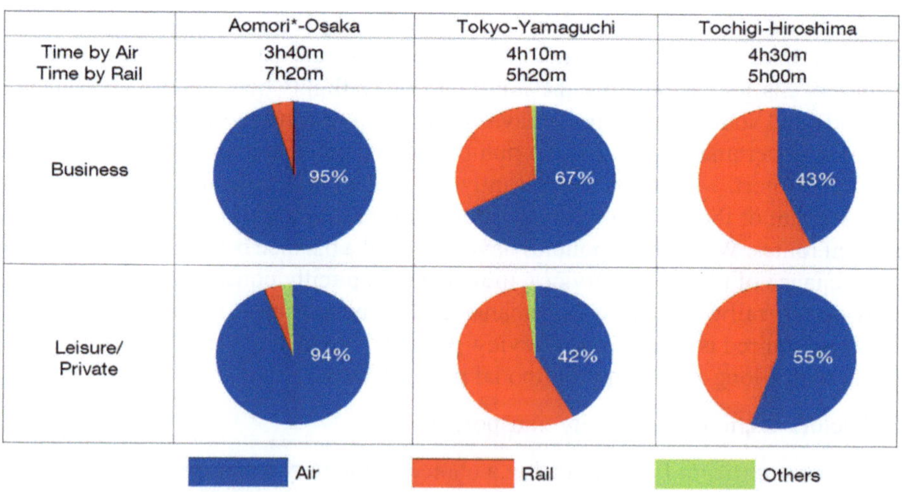

*Without station of Shinkansen in 2005

Fig. 2.23 Modal share by purpose of trip (*without station of Shinkansen in 2005). (Source: MLIT 2008)

Shinkansen was not developed in 2005 and a travel by railways took a longer time than by the air by 2 h 40 min. Other two OD pairs are connected by Shinkansen where the travel time difference is just about 1 h between the two modes, and the railways have relatively larger shares in these OD pairs.

More business travellers prefer to use air transport than leisure travellers because of the difference in value of time between the two groups. However, when the access to airport is inconvenient, a higher railway share of business trips than other trips is often observed. Among 404 OD pairs with enough samples in Fig. 2.23, 145 OD pairs (33 % of the total) show significantly higher shares of air transport in business trips than in private/leisure trips. The counterintuitive result is observed for 22 OD pairs, for all of which travellers must endure long access trips to airports located outside the prefectures.[4] [5]

In Japan, the competitive strength of railway is limited to the HSR corridors, and there is a large variation of ridership between conventional railway and HSR even for similar travel distances. In fact, the figure for the railway share is much influenced by HSR service. Also, the use of different modes for intercity travel depends on the trip purpose, such as business, tourism or private trips (see Fig. 2.24).

2.3.1.4 Intercity Transport Policy in China

During the 2005–2010 period, China accelerated construction of the integrated transport network corridors which were composed of the railways, highways, waterways, civil aviation and pipelines, focusing on "five vertical and five horizontal" transport corridors. A large-scale HSR, the West Rail Link, and interregional railway links were constructed. A national trunk road network and eight western highways were opened to traffic. The rural road system also has been significantly improved; a number of specialised coal, crude oil, iron ore and container terminals have been put into operation; Yangtze River, Pearl River and Beijing–Hangzhou Grand Canal are the backbone of China's inland waterway transport system. In 2010, China owned 175 airports for international and interregional air transport on longer distances.

In 2007, the National Long-term Comprehensive Transportation Plan prepared by the National Development Planning and Reform Commission for intercity transport infrastructures was established, which set national transport policy guidelines for all transport projects of national importance. Many other regional and modal transportation plans followed. In these plans, all cities of different sizes shall be interconnected by a hierarchy of road networks. A motorway system should serve the cities with a population of more than 200,000 inhabitants. The length of national

[4] A resident of Tochigi using Haneda Airport in Tokyo must endure a multimodal access trip to the airport of 1 h 30 min by Shinkansen and conventional rail.

[5] We use 404 OD pairs in total, which satisfy the followings: There are more than 50 samples for both business and leisure/private trips, and there is more than one air transport passenger. We count the total number of departure and arrival of each OD pair in this analysis.

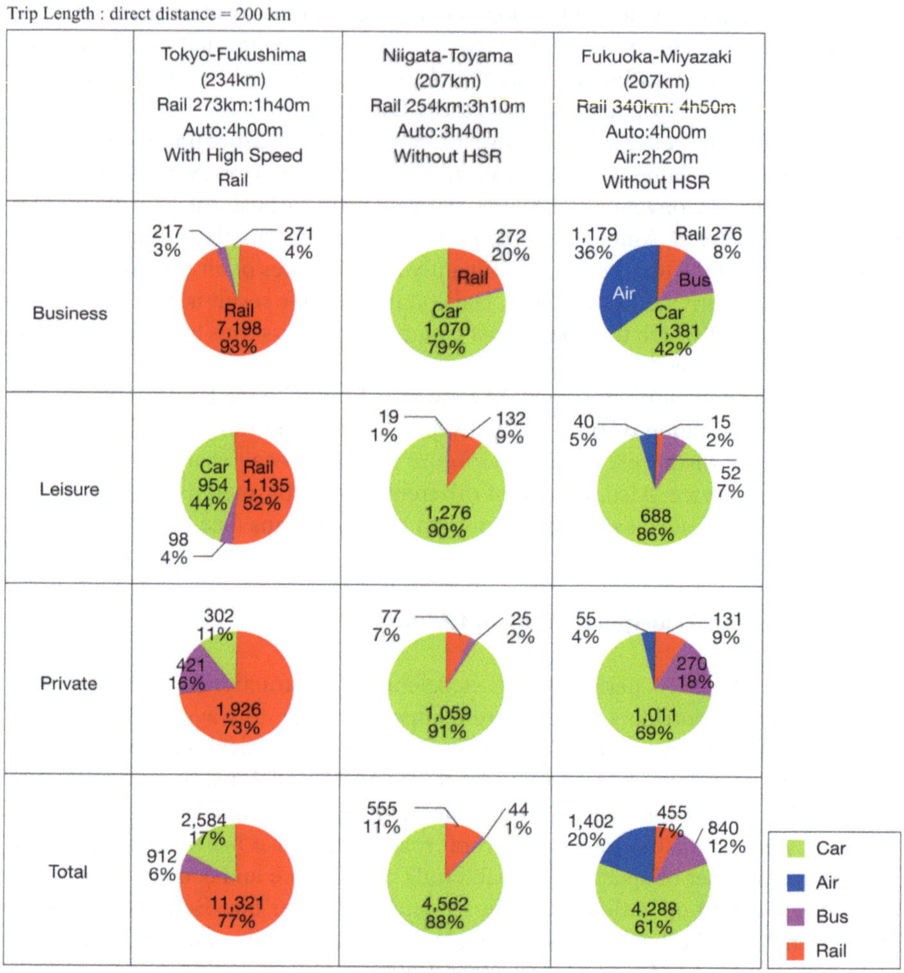

Fig. 2.24 Modal share with or without Shinkansen

railway networks will be extended to 120,000 km, among which HSR networks will contribute around 15,000 km. By 2020, the network density of railway, inland waterway and petrol gas pipeline will be 1.25, 1.35 and 1.25 km per 100 km^2, respectively, while the highway network density will be 31.25 km per 100 km^2.

In China, the state railways have carried 32 % of long-distance passengers in 1978, and this figure has moved down to 5 % of the passengers in 2010. The railways carried 44 % of freight transport in 1978, while in 2010, this share decreased to 11 %. Now railway construction has been the major concern in state infrastructure investment. According to the 12th 5-year plan (2011–2015) prepared by the National Development and Reform Commission, the construction of a 40,000-km national fast railway network will connect all the cities with more than 500,000

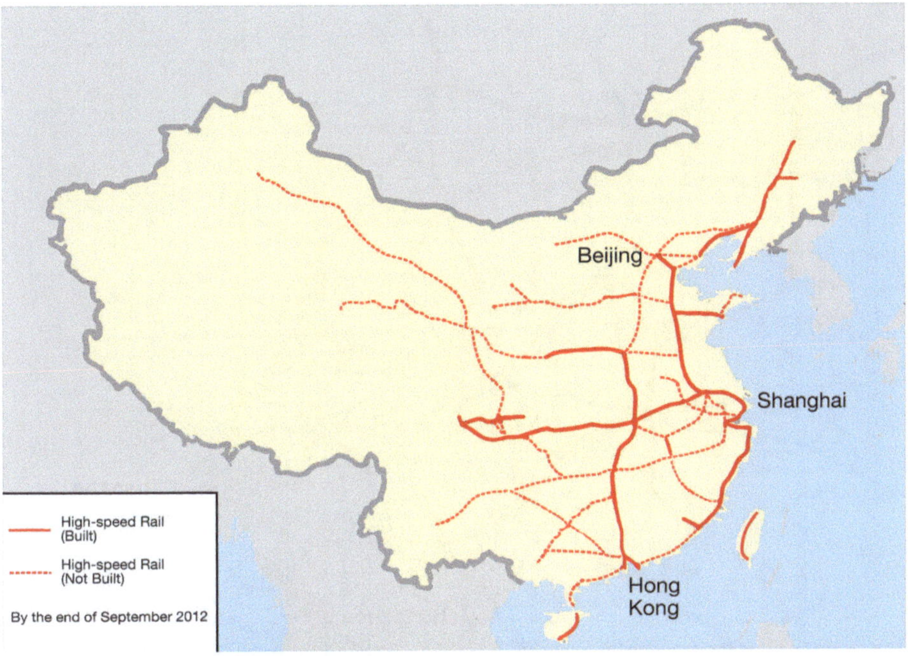

Fig. 2.25 China's long-term railway network plan

inhabitants. Construction of HSR will be speeded up in the plan. The HSR network will be 16,000 km long to connect all the megacities in mainland China (see Fig. 2.25).

The railway and sea transport connection for coal transport from the production areas in the Northwest to the Southeast, where there is a high energy demand, is emphasised. Heavy railway corridors construction will be enhanced to transport coal from the Northwest to the coastal ports and from there to the South by sea, and to the provinces in middle China, such as Hunan, Hubei and Jiangxi, with the capacity of 3 billion t transported on these rail corridors.

More priority will be given to the railway extensions to the West to promote economic development there.

The national railway system will cover major freight collection and distribution areas and every city with over 200,000 inhabitants. In the densely developed areas, such as Yangtze Delta, Pearl River Delta, Bohai Circle, etc., not only the state HSR but also the intercity railway will be encouraged. According to the "Yangtze Delta Region Intercity Railway Network Plan" prepared jointly by Railway Ministry, Shanghai City, Zhejiang province as well as Jiangsu and Anhui provinces, the total railway length in the region will reach 10,400 km, within which the HSR will be 2600 km long (see Fig. 2.26).

Highway construction in China has progressed well in the course of economic development, as in the first phase a good highway system must be provided. This

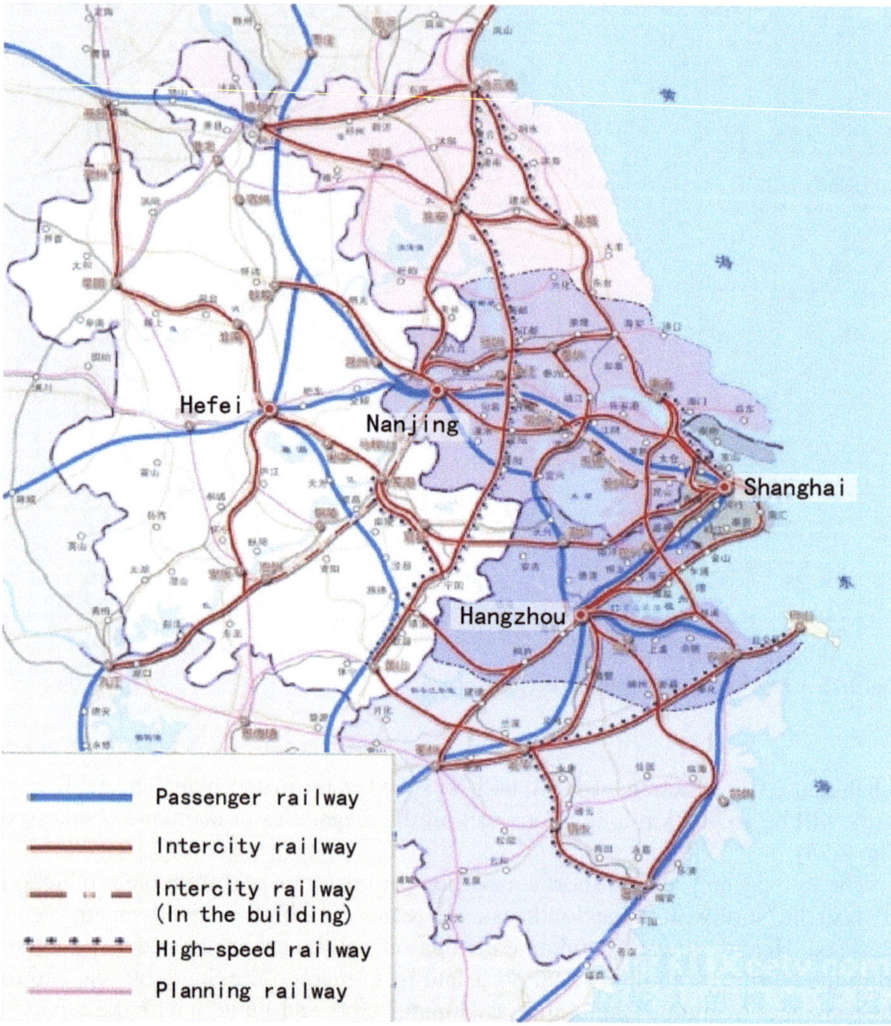

Fig. 2.26 Intercity railway network in the Yangtze Delta region

concept is widely accepted in China. Now highway takes 76 % of the freight transport, while in 1978, it carried only 34 % of the freight. Since the first motorway which was opened to traffic in 1988 for 20 km from Shanghai to Jiading, it has reached 74,000 km. The national motorway system is planned to be completed with 83,000 km length by 2015, connecting all cities with a population over 200,000; by then, coach service will be provided to all towns and over 92 % of all villages. The plan for inland waterways, air and pipelines is also listed as follows:

- Substantially improve the inland waterway transport system through upgrade construction projects of major channels.

- Extend air transport service for 80% of population to have access to air flight service within 100 km.
- Establish interregional transport of crude oil and petrol, natural gas pipeline networks and the capacity of pipeline connections to neighbouring countries. The network length is planned to reach 150,000 km in 2015.

2.3.2 Planning System for Intercity Transport and Spatial Development with Consideration of GHG

2.3.2.1 Intercity Transport in the Planning System and Consideration of GHG in Europe

1. Europe

It was pointed out in Sect. 2.3.1 that the EC develops guidelines—approved by the European Parliament—for the TEN-T network, while the planning competence is still with the single-member countries. However, the member countries have to follow EU law with respect to environmental appraisal.

In 2001, the EU introduced a directive for SEA (Directive 2001/42/EC) for plans and programmes, while environmental impact assessment (EIA) had been introduced already in 1985. By the SEA Directive, which should have been transposed since 2004, each country in the EU was required to conduct SEA at an early stage in developing any major plan, programme or project. This SEA Directive includes impacts on climate as an evaluation item. Each national government has responsibility for methods for assessment tests. The way of implementation differs slightly among the member countries.

It is suggested in documents prepared for the European Parliament to perform a comprehensive climate impact analysis (CIA).[6] This should be a life cycle analysis, which calculates the CO_2 footprint, including infrastructure construction and upstream/downstream processes. This suggestion is still under discussion and not realised yet in any of the member countries.[7]

2. United Kingdom

The planning system related to intercity transport in the UK has changed since the Labour Party came to power in 1997.

Figure 2.27 represents the present (2010) planning system in the UK, which is divided into policy, plan and projects in the fields of transport planning, land-use planning and environmental appraisal. In 2000, LTPs that replaced the Transport Policy and Programme (TPP) were introduced by the Transport Act 2000 and later amended by the Local Transport Act 2008. Instead of an annual bidding round of

---

[6] European Parliament (2013).

[7] See Schade et al. (2013).

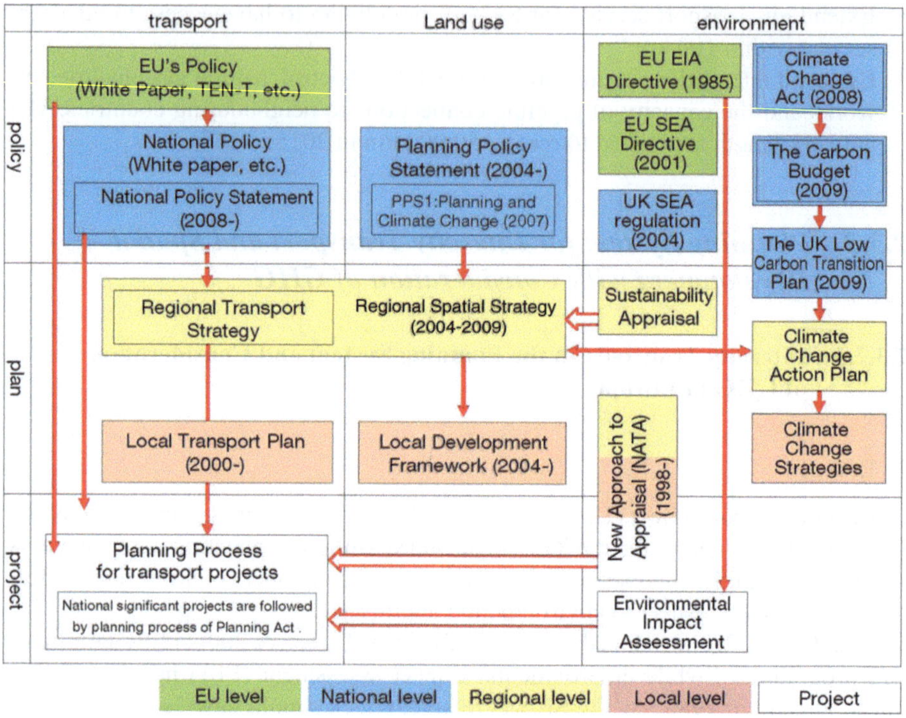

Fig. 2.27 The present planning system related to intercity transport in the UK

TPP, the LTP system is based on 5-year plans. LTP also puts greater emphasis on targets, performance indicators and monitoring. Action on climate change was designated as one of the criteria for assessing the quality of LTPs in the guidance on LTP. Requirements of the criteria have included consideration of the need to reduce GHG emissions and assessing the impacts of policies on CO_2 emissions. The first and second rounds of LTPs have covered 2001–2006 and 2006–2011, respectively. The guidance was published in 2009, which will be applied to the third round of LTPs. This guidance has put more emphasis on the consideration for climate change in response to the "Climate Change Act 2008" and "Low Carbon Transport: A Greener Future".

The system of spatial planning of nationally significant infrastructure projects was reformed by The Planning Act 2008. The Planning and Compulsory Purchase Act 2004 was amended by The Planning Act 2008 to include policies designed to ensure the development and use of land contributes to the mitigation of, and adaptation to, climate change. It introduced new systems such as the Infrastructure Planning Commission and national policy statements, through which the government provides a policy framework to the commission in the planning process of nationally significant infrastructure projects.

In addition, a method of transport appraisal named the New Approach to Transport Appraisal (NATA) was introduced by the white paper: "A New Deal for Transport", and used initially for a set of multimodal studies (MMSs) of areas that have experienced serious transport problems. MMS is an appraisal method which was conducted in areas with serious transport problems regarding multimodal transport. The results of the evaluation of the process of MMS must be reflected in both the Regional Transport Strategy and LTP.

The Climate Change Act was enacted in 2008. A carbon budget and related policies were introduced with the act. An obligation to reduce CO_2 emissions is also imposed on the transport sector. Therefore, transport planning has had a stronger relation to climate change policies than before in the UK.

Based on the EU Directive, the UK provides regulations that are consistent with the institutional planning procedures for the country. In addition, the current process for the plan conforms and corresponds to the SEA guideline that directs to satisfy conditions for the EU Directive. The evaluation items are set depending on the nature of the plan or project; however, the impact on climate change is included as a standard item. The preparation process of the regional spatial strategy (RSS) includes a sustainability appraisal, which satisfies the SEA Directive, in parallel with the planning process. Thus, the RSS also includes an evaluation item for the impact on climate change. The 2007 Planning Policy Statement I: Planning and Climate Change determined that this sustainability appraisal process has included consideration of the impact on climate change. As a result, projects for intercity transport included in the RSS are considered in terms of their impacts on climate change.

For transport plans and projects, guidance on the NATA was published in 2000. NATA presented five criteria for the appraisal of transport investment and policies: integration, safety, economy, environmental impact (including the impact on climate change) and accessibility. And NATA has required making an appraisal summary table (AST), which includes impacts on climate change as one of the evaluation items.

3. France

A new national transport scheme, SNIT, has been introduced by the law of "Grenelle I", mentioned in Sect. 2.2. The first draft of SNIT was submitted in July 2010. After numerous discussions with elected officials and a public consultation held from January 2011 to March 20, 2011, a revised version of SNIT was published in November 2011. One of the four policy axes of the scheme is reducing environmental impact of transport infrastructure. The scheme identifies projects and measures whose implementation is desirable for sustainable development. The estimated amount of budget for those projects in the next 20–30 years is 245 billion € of investment, modal split of which is 71 % for railway, 9 % for river, 2 % for port, 17 % for road and 1 % for air. This means that the scheme clearly intends to put focus on the investment of railway infrastructure such as HSR. The law "Grenelle I" has also decided that adaptability of climate change and energy control should be included in the goals of regional developments for metropolitan areas with populations from 100,000 to 150,000 and that standards for GHG emissions should be introduced in the plan "Schéma de Cohérence Territoriale" (SCOT). SCOT is a spatial plan replacing the former master plan "Schéma Directeur" (SD) under the law "Loi de solidarité et renouvellement urbains" (SRU) in 2000. The objective of SCOT is to indicate a general policy of spatial development and coordinate compatibility between individual plans such as transport and land use. A framework of planning related to intercity transport that is able to consider its impacts on climate change has been specified with SCOT.

The SNIT lays down the following criteria for the choice of transport links (Quinet, E (2010))[8]:

- The net balance of GHG emissions produced or avoided by the project in relation to its cost
- Progress made on other projects and the prospect of saturation on the networks concerned
- Environmental performance (combating noise pollution, severance effects, biodiversity conservation, etc.)
- Multimodal access, economic development, opening up regions and regional development at the various levels
- Improvement of the efficiency, safety and cohesion of the existing transport system
- Fulfilment of the objectives relating to the provision of access for persons with reduced mobility provided for under national legislation

While the SNIT plans following Grenelle I were very optimistic with developing the infrastructure networks for rail, urban PT and inland waterways, the recently published report "Mobilité 21" for the French Ministry of Transport[9] has mirrored the plans with the financial feasibilities after the world economic crisis. Conclusion is that the possibility for new transport infrastructure investments are very limited if the traditional financial instruments are applied in the future (scenario 1). An investment strategy along the lines of Grenelle I is only possible if the budget for transport investments is substantially increased, making extensive use of new financial instruments.

4. Germany

The legal framework for federal transport infrastructure planning foresees a complex interaction between the administration (Ministry of Transport), the federal cabinet, the federal parliament (Bundestag, Bundesrat), the states (Länder) and the municipalities. This follows from the "subsidiarity" principle of division of power, which allocates substantial decision competence to the Länder (states) and municipalities also in the case of federal investment plans. Figure 2.28 illustrates the se-

[8] Quinet, E. (2010), "The Practice of Cost-Benefit Analysis in Transport: The Case of France", OECD/ITF Joint Transport Research Centre Discussion Papers, No. 2010/17, OECD Publishing http://dx.doi.org/10.1787/5km4q8hpgwwd-en

[9] Comission Mobilité 21 (2013).

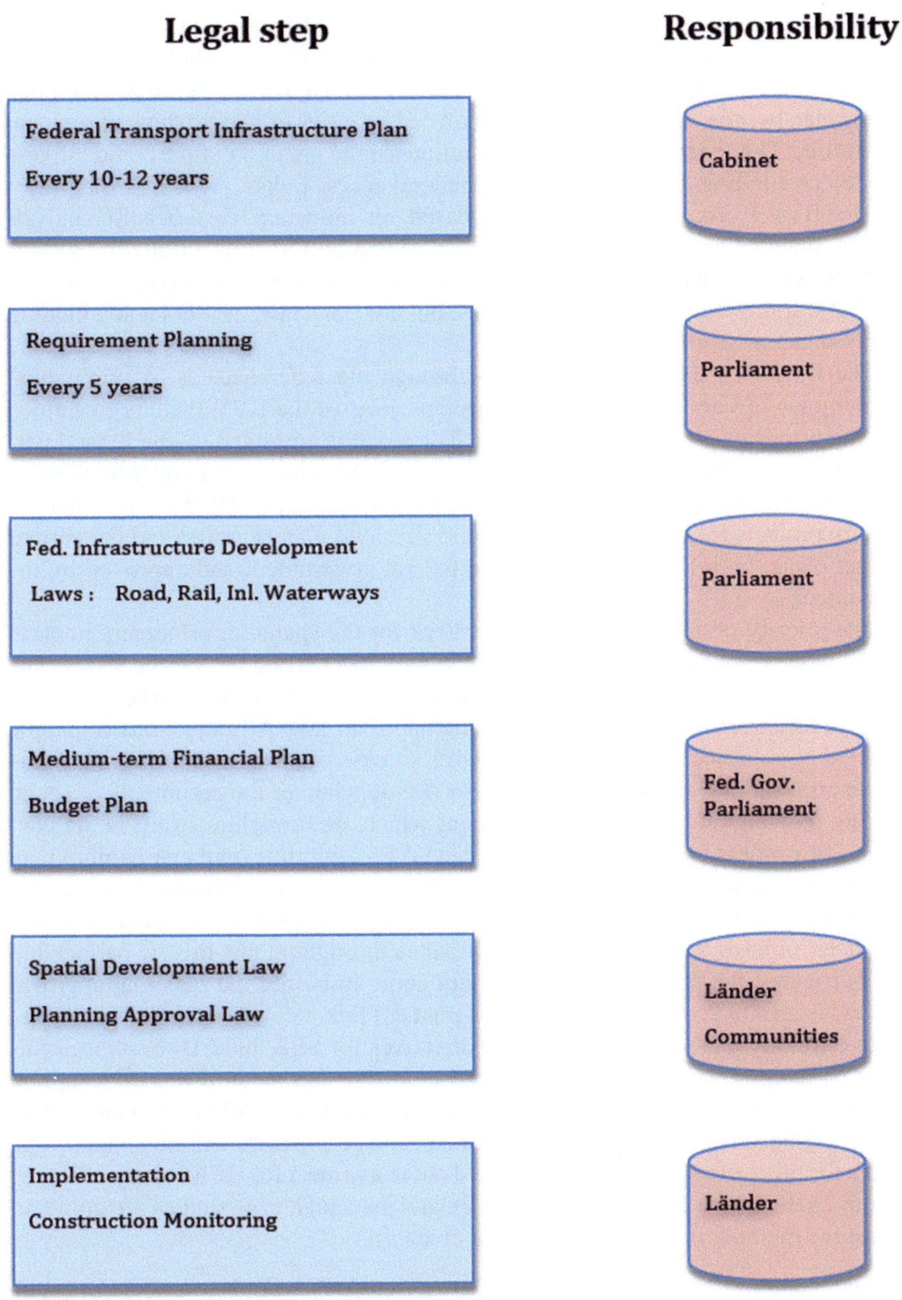

Fig. 2.28 Planning process for federal transport infrastructure projects in Germany

quence of legal steps. The process is started by the Ministry of Transport, which develops the BVWP (federal transport infrastructure plan), including the federal roads, railways and inland waterways. Note that ports and airports are not included in this plan because the latter are widely owned by cities or partnerships of regional authorities. The federal government can influence the planning of ports and airports only by co-funding and the provision of federal access links.

The BVWP assessment scheme is based on monetary cost–benefit analysis (CBA), but also considers the requirements of EIA and SEA according to EU law. GHG emissions are included as one criterion of the environmental assessment. The BVWP is approved by the federal Cabinet, but it is not a law, i.e. not legally binding at this stage.

The BVWP is transposed to legacy through the federal needs (requirements) planning law. Every 5 years, the suggested projects of the BVWP are re-examined and the priorities are eventually revised. The result is approved by the federal parliament and transposed through different laws for the development of federal transport infrastructure modes (road, railway, inland waterways—IWW). The resulting priority projects feed into the medium-term financial plan (5 years) and the current budget plans, both to be decided by the federal government and approved by the parliament.

The priority projects then undergo a check for the spatial development impacts on the base of the spatial development law which is carried out by the states. The final approval of the design and appropriateness of environmental protection measures is taken on the basis of the planning approval law. All objections of people affected by a project have to be considered. In case of violations of private rights and expropriations, the law courts have to decide whether the arguments of social welfare outweigh the private disadvantages which are brought forward in the lawsuits. The process of planning approval can take a long time (eventually more than 4 years), can lead to changes of the project design (guideway alignment, noise protection measures) and can increase the originally planned costs considerably. It furthermore imposes barriers to private co-finance through public–private partnership (PPP) because the private investors cannot come in before the planning approval process is terminated for all sections of a planned link.

In Germany, the requirements of the directives for SEA and EIA have been fulfilled on the project scale, as the standardised evaluation method was only applied on this level. For the next revision of the planning process and the evaluation procedure, to be finalised in 2015, an extended scheme for SEA will be applied. The basics for this extension have been worked out in a project for the Ministry of Transport in 2010 (Bosch & Partner et al. 2010), and presently, case studies are going on to test the method. It includes four major changes:

1. Application of SEA to the comprehensive transportation plan in addition to the single projects
2. Harmonisation with EIA, i.e. avoidance of double assessment
3. Definition of 12 impact areas and appropriate indicators (monetary and nonmonetary), including GHG (area 9)
4. Improved participation of stakeholder groups

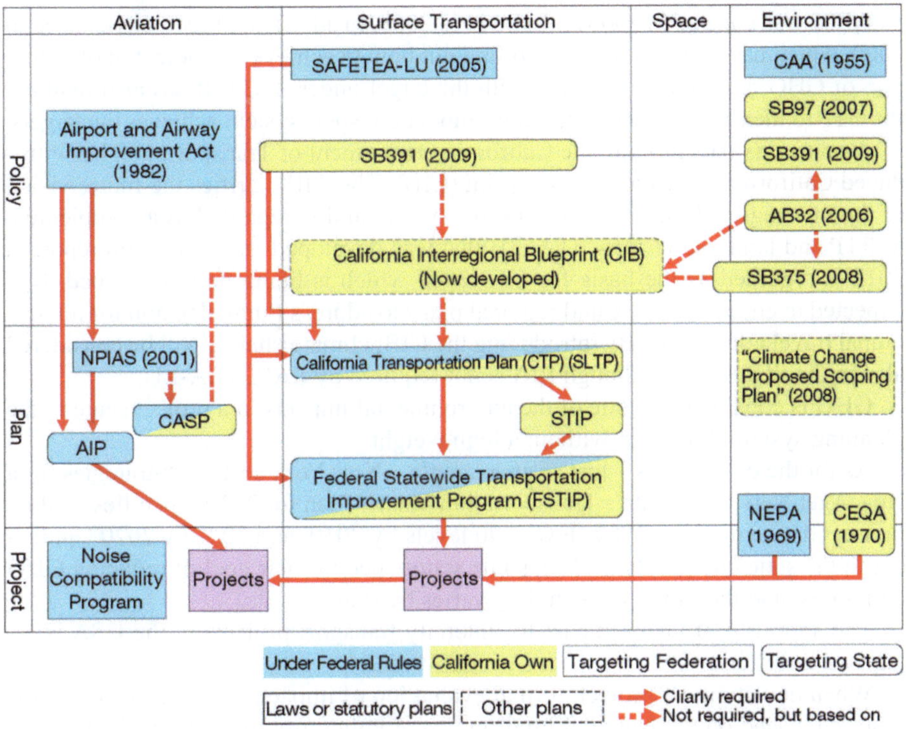

Fig. 2.29 The planning framework and procedure related to intercity transport in the state of California, USA

2.3.2.2 Intercity Transport in the Planning System and Consideration of GHG in the USA

Figure 2.27 shows the planning framework and procedure related to intercity transport in the state of California. In the USA, the "Safe Accountable Flexible Efficient Transportation Equity Act: A Legacy for Users" (SAFETEA-LU) has dictated federal policy for road and railway transport, but has not referred to climate change. This legislation has obliged each state to develop a state-wide long-range transportation plan (SLTP) and transportation improvement program (STIP), which targets road and railway transport primarily and also airports as an integral part of an intermodal transport system for the USA.

California Transportation Plan 2025 (CTP2025) was developed in 2006 as its SLTP. This plan refers to GHG emissions in outline, but does not define a specific goal of reducing GHG emissions. It also refers to measures to reduce GHG emissions, but these only involve improving the fuel efficiency of cars. In addition, CTP2025 proposes HSR as part of the state's plan. However, the main aim of the project is improving intrastate mobility and accessibility. It does not aim for reduction of GHG emissions (Fig. 2.29).

AB32 was enacted in 2006 and SB375 in 2008. In addition, SB391 was enacted in 2009, which has required the CTP to find out how to achieve full potential of reduction of GHG emissions to comply with the target under AB32. It has also required the CTP to identify how the state's multimodal transport system achieves the targets.

Against this background, the California Department of Transportation has introduced California Interregional Blueprint (CIB). The CIB clarifies the future vision of the interregional multimodal transport system and is regarded as a complement to RTP and land-use visions. The CIB that was developed as a mid-term report in 2012 is regarded as the basis for CTP 2040 which is being now developed. It is expected to encourage state and regional plans to address future demand for interregional travel adequately. By introducing the CIB, climate change will be considered also in the SLTP stage although EIA is not required by SAFETEA-LU.

GHG is considered as one of the environmental impacts in a project stage in the planning system, however, without a high weight.

As for the other states—just as an example—New York's SLTP "Strategies for a New Age: New York State's Transportation Master Plan for 2030" specifies "reduction in GHG emissions 5 % below 1990 levels by 2010, and 10 % by 2020", pursuant to the state energy plan. Thus, some states specify their goal of reducing GHG emissions, and therefore, the handling varies by state.

The appraisal of GHG impact for intercity transport projects in the USA is organised as follows:

When developing a project, formal evaluation of impacts on the environment is required, based on the National Environmental Policy Act of 1969 (NEPA), which took effect in 1970. NEPA requires preparation of an environmental impact statement and evaluation of impacts on the environment when a serious impact on the environment is anticipated. The impact is based on ecological, aesthetic, historical, cultural, economic, social and health items. However, the impact on climate change or global warming is not included. Therefore, in practice, the EIS does not consider GHG emissions adequately.

On the other hand, there are cases in which states provide their own formal process for evaluating impacts on the environment, such as California Environment Quality Act (CEQA) in the state of California and State Environment Quality Review in the state of New York. For example, a procedure based on the CEQA is required for projects implemented in the state, whether they are federal projects or not. The CEQA requires that when a serious environmental impact might occur, an environmental impact report must be prepared, evaluating the impact on the environment, but, as with NEPA, the evaluation of GHG impact was not required clearly. (The "environment" is defined as the physical conditions, including land, air, water, minerals, flora, fauna, ambient noise and objects of historical or aesthetic significance in the 2009 CEQA Statute and Guidelines). However, in 2007, Senate Bill No. 97 was passed, which required the Office of Planning and Research to provide a guideline for mitigation or the effects of GHG emissions and their influence by July 2009. By January 2010, the Resource Agency must have inspected and adopted the guidelines. Finally, GHG impact is specified as a target of the evaluation under CEQA. Of course, GHG impacts must be evaluated in EIRs.

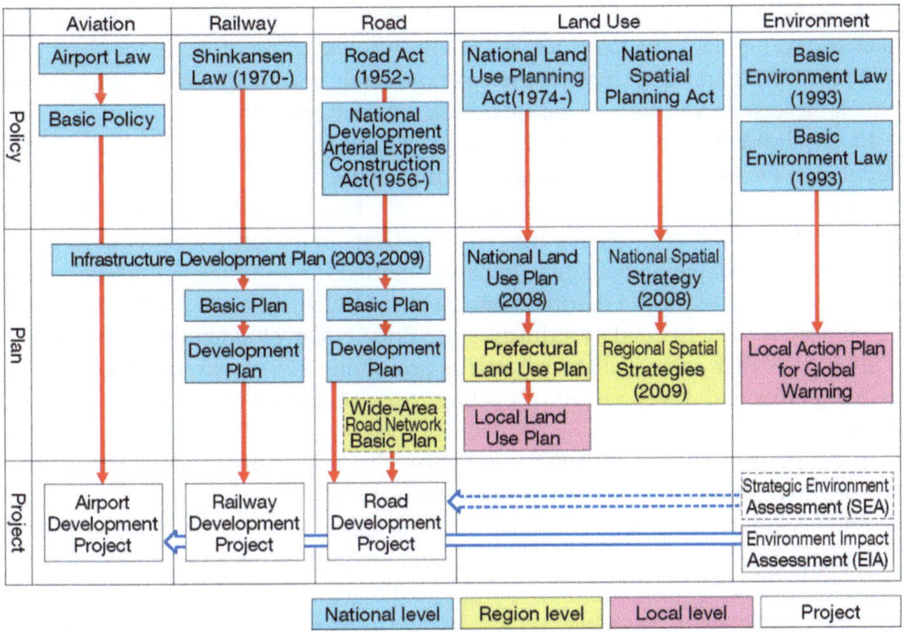

Fig. 2.30 Present planning system related to intercity transport in Japan

2.3.2.3 Intercity Transport in the Planning System and Consideration of GHG in Japan

1. Policy on Intercity Transport and Consideration of the climate change in Japan

Intercity transport policies related to infrastructure development in Japan are provided in the form of legal provisions. There are separate laws for different transport modes such as the Nationwide Shinkansen Railway Development Law, or National Highway Law. However, considerations for the climate change are not specifically mentioned in these laws.

Regarding the transport policies of each mode, the road sector, which accounts for 90 % of the total GHG emissions from transport, has many measures for tackling global warming. In the road sector policy papers, such as "Principal Road Policy by the Infrastructure Development Council" and "Middle Term Road Development Plan proposed by the Road Subcommittee in 2008", a concrete target for GHG emissions and various kinds of countermeasures are indicated (Fig. 2.30).

2. Planning System related to Intercity Transport and Consideration of GHG in Japan

The planning system of transport in Japan is basically divided by modes, such as railways, aviation and roads, and a master plan and development plan for each mode is set on the basis of the relevant basic law. However, the main component of these plans is development and management of infrastructure. Therefore, in the basic law of the respective transport modes, there is no reference to the global warming even though these are provisions of consideration for local environment.

In the Infrastructure Priority Development Plan determined by the cabinet meeting on October 10, 2003, mitigation of global warming is indicated based on the "General Principles of Countermeasures for Dealing with Global Warming". In particular, concrete measures for reducing GHG emissions, such as promoting energy-saving vehicles, managing transport demand, promoting PT usage and a modal shift to railways and maritime transport and tree planting in public spaces are recommended. The Secondary Infrastructure Priority Development Plan, published in 2009, says that policy measures such as promoting ITS, flexible charges of expressway tolls and development of cycling roads should be promoted to meet the targets of the Kyoto Protocol.

In the field of spatial planning, a national plan and regional plans (for extensive regions) are to be prepared based on the National Spatial Planning Act. In the National Spatial Strategy, "Promotion of Preventing Damage Caused by Global Warming", given in Chap. 7, and "Basic Countermeasures for Environmental Preservation and Landscape" are discussed. In particular, it describes countermeasures that make the energy demand/supply structure more environmentally friendly. In each of the RSSs settled on, based on the National Spatial Strategy, plenty of descriptions related to global warming or climate change can be found, as shown in Table 2.2.

3. Appraisal of GHG impact in the process for intercity transport planning in Japan

Regarding procedures, EIA has been conducted in the stage of project planning in Japan. The Ministry of the Environment of Japan published SEA guidelines in 2007. The Environmental Impact Law of Japan was revised in 2011 to cover the SEA procedure for the project development stage, which assesses its location and scale.

2.3.2.4 Intercity Transport in the Planning System and Consideration of GHG in China

China's National Development and Reform Commission is responsible for the medium- and long-term comprehensive transport network plan. In 2007, the commission issued the most recent medium- and long-term plan for 2020. It is stated in this plan that there is not enough capacity to meet the further development of transport demand, and China will still face bottlenecks and structural constraints in the intercity transport system. It is essential to establish a convenient, unimpeded, high-efficiency, safe and well-coordinated multimodal transport system, taking full advantage of the highway, railway, waterway, pipeline and air transport system as a whole to meet the transport demand from social and economic development at the lowest environmental cost and resource consumption. This plan provides the fundamental requirements to other transport sectors and includes the key intercity transport projects.

Table 2.2 Descriptions related to climate change in the regional plans

	Transport policy measures							Other policy measures			
	Ring road Road network	Mitigating congestion	Eco car, eco drive	Eco public transport	Public transport Bicycle	Freight efficiency Green freight, modal	Port development	Urban structure (ex. compact city)	Urban afforestation Forest	Energy saving Alternative energy	Disaster prevention caused by climate
Tohoku	O	O	O		O	O			O	O	
Metropolitan	O	O	O		O	O		O	O	O	O
Hokuriku	O	O	O	O	O		O		O	O	O
Chubu	O	O	O	O	O	O		O		O	
Kinki	O	O	O	O	O	O		O	O	O	
Chugoku		O			O	O		O	O	O	O
Shikoku		O				O			O	O	O
Kyushu	O	O			O		O			O	O

In 2012, China's National Development and Reform Commission updated the medium- and long-term plan by the 12th 5-year Comprehensive Transport Plan (2011–2015). Besides the infrastructure construction project plan, environment protection plays an important role among the goals defined:

Recycle economy This means to vigorously develop recycling processes in the economy and promote the construction of green transportation systems, increasing the efforts for energy saving in transport to control GHG emissions in the transport sector to improve the sustainability of the overall transport system.

Resources saving This includes saving resources in all aspects of transport system planning, construction, operation, maintenance and also in intensive use of land, coastline, airspace and other resources to improve the level of resource utilisation. Accelerate the development of rail transport, waterways and other resource saving-oriented modes of transport to improve the efficiency of resource use. Strengthen recycling of waste materials and other renewable resources use in the transportation infrastructure construction.

Energy efficiency Under energy efficiency, strengthen R & D work to promote the use of energy-saving new technologies, new processes and new equipment, especially to improve energy-saving technologies applied in rolling stock, aircraft, ship and port. Raise the proportion of railway electrification rate and encourage the use of electricity power-driven port-handling facilities and eliminate energy-intensive transport equipment and processes. Improve transport management level to improve energy consumption efficiency.

Environmental protection Under environmental protection, enhance environmental awareness in transport planning stages and strengthen EIA in transportation infrastructure construction. Implement at dynamic monitoring of the environmental impact during the construction process. Encourage the use of clean and environmentally friendly transport technologies and the equipment to reduce emissions of pollutants and carbon dioxide as well as effective control of noise pollution.

The Ministry of Transport published the 12th Five-Year Transport Development Plan in 2011, when railway was separated from their responsibility. In the plan, five categories of targets were listed for the three modes of highway, waterway and air transport, in which green transportation target is related with environment protection (Table 2.3).

Table 2.3 Green transport targets in 2015 in China (% reduction of CO_2)

Mode	Indicator	Target in 2012
Highway	CO_2/passenger-kilometre or tonne-kilometre reduction (%, base year 2005)	11
	Highway land area/vehicle carried reduction (%, base year 2010)	5
Waterway	CO_2/tonne-kilometre reduction (%, base year 2005)	16
	Tonnes processed/berth-kilometre increase (%, base year 2010)	5
Air	Air flight energy consumption and CO_2 emission reduction (%, base year 2005)	>3
Total	TSP and COD/tonne-kilometre reduction (%, base year 2010)	20

The former Ministry of Railways has also issued a medium- and long-term railway network plan according to the requirements from the Comprehensive Transport Network Plan in 2008. It sets the goal of railway network development. The ratio of electrification will be increased up to 60 %. In this plan, it is hoped that with the increasing capacity of railways for passenger, there will be more space for freight transport in railways. Energy saving and environment protection are listed as the principles in the plan. In the previous 12th Five-year Development Plan of Railways prepared by railway ministry, it was mentioned that by 2015 the CO_2 emissions would be reduced by 5 % compared to 2010.

2.4 Measures to Address Global Warming in the Transport Sector

In Chap. 1, the possible measures to address global warming in the transport sector have been described and structured by the CUTE matrix (Table 1.6). This section aims at discussing the most important measures in some detail, starting with technology (Sect. 2.4.1), followed by information technology (Sect. 2.4.2), regulatory instruments (Sect. 2.4.3), economic instruments (Sect. 2.4.4) and necessary changes in the energy sector (Sect. 2.4.5).

2.4.1 Technology Development to Address Global Warming

2.4.1.1 Energy-Saving Technology

In technologies necessary for the energy saving of a traffic mode, an improvement of engines and lightening of the body and airframe contributes to improve the fuel cost efficiency. The decrease of the frictional drag when running at higher speeds is an important technology for efficiency improvement in intercity transportation. Figure 2.31 shows the progress in energy efficiency for each mode. Aircraft, Shinkansen and cars, all have improved their energy efficiency by about 50 % in the past 30 years, and technological development for energy saving has been carried out for each mode. The percentage of improvement is the highest with passenger cars, but starting from the highest level of CO_2 emissions.

Reflecting recent diversification of air-transport demand, and the growth of LCCs in Europe and the USA, passenger airlines have shifted to medium or small planes on small and medium transport distances and energy-saving two-engine planes on longer distances. This is symbolised by the appearance of the Boeing 787, and the forthcoming introduction of the Airbus A350. The new aircrafts are expected to improve energy efficiency about 20 % compared with existing medium-sized B767.

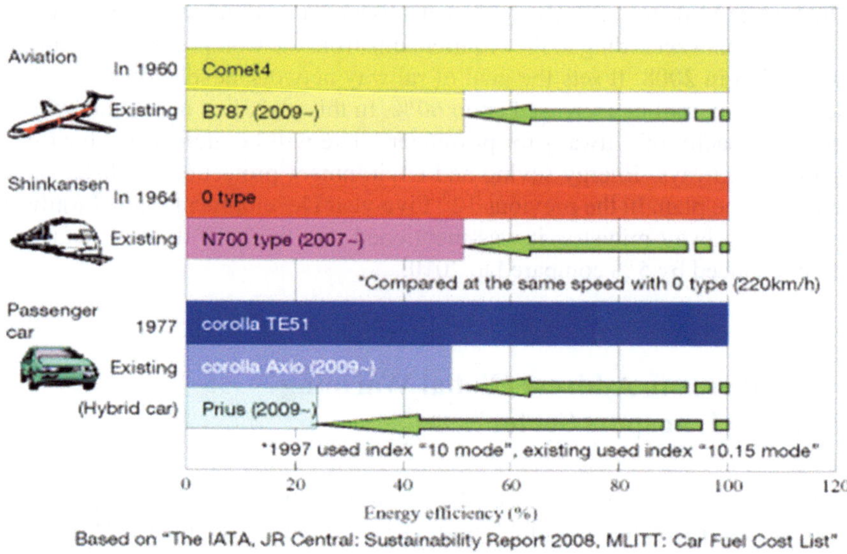

Fig. 2.31 Change in motive power energy efficiency according to mode

The development of small planes (regional jets, RJ) is a remarkable trend in recent years. Not only Japan but also China (the ARJ21) and Russia (Super-Sukhoi jet) are developing their RJs to enter the market dominated by Canadian (CRJ) and Brazilian (ERJ) jets and are achieving an increase in the number of seats, extension of the cruising range and improvement of fuel efficiency at the same time. Mitsubishi Heavy Industries is developing the next-generation regional jet (MRJ) as the first regional jet of Japan, and the MRJs were introduced into All Nippon Airways in 2013 after the first flight in 2011. Fuel cost improvement and CO_2 reduction by about 20 % or more are expected compared with a past isomorphic jet plane.

In the railway sector, the Japanese Shinkansen trains have been developed and modernised a few times since 1964. The energy consumption of the newest train is 50 % less than that of the earliest 0-type at the same speed, and 32 % less if the newest one is operated at 270 km/h (see Fig. 2.32). Although the railways are known as a transport mode with low environmental burden, further continuous technological development is necessary here as well as in the aviation sector.

Such improvements in energy efficiency were realised as a result of combining technologies in various fields such as machine engineering, electrical engineering, material engineering and fluid dynamics. Examples are the lightening of airframes and vehicles based on composite materials and material processing technology, surface smoothing and shape improvement of airframes and vehicles or regenerative braking technology that recycles rejected thermal energy (or kinetic energy).

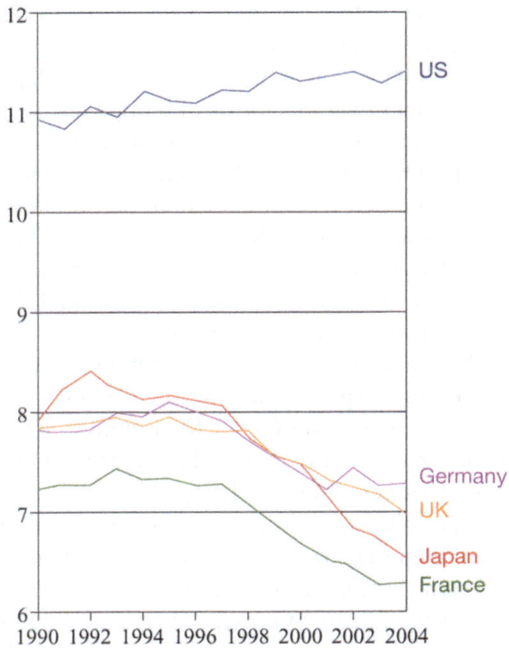

Fig. 2.32 Trend in fuel intensity of new car/SUV (litres of petrol equivalent per 100 vehicle-kilometres). (Source: OECD/ITF (2008)[10])

2.4.1.2 Decarbonising Technology for Next-Generation Vehicles and Airplanes

Since the Kyoto Protocol, technology developments towards lower carbon footprint (CFP) have been promoted, temporarily driven by increasing oil prices. In the railway sector, JR East has begun, since 2007, operation of diesel-hybrid vehicles running on electricity for mountain trains for the first time in the world. The energy efficiency of these vehicles is 20 % higher than that of the old diesel vehicles, and harmful materials in the exhaust gases, such as NO_x and black smoke, are reduced by 60 %. Hydrogen hybrid technology is also being developed for zero-emission vehicles, and driving tests are being carried out. A test run at 100 km/h has already been carried out in the test yard. Diesel-electric engines are used on non-electrified lines, which are about 40 % of all railway routes in Japan, and these are chiefly operated on local railways. A further effect of the GHG reduction might be expected when these hybrid trains are introduced to local routes, leading to the activation of the local railways in consideration of global warming.

In the aircraft sector, since 2008, the Boeing and Airbus companies have executed test flights with passenger planes using biofuel, in cooperation with airlines of various countries. A test flight was carried out with Virgin Atlantic Airways with a jumbo passenger plane (B747–400) in February 2008, and Air New Zealand

[10] OECD/ITF (2008), Greenhouse Gas Reduction Strategy in the Transport Sector: Preliminary Report http://www.internationaltransportforum.org/pub/pdf/08ghg.pdf

succeeded with a test flight using a 50–50 blend of fuel oil and a second-generation biofuel in December 2008. The test flight using a different biofuel was carried out in 2009 with Continental Airlines and Japan Airlines. Many other airlines have succeeded in various forms of biofuel test flights. Qatar Airways succeeded with a commercial flight using gas-to-liquid (GTL) fuel made from a natural gas in October 2009. It attracted attention as a technology that lowered dependency on oil.

Technological development of decarbonisation by diversification of fuel source is being advanced in the airline industry. The International Air Transport Association (IATA) announced the road map of the introduction of the hydrogen fuel and the fuel cell in 2008. Moreover, German DLR succeeded in a test flight with hydrogen fuel cells in a small plane in 2009.

Moreover, the next-generation transport in Japan—a linear motor car (Maglev)—is worthy of special mention. JR Central announced the Linear Chuo Shinkansen plan to connect the Osaka Metropolitan Area with the Tokyo Metropolitan Area with a magnetically levitated train (Maglev) operating at 500 km/h. It is planned to open between Tokyo and Nagoya in 2027, and to be extended to Osaka by 2045. A super-high-speed Maglev system is already in operation in Shanghai which links the Pudong District with the airport Pudong. The Japanese development applies a different Maglev technology by using the repulsion power and the attraction power of superconducting magnets. In running tests along the Yamanashi test line, the fastest speed in the world for a railway, 581 km/h, was recorded. That is a very-high-speed mode, but its CO_2 emission per capita is about one third of airplanes, given the current composition of electricity generation sources (Tokyo-Osaka). Because Tokyo and Osaka will be connected in 67 min after operation is begun, the world's largest economic region, with a population of 64 million living within a travel distance of 1 h, would be created. The environmental efficiency and economic viability will be discussed in Chap. 4.

2.4.2 Information Technology

Technological developments to support the efficient services of information and telecommunication systems are being advanced remarkably. As an example, airlines aim to establish efficient operation by devising flight path and operation methods, such as linearisation of flight routes by introduction of area navigation (RNAV), continuous descent arrival (CDA) and dynamic optimisation of sea route (UPR, DARPS). The ICAO created the global air traffic management (ATM) operation concept in 2003. ATM is intended as integrated, safe and efficient management of air traffic, and the organisation describes how to operate ATM in the future in order to address global warming. Based on the concept, the approaches have started for the improvement of the ATM, which are named NextGen (Next Generation Air Transportation System) in the USA, SESAR (Single European Sky ATM Research) in Europe and CARATS (Collaborative Actions for Renovation of Air Traffic Systems) in Japan.

Modern rail control system allows for flexibly moving safety blocks between trains and for a more continuous and energy-saving movement. Furthermore, bottleneck situations can be managed more efficiently as, for instance, allowing two passenger trains to approach a railway station in a synchronised way and to stop at the same platform. Freight trains can be guided more flexibly in the network to bypass congested nodes. In countries, in which the railway network is densely meshed and strongly interconnected, it is most important to inform the passengers as soon as possible in case of a disturbance of train schedules. Contrasting Japan, where the main intercity traffic is flowing alongside one corridor, in Europe, the trains often have to be synchronised at particular connection stations such that people can change the trains without much time losses and inconvenience (e.g. change on the same platform to the opposite train car).

In road traffic, the route guidance systems have established a functioning market since several years. The recent developments go towards dynamic guidance systems, which take the actual traffic situation into consideration and react quickly to congestion or accidents. Cars and trucks are developing into information centres and include all kinds of assistant systems to reduce stress for the drivers and help to reduce energy consumption. As the car is often a necessary element of intermodal transport chains, its more efficient use can contribute substantially to save energy and CO_2.

2.4.3 Regulatory Instruments

Regulation of technology and behaviour is the standard instrument of environmental policy. In the road sector, this includes:

- Emission standards for air pollution (the US, Californian, Japanese, European limit values for air pollutant emissions and noise)
- Emission standards for CO_2 (CAFE in the US, EU and Japanese standards)
- Standards for local concentration of air pollution
- Safety standards (crash zones, airbags, accident equipment)
- Obligatory car checks

The railway industry is one of the most strongly regulated industries of the world. This concerns in the first instance:

- Safety regulations
- Emission standards for diesel locomotives (in the EU: standard in analogy to truck emission limits)
- Technological regulations
- Service regulations

With respect to CO_2 efficiency, there is not particular regulation necessary because the railway companies are intrinsically interested in saving energy to reduce operating costs. In some cases (e.g. Deutsche Bahn AG in Germany), the rail company sets internal targets for CO_2 reduction to present itself as a voluntary top runner to improve on image and political support.

1. Regulation on automobile fuel efficiency

The first oil crisis beginning with the fourth Mideast War in 1973 disrupted the world economy, and the influence was especially large for Japan, which was dependent for about 80 % of its oil on the Mideastern countries. Also, it was a turning point for broad development of energy saving in various industries, and the automobile industry was required to achieve further improvements in fuel efficiency at the same time that it was addressing the air pollution issue.

In Japan, fuel efficiency has been drastically improved since 1998 because of the top-runner regulation, which was implemented in 1998 by updating the Energy Saving Law established in 1979. The top-runner regulation sets the standard for the next period above the efficiency level of the most energy efficient product in the current period. The standard is updated every 5 years for automobiles, and succeeds in promoting continuous technological development. As a result of those developments, the target for fuel efficiency set in 1999 was fulfilled by all Japanese automobile companies by 2007, and further development towards the updated target for 2015, set in 2007, is expected.

Further regulations applied in the USA and in Europe address the internalisation of external costs of transport in a more general way and will be explained in Chap. 3.

2.4.4 Economic Instruments

1. Environmental taxes

Environmental taxes can be levied either on the magnitude of pollutant emissions or on proxy indicators. With respect to CO_2, the energy or fuel consumption is traditionally used as proxies. In the EU, such taxation policies are regulated by the energy taxation directive, which sets lower limits for energy consumption taxes. This also applies to fuel consumption of passenger cars and trucks. The presently prevailing taxation scheme, which has been introduced in 2003, resulted from a political compromise and is not based on environmental logic. Diesel consumption is taxed much lower than gasoline, which has motivated the industry, in particular in Germany, to produce high-powered diesel-driven premium cars, which is counterproductive to the CO_2 reduction issue. The proposal of the EU Commission from 2011 therefore tried to bring in more environmental rationality by linking the tax level to energy density of fuel and CO_2 production of its use. Because of the strong resistance of some powerful European automobile manufacturers, this proposal is still pending.

In Germany, a particular environmental tax is added to the energy tax on fuel consumption; however, the revenues go to the general budget and are not earmarked for environmental investments. While the tax rates appear high in an international comparison, one has to keep in mind that there are many exemptions, e.g. for ag-

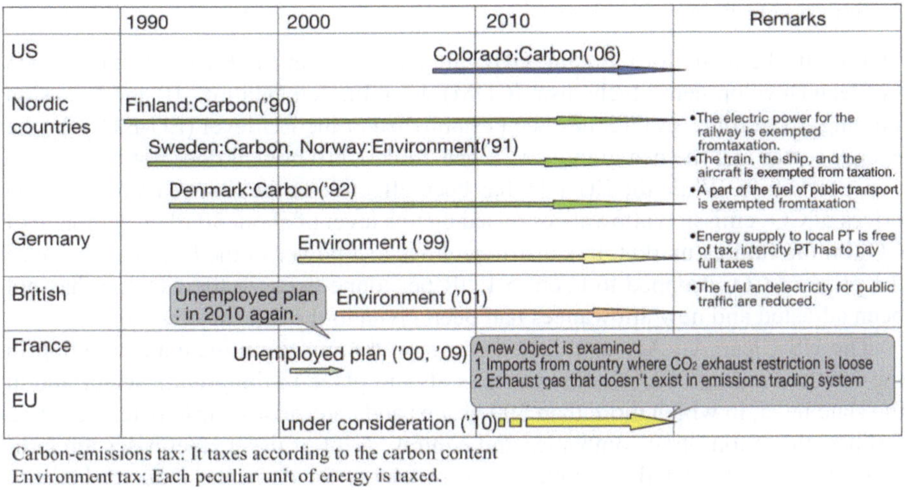

Fig. 2.33 International trend of environment tax

riculture and parts of the industry. Local public transit is also free of energy tax, while intercity rail transport is fully taxed (also with VAT). Figure 2.33 shows that the characteristics of environmental taxation are most different among the European countries. This underlines that—although a common framework for taxation exists with the EU Directive—the actual situation is very heterogeneous and will hardly be changes in the future because all fiscal policy actions have to be decided unanimously in the EU.

In some countries, a part of the revenue is used for capital formation to implement energy policy and CO_2 reduction measures, as, for instance, in Britain and Italy. In Britain, about 80 % goes to reduce the amount of the pension premium load of segment of industry, and a part of the remainder is allotted as capital for executing energy efficiency measures. Moreover, the carbon trust company, which is an independent non-profit corporation, has been established, and this carbon trust company is playing a central role in guideline decisions regarding CFP in Britain, as described later. In the countries where the environment tax has been introduced, the environment tax system combines various preferential treatments that take the environmental policy and economic impacts into consideration. As for the environment tax of Britain (Climate Change Levy), to maintain the global competitiveness of the home industry, a tax reduction of 80 % is admitted for industries and companies that agree to the Climate Change Agreements (CCAs) with the government to accomplish the goal of abatement of emissions and energy efficiency improvement. Moreover, in some EU countries, reduced tax rates and an exemption for electric power that uses renewable energy have been implemented. There are reduced rates and exemptions for electric power and energy used for PT in some but not in all the EU countries and not for all segments of the transport sector.

2. Emission trading

Market mechanisms for reducing GHG are suggested in the Kyoto Protocol, such as Clean Development Mechanism (CDM), Joint Implementation (JI) and Emission Trading. In Europe, an ETS has been established at the EU level (EU-ETS) for the energy sector and the heavy industry since 2005. Until now, it does not include the transport sector. While the EU-ETS has been effective in the first years because the prices of CO_2 emission allowances raised up to a level of about 30 EUR per tonne of CO_2 on the futures market, this instrument has lost power in the last years because the price of CO_2 dropped to about 5 EUR per tonne because the CO_2 cap has not been adjusted and new allowances had been given to the industry free of cost.

The US Clean Air Act established emission trading to mitigate air pollution in the 1970s. Trading of CO_2 allowances is also in place in the private sector and at the state level, in which more than 300 groups and enterprises participate, including airlines, the automotive companies and Amtrak. A lot of these enterprises are indirectly the objects of federal government measures for domestic industry because they participate in the "Climate Leader Program" of the United States Environmental Protection Agency (EPA), which began in 2002 and sets the target reduction of GHG emissions. At the state level, the Western Climate Initiative (WCI), in which seven states in the Western USA and four Canadian provinces participate, made a draft of an emissions trading system public in 2008. The Midwestern Greenhouse Gas Reduction Accord (MGGRA), in which six Midwestern states participate, also made their draft public in 2009. And, they made their scheme of cap-and-trade type emissions trading public in 2010. The Regional Greenhouse Gas Initiative (RGGI) by ten northeastern states already began the system in 2009. However, though an emissions trading system bill was passed in the House of Representatives at the federal level in the USA in 2009, discussion in the Senate was abandoned in 2010, and the possibility of its future introduction is unclear. In Japan, since 2008, the government and industry have been trying it jointly through the voluntary participation of enterprises that do not make the upper bound of the amount of the emissions an obligation. On the other hand, the approach in each municipality has already been started. The Tokyo Metropolitan Government in 2010 began emissions trading for offices that use energy equivalent to crude oil conversion of 1500 kl or more (Fig. 2.34).

3. Various measures

Britain announced the "Carbon Budget" with the approval of the Climate Change Act 2008. The carbon budget shows the upper bound of the greenhouse emission burden that Britain emits every 5 years. It was introduced to achieve the long-term goal of a 34% reduction by 2020 and an 80% reduction by 2050 compared with 1990 announced by the climate change method.

Attention is also paid to the concept of CFP, through which GHG reduction can be made transparent to everyone. This approach displays the amount of CO_2 emissions and makes the amount of GHG exhausted during the life cycle of the commodity or service visible. Britain decided to implement the world's first guidelines

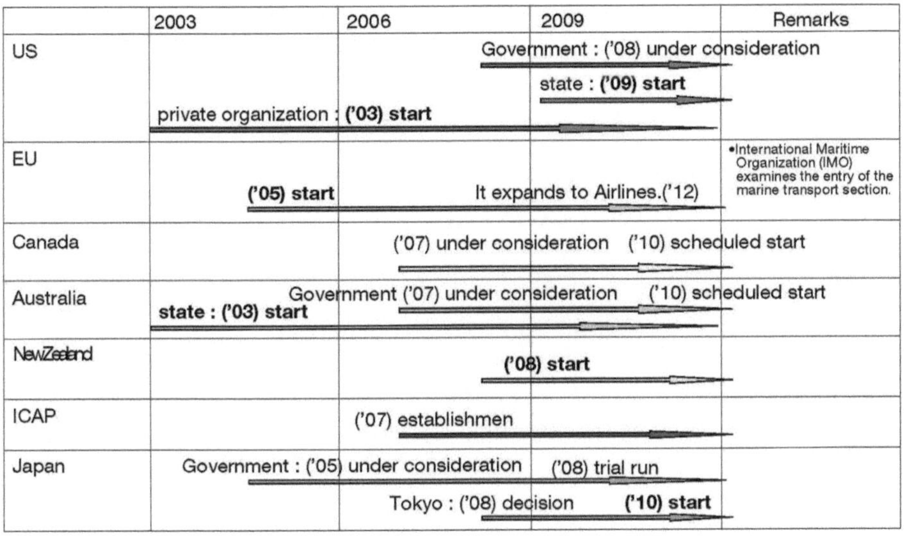

Fig. 2.34 International trend of emissions trading

of this kind at the national government level in 2008. In Japan, the Ministry of Economy, Trade and Industry has been trying such a system in cooperation with other concerned ministries since 2009. The purpose of this system is to further reduce GHG through cooperation among the enterprises that compose the supply chain, in order to change consumers' behaviour. The CFP display has already been executed now in Germany and France, and Canada has settled on the guideline.

Similarly, travel eco-points are used as an approach of "make visible" in Kochi Prefecture in Japan. This approach displays, on the user's personal computer or phone handset, the amount of CO_2 reduced by using the train or bus with an IC card rather than a private car. The reduced amount of CO_2 is added up as eco-points, and all eco-points of users during the year are used as a measure for controlling global warming activity of Kochi Prefecture. The prefecture began to operate this system in 2009. Not only can CO_2 be reduced by changing the thinking of residents, but it is also expected that utilisation of regional PT can be promoted by this approach.

2.4.5 Necessary Changes in the Energy Sector

a. World Energy Outlook of the IEA

In the World Energy Outlook (WEO), the International Energy Agency (IEA 2013) has presented three scenarios for the development of energy demand and its supply by different sources from 2011 to 2035. The "Current Policies Scenario" projects the present policy trends to the future. The "New Policies Scenario"—which is the

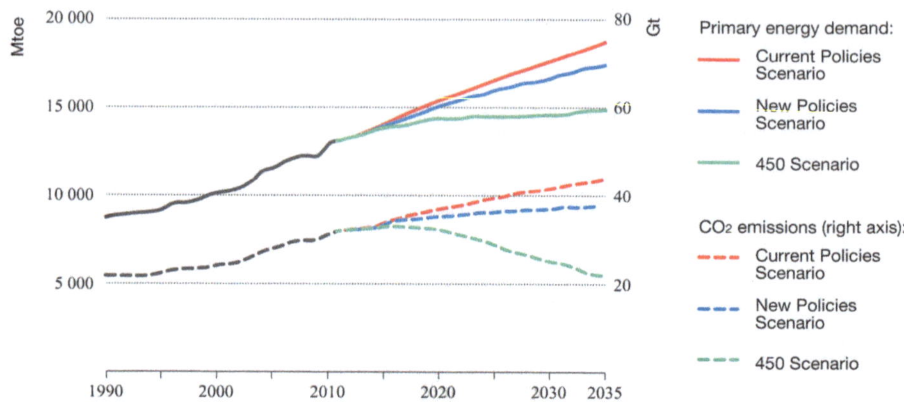

Fig. 2.35 Development of world primary energy demand and GHG emissions until 2035. (Source: IEA 2013)

central scenario of IEA—takes account of declared policy intentions and presumes that these are implemented. The "450 Scenario" describes an energy pathway compatible with a 50% chance of limiting the long-term increase in average global temperature to 2 °C. All scenarios predict an increase of world energy demand at different orders of magnitude: 43% (Current Policies), 33% (New Policies) and 14% (450 Scenario). This results in a change of GHG emissions by 38; 19; −31%, between 2011 and 2035. Even under the assumptions of the 450 Scenario, the drastic decrease of GHG production would only achieve the 1990 levels but is still far from the Intergovernmental Panel on Climate Change (IPCC) reduction target of 50% compared with 1990 levels worldwide and 80% for Organisation for Economic Co-operation and Development (OECD) countries (Fig. 2.35).

Figure 2.36 gives the reason for the substantial increase of world energy demand. While energy demand in Europe or Japan will be stagnating and increase modestly in North America, the emerging and developing economies in Asia, Africa and Latin America will drastically expand their energy consumption. This is partly caused by population growth in these areas and partly by the assumption that the economies will develop more prosperously, following growth champions like China and India.

A decoupling of GHG emissions from population and economic growth would presuppose that regenerative energy sources substitute the fossil sources. Coal and oil would have to be replaced by wind, solar, hydro and biomass sources. The IEA (2013a) gives a numerical example for the dramatic change of the energy supply structure which would be necessary in the next two decades to move to the sustainable "450 path". Against the background of the present policy debate, which has made transparent the egoistic positions of the parties at the World Climate Conference in Warsaw in November 2013, it seems highly improbable that such a leapfrog towards regenerative energy sources will happen.

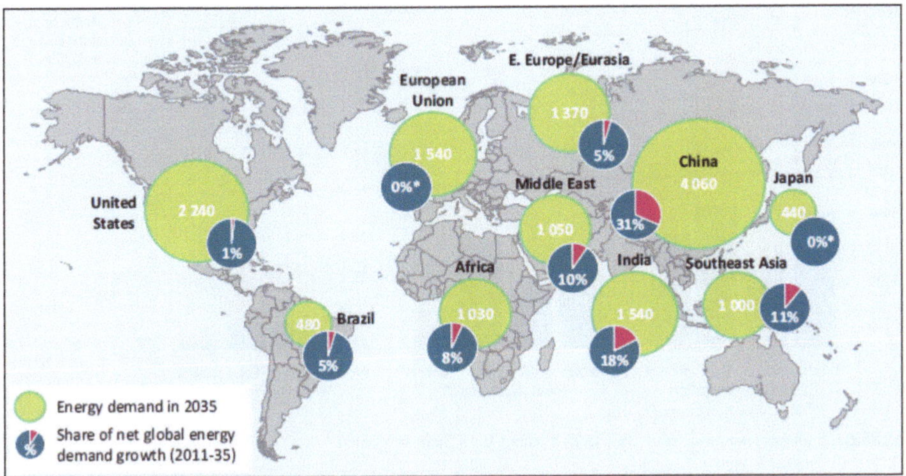

This map is without prejudice to the status of or sovereignty over any territory, to the delimitation of international frontiers and boundaries and to the name of any territory, city or area.

* These regions experience a decline in primary energy demand in 2035 relative to 2011.

Fig. 2.36 Energy demand in the regions of the world in 2035 and growth from 2011 to 2035. (This is without prejudice to the status of or sovereignty over any territory, to the delimitation of international frontiers and boundaries and to the name of any territory, city or area. *These regions experience a decline in primary energy demand in 2035 relative to 2011). (Source: IEA 2013)

Figure 2.37 shows the results from a "New Policies Scenario" which look already very optimistic compared with the present policy trend. Nevertheless, the most problematic fossil energy sources, coal and oil, will contribute the largest shares to the total energy supply. Renewable energy shows a steep upturn and will contribute 18 % of the total energy supply in 2035, starting from 11 % in 2011.

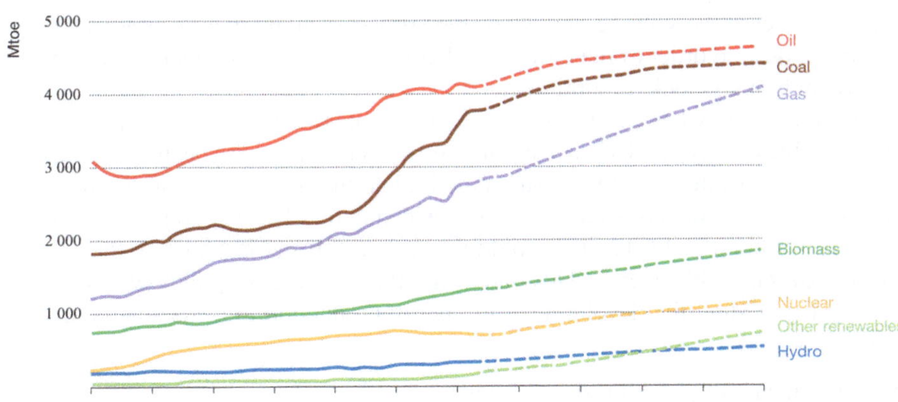

Fig. 2.37 Development of fossil sources and renewables until 2035. (Source: IEA 2013)

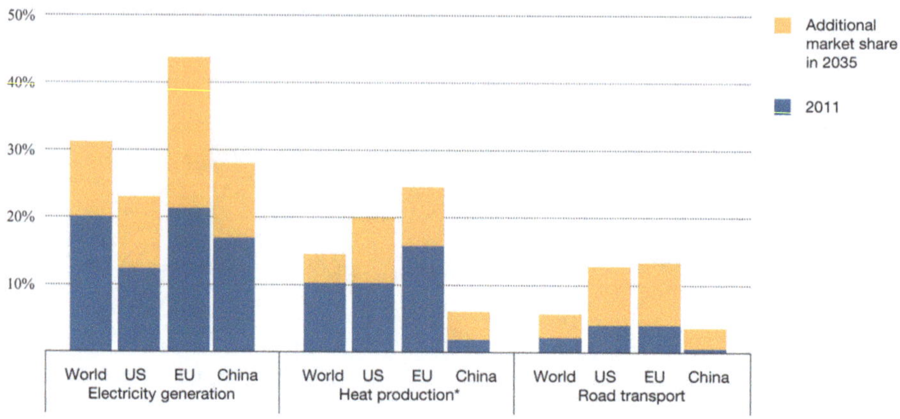

* Excludes traditional biomass. Note: US = United States; EU = European Union.

Fig. 2.38 Contribution of renewable energy to total energy demand. (Source: IEA 2013)

When it comes to evaluate the prospects of renewable energy sources, one has to consider two countervailing aspects. First of all, biomass plays a big role in the WEO and is by far the dominant renewable energy source. It plays a big role for changing the CO_2 footprint of road and air transport. The EU Renewable Energy Directive had foreseen a blending of fuel with at least 10 % of biofuels, but presently, there is discussion about a reduction of the cap to 6 %. Reason is that biomass of the first generation touches other sustainability issues, namely the feedstock production contributing to deforestation and crowding out of food production. Biomass of the second generation (from agricultural residues or dedicated energy crops like palm oil) for supplying cellulosic biofuels is not available at a sufficient volume and the research on algae farming has until now not reached a state which could give hope to increase the volume of sustainable advanced biomass in a foreseeable future in the desired order of magnitude. If there is no progress in this field, the WEO figures on biomass look too optimistic.

Secondly, policy has by far not exploited the potential to support the production of renewable energy. In 2012, a sum total of US$ 111 billion has been spent on subsidies for renewables, which appears quite high at the first glance. But in the same year, US$ 544 billion has been given for subsidising fossil energy sources. Inverting this relationship of subsidisation would foster research and improve on the market chances of renewable energy. This underlines that a change of policy worldwide is the key for guiding the development of energy production to a sustainable trajectory.

b. Consequences for the transport sector

Figure 2.38 is taken from the "New Policies Scenario" of WEO and depicts the additional market share of renewable energy in electricity generation, heat production and road transport—by far the biggest energy consumer of the transport sector. It shows that even a tripling of renewable energy in the transport sector would only lead to a share of energy used in the transport sector of about 6 % worldwide. In the USA and the EU, this share could reach about 13 % and would mainly stem from blending biofuels and using electricity produced by renewable sources. While huge efforts, including subsidisation, are necessary to achieve this change, the result is still modest and represents only a first small step towards a zero-carbon world in transport.

Automobiles have only few options such as electric vehicle (EV), fuel cells, hybrid engines and biofuels. Although much progress has been made with developing e-driven automobiles, a major breakthrough of EV and hydrogen-based technologies cannot be expected before the problems of producing regenerative energy are solved. The most promising technology for mass production is hybrid propulsion combining fuel combustion engines with electrical power trains. This can lead to a reduction of energy consumption, in particular for large cars, and bring down the average fuel consumption of a new car fleet by about 30 %. For road freight transport, the potential of EV is very limited; only in urban areas and for light-duty vehicles, this might be a future option. Therefore, the main strategies have to be based on hybrid propulsion and biofuels.

Ships can also attempt diversification of energy sources, even including wind power. But the main instruments for reducing energy consumption is increased efficiency of engines and lower operating speeds. A further big challenge consists in changing to clean fuel and to tighten emission standards. Starting in 2014, ships fuelled only with heavy oil will not be allowed to sail Europe's North and Baltic seas after the International Maritime Organisation (IMO) has declared these seas an "Emission Control Area". If such regulations stimulate the maritime shipping industry to use liquid gas, this would also have a positive impact on maritime CO_2 production.

For airlines, biofuel is at present the main prospect because of many technological constraints, and they have few options. However, at present, the total production volume of biofuels of the second generation would have to be allocated to aviation to help this sector to pivot into a sustainable development path. The airline manufacturers have invested much technical intelligence into the reduction of energy consumption of airplanes. Nevertheless, the overall result will be limited to 20–30 % for airplanes of the new generation, a progress which will quickly be eaten up by the growth of air transport.

Since the railway system is the first instance run by electricity and hence CO_2 is emitted at the source of electricity generation, railways have more options to achieve a zero-carbon system than other modes. The railways, including Maglev, can contribute in two ways to a technologically based CO_2 reduction: Firstly, they can feed in electricity from the energy supplier with the least CFP as, for instance, hydro or wind energy. Secondly, they can use rolling stock technology which minimises energy consumption through energy recycling, aerodynamics and, in the future, by solar roofs. For example, in Germany, Deutsche Bahn AG plans to increase the share of renewable energy from 24 % (2012) to 40 % (2030). This includes "green service products" such as zero-carbon tickets for bonus card ("BahnCard") owners and ordered zero-carbon traction services for freight transport. This exceeds by far the forecasted share renewable energy in road transport, which is estimated at about 12 % by IEA under the "New Policy" scenario for 2035.

One can draw the conclusion from these facts that the shift strategy from road and air to rail will be the most carbon efficient strategy in the coming decades per unit of transport (passenger-km, tonne-km). While there is undoubtedly big progress with car and air transport technology, it is by far not enough to achieve the carbon performance of the railways on relationships for which railways can expect a high rate of occupancy.

Chapter 3
Climate Change Impacts of Intercity Transport in the Context of External Costs and Their Internalisation

Werner Rothengatter, Yoshitsugu Hayashi, Koichi Fujisaki, Hirokazu Kato, Takaaki Okuda and Naoki Shibahara

3.1 Introduction

Although the climate change issues of intercity transport are in the focus of this book, it is necessary to discuss their impacts in the context of external costs of transport and their internalisation. The reason is that the external costs of transport—e.g. stemming from noise, air pollution or uninsured accident costs—give rise to applying a number of policy instruments for their internalisation. Examples are taxes, charges, regulation or infrastructure provision for environmentally more friendly transport modes. Instruments which aim at reducing greenhouse gases (GHG) of transport therefore have to be coordinated with instruments which give incentives to lower other external costs. Otherwise, problems can occur with addressing the right lever points of instruments and triggering them optimally, for instance with finding the best balance of ETSs and fuel taxation. Also, the acceptance of stakeholder

W. Rothengatter (✉)
Karlsruher Institut für Technologie (KIT), Karlsruher, Germany
e-mail: werner.rothengatter@kit.edu

Y. Hayashi
Graduate School of Environmental Studies, Nagoya University, Nagoya, Japan
e-mail: yhayashi@genv.nagoya-u.ac.jp

H. Kato
Graduate School of Environmental Studies, Nagoya University, Nagoya, Japan
e-mail: kato@genv.nagoya-u.ac.jp

K. Fujisaki
(Formerly) Institution for Transport Policy Studies (ITPS), Tokyo, Japan
e-mail: fujisakiyk@gmail.com

T. Okuda
Graduate School in Business Administration, Nanzan University, Nagoya, Japan
e-mail: okuda1@nanzan-u.ac.jp

N. Shibahara
Japan Environmental Management Association for Industry, Tokyo, Japan
e-mail: shibahara@jemai.or.jp

Y. Hayashi et al. (eds.), *Intercity Transport and Climate Change*,
Transportation Research, Economics and Policy 15, DOI 10.1007/978-3-319-06523-6_3

groups plays a role and might countervail environmental policy programs in which stakeholders feel to be treated unfair because of an overkill of particular instruments (in most cases taxes). Section 3.2 introduces the characteristics of external costs, their relevance with respect to disturbing efficient allocations in the transport sector and the orders of magnitude for the different types of externalities. International examples are presented for the various ways of internalising external costs and the particular impacts of GHG-reducing instruments which partly can be complementary but partly can also be conflicting with other instruments.

Section 3.3 aims to illustrate that external costs and GHG emissions have a long-term impact and require a life-cycle approach. While many studies on external costs and GHG emissions of transport are static in nature and restricted to a one-period analysis, this section shows how to estimate the life-cycle CO_2 emissions of intercity rail versus air transport. The result of this analysis is in particular relevant for policy measures which aim at a long-term control of GHG emissions through supporting low-emission transport modes. The latter requires long-term irreversible infrastructure investments in terms of high-speed rail (HSR) tracks and stations or airport facilities accommodating short-distance air transport. This is considered by applying a life-cycle assessment (LCA) approach which integrates all GHG emissions of a transport system including the infrastructure provision.

Section 3.4 focuses on economic instruments to reduce external costs and carbon footprint, in particular environmental taxation and charging. As the proper dimensioning of such instruments is of utmost importance, it will be shown how the appropriate magnitude of taxation can be derived from different economic approaches such as Pigou or Baumol–Oates pricing. It will turn out that Baumol–Oates pricing is an appropriate instrument to control the target achievements of emission reductions and is in particular preferable if the government decides on explicit reduction targets for external costs and GHG emissions.

3.2 External Cost of Transport

3.2.1 Contents of this Section

Transport is a necessary element of economic activities, as soon as the latter take place at different locations in space. It provides a bridging of space for sourcing, distribution and industrial exchange (freight transport) as well as for business travel, commuting, tourist activities or leisure (passenger transport).

While transport, in principle, is a positive economic activity as it contributes to consumption and production efficiency, there may occur negative effects on people, nature or production processes. When such negative effects are not processed through market forces, i.e. are free of cost for the producer, they represent external effects and require state intervention. In general, car traffic and aviation are causing the highest external costs per unit of travel (per passenger-kilometer (pass-km)).

In the following, the terms "external effects" and "external costs" are defined more precisely and reflected against the background of relevant theories (Sect. 3.2.2). In Sect. 3.2.3, we go into some details of identified external costs and analyse their nature, the possibilities of economic valuation and the options for internalisation. Section 3.2.4 is dedicated to the climate change problem, which deserves special treatment because it can be regarded the biggest challenge after the industrial revolution. Section 3.2.5 presents some comprehensive approaches for empirical measurement of external costs of transport, as they have been applied in European studies, and gives a summary of findings for the European case. Section 3.2.6 gives the conclusions and recommendations for the treatment of external costs of transport in Japan.

3.2.2 Properties of and Theories on External Effects

3.2.2.1 Properties of External Effects and External Costs

External effects have the following properties (see Rothengatter 1999, 2000):

- They occur through an unplanned interaction among agents
- They work outside the market, i.e. they have no market price
- They affect the dynamic efficiency of the economy negatively, i.e. they induce the wrong economic incentives

The first property means that at least the party that is negatively affected has not planned this effect. In many cases, the affected party is not aware of the causation and the working mechanism of the effect. The second property explains why the external effects may be market distorting because the producer of external costs will not get a bill for the cost and the producer of external benefits will not be able to send an invoice to the benefiting parties. To understand the third property, it is necessary to introduce the concept of dynamic efficiency. Dynamic efficiency, according to the ideas of the Nobel Prize laureates D.C. North (1990) and J. E. Stiglitz (1990), means that the economy is able to adjust to the sustainable direction in the medium- or long-term run. Many economic effects are not planned and work outside the market without generating market failures (e.g. implicit contracts as for instance mutual aid). Only if interactions outside the market cause wrong incentive effects which disturb the adjustment of the economy to a long-term sustainable direction, then they imply external costs and require state actions.

3.2.2.2 Pigou Theory

This approach (Pigou 1924) starts from the proposition that the individual agents can observe their (individual) average costs of travelling and plan their route or mode choices rationally on the basis of this information. In a network system, this

Fig. 3.1 Pigou approach to quantifying external costs

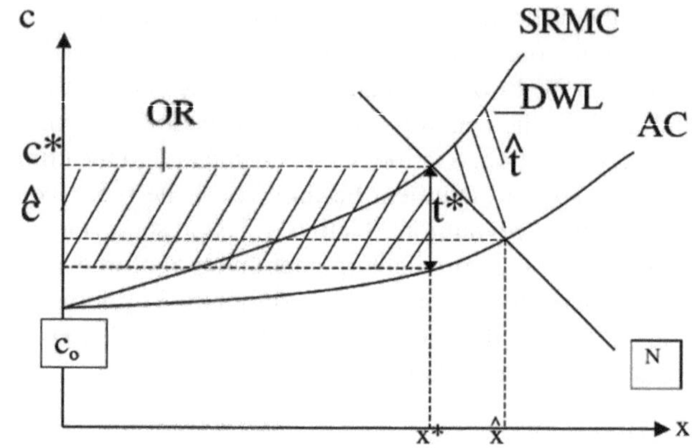

leads to the so-called user equilibrium situation. This means with respect to route choices that all routes used between an origin and a destination show the same (generalised) average cost. Therefore, no user has an incentive to change the route decision taken. The social optimum, however, would presuppose that the (generalised) marginal costs for every used route would be the same, such that no user can improve his position without making another user worse off. These situations correspond to the Wardrop principles of transport engineering (Wardrop 1952) or the Nash versus Pareto equilibria in economic theory.

The corresponding situations are presented in the Marshall diagram of Fig. 3.1 We pick out only one route and analyse the cost and the demand side by constructing an analogy to the partial market theory of A. Marshall (1920). If there are only two route alternatives used, then the diagram explains the total market situation, because the second market is just the complement of the first. If there are more than two routes and eventually further choices (mode, destination, travel time), then the diagram only gives a partial view of the problem. It helps to understand the situation, but a total-optimisation approach would be required to determine the user equilibrium and the systems optimum.[1]

In Fig. 3.2, SRMC denotes the short-run marginal costs, AC the average costs and N the demand curve. If the individuals maximise their utility on the basis of the perceived average costs, then the activity volume will be x^. The systems optimum would be x^* and the welfare loss associated with the user equilibrium is "dead weight loss" (DWL). A Pigou tax t^* in terms of the difference between marginal and average costs would internalise the externality and lead to the social optimum x^*. It is essential that t^* is computed for the activity level x^*, which cannot be observed, rather than follows from an optimality calculus. If the Pigou tax would be calculated

[1] We should emphasise the limited explanatory power of the Pigou diagram (Fig. 2.1) because many economic approaches start from this simple platform and do not consider the failures occurring when the set of choices for the agents is much richer. As to the network-based optimisation approaches to model user and systems optimum, see LeBlanc and Rothengatter (1982).

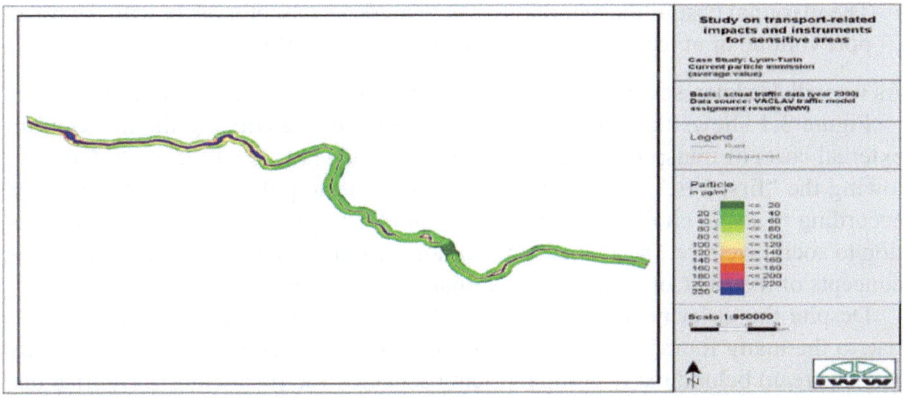

Fig. 3.2 Lyon–Turino corridor exhibiting particle concentrations stemming from road traffic. (Source: IWW 2003)

based on the observed level of activity $x^\wedge$, the result would be $t^\wedge$ which would be a much too high dose for the internalisation strategy.[2]

Pigou has started this exercise with the external costs of congestion. Such costs occur if a user does not consider the impacts, which he/she is imposing on others, in terms of time losses or additional operating costs. These external costs are identical to welfare losses, i.e. the DWL. They will vanish once the optimal Pigou tax is implemented.[3]

The Pigou scheme is not restricted to congestion externalities. Other types of externalities can be added to the marginal cost function SRMC, which then will represent a social marginal cost function. This also includes the marginal costs of climate change as will be shown later. The rest of the Pigou procedure is left unchanged. The optimal Pigou tax (or optimal externality charge) is calculated by taking the difference between the (individually perceived) average cost and the social marginal costs, at the system's optimal traffic volume x^*.

Again, the graphical representation in the second dimension looks simple and can be found in every textbook on welfare economics or optimal pricing. But looking at the theoretical background, we must be aware that this simple adding of marginal cost functions is only consistent if two conditions hold[4]:

- The externality has to relate to traffic volume and has to be quantified as an additional social cost when adding one unit of transport activity.

---

[2] This is mentioned here because many empirical quantifications of the Pigou tax, in particular in the transport sector, do not take this risk of failure into account.

[3] In many empirical applications, the external marginal costs of congestion are measured by taking the difference between the marginal costs and the cost under free flow conditions, c_0.

It follows immediately that the congestion externality is overestimated by far in this case (see the UNITE study for the EU Commission). INFRAS and IWW (2004) has shown that the magnitude of overestimation can reach a factor of 7.

[4] We emphasised these conditions because they are violated in many practical applications.

- The marginal cost of the externality has to increase monotonously with the transport volume or at least to be a constant per unit of traffic.

As we will see later, several external costs of transport do not meet these conditions.

Figure 3.1 shows the simple textbook exercise for calculating the mark-up for external costs of transport. Optimal charge is *t** and leads to a revenue of OR. Following the "first-best" welfare theory OR goes to the public budget and is spent according to social choice rules, i.e. for the public projects with maximal contribution to social welfare. In practical applications, for instance in the transport pricing concepts of the EU Commission, an earmarking of OR is considered.

Despite the strict neoclassical world of unrealistic assumptions—which gives rise to the many footnotes and side remarks—the Pigou concept is the main theoretical stream behind many optimal pricing schemes and treatments of external diseconomies. The applications are often characterised by pragmatic simplifications to overcome the problems of unrealistic assumptions, as for instance cancelling effects which do not fit into the scheme, linearisation of non-convex functions or introducing linear mark-ups for effects that are not captured by the theoretical approach (e.g. noise, nature and landscape).

3.2.2.3 Coase Theory

Coase has established an alternative theory of external effects in his famous article "The Problem of Social Cost" (1960). His findings are:

1. Externalities occur because property rights for a jointly used resource are not well defined (e.g. air, water, soil).
2. Externalities are reciprocal, i.e. the production activity of the "perpetrator" is harmful for the "victim", but reducing the production activity will be harmful for the "perpetrator".
3. If transaction costs are negligible, an efficient allocation of resources will be obtained through a clear allocation of property rights, regardless of how the property rights are assigned (to the "perpetrator" or to the "victim").
4. If transaction costs are not negligible, it has to be checked whether the benefits from reducing the harmful activity are higher than the losses.

Coase has explained the important fourth proposition of his theorem in the following way (Coase 1960, p. 27):

> The problem which we face in dealing with actions which have harmful effects is not simply one of restraining those responsible for them. What has to be decided is whether the gain from preventing the harm is greater than the loss which would be suffered elsewhere as a result of stopping the action which produces the harm.

This was misunderstood for a couple of years as a postulation of public inactivity in case of environmental external diseconomies.[5] But this is not the heart of the Coase theorem. The theorem postulates, indeed, to examine whether an externality is

[5] Remember that Ronald Coase was a professor of the University of Chicago, which in the 1970s started to re-establish the neoclassical economic theory by integrating modern attributes. This led

really detrimental to dynamic efficiency (see the definition of externalities above). To understand the main idea, let us cluster the interactions including the transport sector that may cause externalities into three categories:

1. Interactions within a transport sector (e.g. congestion)
2. Interactions between the transport sector and industry sectors (e.g. traffic noise and vibrations disturbing production)
3. Interactions between the transport sector and the stock of non-renewable resources (human resources, environmental resources)

Starting with the first interaction phenomenon, we immediately see that the congestion externalities are external to the individual users but internal to the road sector.[6] Nevertheless, there is a negative impact for society because road transport is inefficient as long as congestion externalities occur. Whether or not a Pigou scheme is appropriate to remove the source of inefficiency is dependent on the impacts of this scheme. It may impose high transaction costs, lead to undesired diversion effects (if good alternatives like public transport are missing) or change the land use towards a suboptimal direction. A Coase-check means that some sort of a cost–benefit analysis has to be carried out to calculate the balance of such positive and negative results. This check is the essential part missing in the Pigou concept.

Interactions between transport and the industrial sectors are usually based on market-conforming arrangements, expressed by pricing and explicit contracts. But there may be also interactions outside the market, for instance the delivery activity of a haulage firm for retail company *A* may disturb retailer *B*. On a larger scale, there may be many such interactions that work multilaterally. Therefore, there is little net benefit in removing such types of externalities because the transaction costs for an internalisation scheme would be very high.

Interactions between the transport sector and the sector of non-renewable resources are different from the above examples because non-renewable resources cannot simply be traded off against other material goods. Environmental resources (including the climate resource) and human resources are limited in this sense and it is a challenge for society to prevent these resources from being exploited by production and consumption activities. From this follows that society has to control the ownership of property rights and the rules for using non-renewables for production or consumption activities. This is the challenge behind the issue of "internalisation of externalities" and goes beyond the "setting prices right" recommendation of the economic main stream.

3.2.2.4 Theory of Ecological Economics

The environmental economics approach starts from the definition of safe minimum values. A clear distinction is made between externalities, which generate some

to the revival of supply-side economics, which places its main emphasis on the functioning of production and distribution as well as the minimisation of state interventions ("Chicago School").

[6] Note that external costs of congestion only consist of additional time losses and operating costs. Other externalities like climate effects are treated as their own category.

inconvenience and utility losses of the affected parties, and externalities, which cause risk for human life and for nature (e.g. for rare species, biodiversity or climate change). While the first category of externalities can be traded off against monetary compensation, the second category is not tradable. The state has the task of setting clear regulation standards, perhaps supported by incentive compatible price signals, to ensure that the safe minimum standards are not overrun.

Examples of the protection of human resources are limit values set for noise emissions or for concentration of air pollution (Directive 1999/30 EC for particulate matter and NO_x). Rare species of flora and fauna are protected by the World Wildlife Fund (WWF) or the Habitats Directive of the EU (92/43 EC). They impose mandatory constraints on transport planning (see for instance, the Oeresund Project and the special treatments of problems like breeding grounds of birds or the stabilising of water flow current between the North Sea and the Baltic Sea) .

CO_2 concentration in the atmosphere affects the living conditions of future generations and should be controlled by international agreements (Post Kyoto Agreements). The Intergovernmental Panel on Climate Change (IPCC) has proposed to limit CO_2 concentration in such a way that the world temperature will not increase beyond 2 °C until the end of this century. The protection of people, flora and fauna is a challenge as such to set limit values on CO_2 emissions and to control them internationally. It can be useful to underline the above precaution argument by economic calculations as was done in the Stern Review (2006). The political decision makers have been substantially influenced by Stern's argument that starting early with effective mitigation measures will yield much higher benefits in terms of avoided future damages than the costs of the investment.

Several economic methods have been developed following this baseline. The best-known and practically relevant approach of this family is the "standard prices" approach, developed by Baumol and Oates (1971)[7]. It starts from a target level for the reduction of the externality considered (Baumol and Oates limited their approach to environmental externalities) and derives the prices, which are necessary to achieve the target level, from the demand functions. This method can be combined with a benefit–cost assessment, which is in particular necessary in the case of multiple externalities and multiple instruments for their internalisation. In the latter case, the pricing scheme is an intrinsic part of a package of instruments, which is optimised with respect to effectiveness and limitation of undesired side effects.

The *multiple externalities/multiple instruments approach* combines the Baumol/Oates prices with further instruments such as emission trading, regulation or emission labelling. In the long run also infrastructure investment and land use optimisation can be powerful instruments. This gives up the Pigou idea that all externalities can be internalised by one instrument, i.e. social marginal cost pricing, and suggests a bundle of instruments composed on the basis of the expected demand reactions in the transport sector considered and the secondary impacts on other sectors of the economy (e.g. land use).

[7] This pricing approach is applied in Sect. 3.4, where it is explained in more detail.

The big advantage of the *multiple externalities/multiple instruments approach* is that every externality can be treated separately to identify which instrument is best to achieve the desired reduction. In particular, effects that are not directly dependent on traffic volume or effects which show decreasing marginal costs can be treated by this approach. Examples are effects on nature and landscape (dependent on the infrastructure provision), upstream–downstream effects (dependent on fuel, vehicle and infrastructure provision[8]) or noise impacts (diminishing per vehicle with increasing traffic). In this context, least-cost calculations can help to find the economically best triggers for reduction policies.

3.2.3 External Costs of Transport

3.2.3.1 Classification

Transport activities are generating a high magnitude of external effects. Estimations come to a share of up to 7 % of gross domestic product (GDP) (INFRAS and IWW 1995, 2004). In other words: Spatial movement of people and goods is substantially subsidised by society by covering the costs of non-market impacts. Therefore, the European Commission has decided to take these effects into account, when assessing investment and pricing strategies in the transport sector. According to the study of INFRAS and IWW (2004), which is one of the baselines for the Handbook on internalisation of external costs of transport (2008), elaborated for the European Commission, external costs can be classified into the following categories:

1. ***Congestion costs***
   Congestion externalities occur because the agents do not take into account the impacts of their transport activities on other transport users. This is a typical phenomenon of interactions among agents planning their activities individually. Contrasting the other externalities, they are in the first instance intra-sectoral, i.e. they are covered by the transport users as a whole, but wrongly allocated individually. Therefore, congestion externalities give rise to introduction of management and charging systems, but they are not relevant to an intermodal comparison of external costs.
2. ***Accidents***
   This external cost category includes all social costs of accidents, which are not recovered by insurance premiums/indemnities. In particular, a part of the human costs must be covered by society, as for instance lost production and cost of suffering and pain.
3. ***Noise***
   Noise above a certain level causes health problems and associated costs of medical treatment and production losses.

[8] Section 3.2 shows this in a life-cycle context.

4. ***Air pollution***
   Pollutant emissions like sulphur dioxide, hydrocarbonates, nitrogen oxides or particulate matter cause considerable costs of lost production, reduced consumption utility, impacts on nature and human health/life.
5. ***Climate change***
   Climate change causes a number of impacts on nature, as for instance melting of glaciers and ice caps, rising of sea levels or increased natural disaster events, which lead to high costs of abatement.
6. ***Nature and landscape***
   Infrastructure and transport activities may cause impacts on biodiversity, separation effects and landscape (effects on tourism).
7. ***Urban separation effects***
   Infrastructure and transport activities may separate urban ensembles.
8. ***Upstream and downstream processes***
   Production of vehicles, provision of fuel and construction of the infrastructure generate externalities in the upstream of transport production while the disposal of used vehicles leads to downstream external costs.

3.2.3.2 Quantification and Evaluation Methods

Quantification Methods

The quantitative analysis of impacts can use a top-down or a bottom-up analysis. The top-down analysis will generate values consistent with the macroscopic observations and the associated indicators. It usually leads to the calculation of average figures for the externalities (total costs of impacts divided by the number of traffic activities (e.g. pass-km or t-km).

While aggregate analysis can be done on the basis of clusters of infrastructure and population, the bottom-up approach starts from an impact path analysis which presupposes a more detailed modelling of transport, the environment, the land cover data and the affected population. INFRAS/IWW have applied network analysis, using transport supply and demand modelling and combined this with detailed geographic information systems (GIS) modelling to model diffusion, concentration and exposed population exactly enough. Impact path analysis requires a high spatial resolution and is therefore in general restricted to corridors (see Fig. 3.2).

Evaluation Methods

The widely applied evaluation methods for external costs of transport are:

- Damage cost approach: Future costs of damages are estimated and discounted to determine the present values of damage costs. This is often applied for: cost of accidents, air pollution and noise.

- Willingness-to-pay approach: Potentially affected people are asked for their probable responses to the externality (stated preference), or the decisions of agents are observed and the willingness to pay is calculated backwards (revealed preference). This is often applied for: cost of accidents, air pollution and noise.
- Hedonic pricing: Demand functions are constructed which include the externality—besides prices and budget constraints. Then, the willingness to pay can be derived from the econometrically tested demand functions. This is often applied for: noise, air pollutionin urban areas.
- Avoidance cost approach: A standard is set for the reduction of the externality and the least costs are calculated to reach this standard. This is often applied for climate costs and disturbance of nature and landscape.
- Opportunity cost approach: Safe minimum values are defined and an optimisation calculus is performed to maximise welfare (benefits) subject to the constraints set. The economic values follow from the dual variables of the optimisation program. This is applied through least-cost approaches in safe minimum values calculation (see the example in Sect. 3.3.4).

3.2.3.3 Measurement, Valuation and Internalisation by Type of Effects

In this section, we treat all externalities listed above except for climate change. As the issues of climate change are on the political agenda on all policy levels, they are discussed in Chap. 4.

Congestion Cost

The measurement of marginal external congestion costs, their valuation and internalisation follow directly from Fig. 3.1. The optimal value of the internalisation charge is measured as the difference between social marginal and private average costs at the point of optimal capacity loading. While the theoretical concept seems simple and intuitive, its practical implementation is very difficult and its use for deriving an optimal charging scheme for the use of scarce infrastructure includes a number of caveats:

Problem 1: Proper quantification of speed-flow relationships.
Problem 2: Identification of the optimal point of capacity loading.
Problem 3: Neglecting impacts on land use.
Problem 4: Partial application to network parts or selected traffic categories.
Problem 5: Use of revenues.

Cost of Noise

a. *Nature of the effect and measurement*
Transport noise causes annoyance and health risks. Noise is usually measured in the decibel (dB) scale, which is logarithmic, reflecting the logarithmic manner in which the human ear responds to sound pressure. Since the human ear is also more sensitive at some frequencies than at others, a frequency weighting is applied to measurements and calculations. The most common frequency weighting is the "A weighting", hence the use of dB(A).

The logarithmic nature of noise is also reflected in the relationship between noise and traffic volume. By halving or doubling the amount of traffic, the noise level will be changed by 3 dB, irrespective of the existing flow. This means that an increase of traffic volume from 50–100 vehicles per hour will result in the same increase in the noise level (3 dB) as a doubling of the transport volume from 500–1000 vehicles per hour (Handbook 2008).

People can feel annoyed already at low noise levels, depending on their sensitivity to noise.

Health damage can occur with levels of noise higher than 55 dB(A) at night and 65 dB(A) in the daytime in a relevant order of magnitude. Reduced sleep quality, nervous stress reactions and change of heartbeat frequency are usual responses of the human body to noise. The risk of cardiovascular diseases (heart and blood circulation) increases.

b. *Evaluation*
The negative impacts of noise on human health result in various types of costs, like medical costs, costs of productivity loss and the costs of increased mortality. The risk of serious health impacts can be quantified by evaluating field studies, which

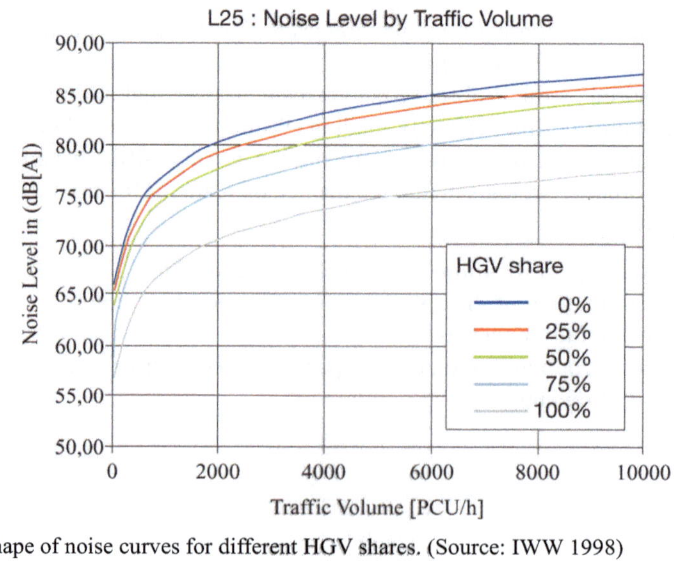

Fig. 3.3 Shape of noise curves for different HGV shares. (Source: IWW 1998)

have been carried out in Switzerland or in the USA. Economic costs are quantified through the costs of medical treatment plus the (opportunity) cost of labour ("years of work lost").

The annoyance impacts of noise are reflected in the willingness to pay for lower noise levels, as for instance expressed by land prices or the room rates of hotels. Such values can be derived from hedonic pricing studies (revealed preference approach) or willingness-to-pay questioning (stated preference approach).

The cost of one additional unit of noise goes up with the noise level. But as the dB(A) curve is logarithmic, the final shape of the noise *cost* curve depends on the steepness of the noise per unit profile. According to many noise studies, the final outcome is a logarithmic curve again. This implies that the marginal costs of traffic noise are decreasing with increasing traffic volume.

c. *Internalisation*

Due to the logarithmic nature of the relationship between noise costs and traffic volume, an internalisation through marginal cost allocation does not make sense, economically. Two alternative internalisation approaches are possible: First, an "average marginal cost value" is calculated and used as a proxy of the marginal cost of noise. Secondly, noise is not internalised through a marginal cost charge rather than through a regulation scheme, noise compensation payments or noise specific obligatory investments. Examples for regulations are speed limits or technical standards. Noise compensation payments in several EU countries are obligatory by law if new roads or starting/landing tracks are realised which extend the disturbed area. Technical standards apply for trucks or buses (capsulated engines) and are also discussed for railway freight wagons (e.g. silent brakes). For new tracks, often noise walls are obligatory.

Air Pollution

a. *Nature of effect and measurement*

Transport-related air pollution causes damages to humans, biosphere, soil, water, buildings and materials. The most important pollutants are the following:

- Particulate matter: PM_{10}, $PM_{2.5}$
- Nitrogen oxides: NO_x, NO_2
- Sulphur oxide: SO_2
- Ozone: O_3
- Volatile organic compounds: VOC

Studies on air pollution costs generally cover the following impact categories (Handbook 2008):

- *Health costs:* Impacts on human health due to the aspiration of fine particles ($PM_{2.5}$/PM_{10}, other air pollutants). Exhaust emission particles are here considered as the most important pollutant. In addition, ozone (O_3) affects human health.

- *Building and material damages:* Impacts to buildings and materials from air pollutants. Mainly two effects are of importance: soiling of building surfaces/facades mainly through particles and dust. The second, more important impact on facades and materials is the degradation through corrosive processes due to acid air pollutants like NO*x* and SO_2.
- *Crop losses in agriculture and impacts on the biosphere:* Crops as well as forests and other ecosystems are damaged by acid deposition, ozone exposition and SO_2.
- *Impacts on biodiversity and ecosystems (soil and water/groundwater):* The impacts on soil and groundwater are mainly caused by eutrophication and acidification due to the deposition of nitrogen oxides as well as contamination with heavy metals (from tyre wear and tear).

The impact pathway approach is an appropriate procedure for measurement and evaluation. This is depicted by Fig. 3.4

The process starts by estimating or forecasting the volume of traffic activity and assigning the corresponding emissions on the basis of vehicle movement cycles. From this, concentration levels are derived using emission concentration models.

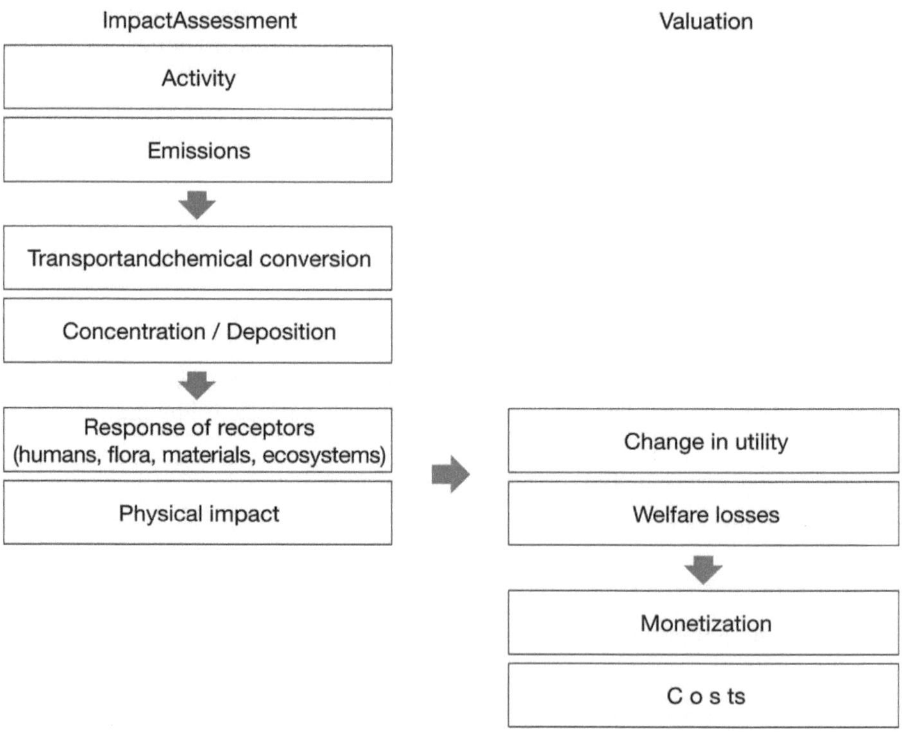

Fig. 3.4 Impact pathway approach for quantification and valuation of air pollution costs. (Source: HEATCO 2006)

Table 3.1 Monetary values for economic valuation in Europe. (Source: HEATCO 2006)

Impact	€ per unit (2002)
Human health, effects in respective units	
Acute mortality—years of life lost due to acute exposure	60,500
Chronic mortality—years of life lost (YOLL) due to chronic exposure	40,300
New cases of chronic bronchitis	153,000
Hospital admissions (respiratory and attributable emergency cardiac)	1,900
Restricted activity days	76
Minor restricted activity days; cough days; symptom days (lower respiratory symptoms including cough); days of lower respiratory symptoms, including cough, in children in the general population, i.e. extra symptoms days	31
Days of bronchodilator usage	1.0

The concentration data then are mapped with the exposed population, buildings and nature to model the human and physical impacts.

b. *Evaluation*
Meanwhile there is a good database on the human and physical damage costs of air pollution following from the different sources of air pollution. The above table gives average values for economic valuations in Europe (HEATCO 2006) (Table 3.1).

c. *Internalisation*
Most of the air pollution impacts increase with increasing traffic volume. Therefore, internalisation through a kilometre-based charge can be an appropriate measure. The effect on traffic behaviour is dependent on the size of the mark-up on the infrastructure charge. If the charge is relatively low, then there will be little incentive to change to green technology. Therefore, charging has to be accompanied by standard setting and regulation, as is done in Europe through the Euro standards for air pollution. At present, Euro 6 is obligatory in Europe for trucks (since January 2013) and for cars since January 2014. The German motorway charging scheme for heavy goods vehicle (HGV) is differentiated according to these Euro standards, including already the new norm for Euro 6, which has created strong incentives for changing to cleaner technology.

Accidents

a. *Nature of effect and measurement*
External accident costs are that part of the costs of accidents which is not recovered by private agents involved and their insurance system. The most relevant cost categories are material damages, administrative costs, medical costs and production losses, plus a social risk value which reflects human pain, grief and suffering. The basics of measurement stem from accident statistics and traffic engineering, which generate probability functions for the type of incidents and their significant drivers. In general, traffic volume is only one of the drivers, while others can dominate, as for instance weather conditions, alcohol, age of drivers or mix of speeds.

A particular problem for evaluating road accidents is related to the concept of risk elasticity. This refers to the relationship between accident risk and traffic volume. In many cases, the risk of a particular type of accident, such as accidents with fatalities, is decreasing with increasing traffic volume for certain intervals of traffic volume. This means that elasticity may come out as negative. This holds for example for motorways and urban expressways where speed goes down with higher traffic volume and the risk of fatal accidents is reduced as well (while the risk of material damage increases).

The accident risks in other traffic modes such as railways or aviation follow completely different causality. The risks for passengers in these modes are lower by orders of magnitude on the average (lower than the factor 1/10 in most countries). However, most of the accidents which occur are spectacular and widely perceived events which are discussed in the media over a long time. Therefore, the subjective perception of risk with respect to rail or air travel is much higher than the objective risk figures.

b. *Evaluation*

The evaluation task consists of two major steps: First of all, it has to be decided which parts of the accidents or accident risks are private and which parts are social. This problem is easily solved in top-down approaches by assuming that all costs of material damages, administrative activity and medical treatment are covered by insurance and are therefore private. The social part is production losses and the social risk value (as the preference of society to preserve health and lives). In bottom-up approaches, the social risk is analysed further and differentiated according to private risk behaviour. As a consequence, the latter approach leads to much lower values of social accident risk.

The individual impacts of accident events can be evaluated with their monetary consequences. The studies on accident costs, such as INFRAS and IWW (2004) or HEATCO (2005) give very detailed classifications of impact type. (A rough classification would be by only fatality, severe injury, slight injury and material damage.) Evaluating the value of lost lives is the most difficult exercise for economists, which is in general performed by estimating the value of future production and the additional social value of pain, grief and suffering (by willingness to pay or revealed preference estimations, eventually derived from traffic safety programs of the state).

c. *Internalisation*

It follows immediately from the measurement problems and the shape of accident costs curves (remember the occurrence of negative elasticities) that accident cost curves are not monotonously upward sloping, as it would be necessary to apply the marginal social cost approach of internalisation. So, many authors suggest applying different instruments rather than marginal costs for internalisation. Possibilities are:

- Mark-ups on insurance premiums (which are differentiated by the insurance companies according to risk categories)
- Regulation of vehicle technology (active and passive safety measures)
- Speed regulations
- Stricter regulation of alcohol consumption
- Driver education

It can immediately be seen that such measures address the drivers of accident risk in a much more direct manner than can marginal cost pricing. But despite the measures above, which are taken with different intensities in the countries, rail and air transport are still much safer than road transport. This big difference, which continues, gives rise to thought about a further tightening of safety measures for road transport as well as a subsidisation of safe transport modes because of their savings of social costs. Incidentally, a big part of congestion costs (about 25 %) on roads is caused by accidents and would be reduced by improving safety.

Impacts on Nature and Landscape

a. *Nature of effect and measurement*
We summarise several effect categories under this item such as (1) the impacts on ground sealing, ecosystems, habitats and biodiversity and (2) the impacts on soil and water pollution.

1. Land taking, ground sealing, ecosystems, habitats and biodiversity
These effects in general relate to the infrastructure provision and must be treated in cost–benefit analysis for new or upgraded infrastructure. They can differ widely by type of region (e.g. visual intrusion in a tourist area compared within industrial areas). Such effects can be a reason for taxation or regulation of particular transport modes, which consume more of these natural resources. But in principle, they also can be added to the material infrastructure costs and allocated on an average cost basis.

2. Soil and water pollution
Soil and water pollution is in general dependent on the level of activities and the technology used. Therefore, this type of external costs can be treated like the costs of air pollution.

b. *Evaluation*
The estimation procedures are (Handbook 2008):

- Repair cost approach for ground sealing and other impacts on ecosystems (disturbance of animals and their biotopes by noise or barrier effects, visual disturbance, etc.; INFRAS and IWW 2000, 2004). Standard price approach for quantifying the negative effects of airborne emissions on ecosystems and biodiversity (through acidification and eutrophication; ExternE 1999; NewExt 2004).
- Two-stage approach for quantifying biodiversity losses: (a) repair costs for reduced species diversity due to *land use change* and (b) repair costs for negative effects of *airborne emissions* on ecosystems and biodiversity (through acidification and eutrophication; NEEDS 2005).
- Two-stage approach for habitat loss and fragmentation: (a) compensation costs for *habitat loss* due to transport infrastructure (creating a compensatory ecosystem) and (b) compensation cost approach for *habitat fragmentation* (OSD 2003).

c. *Internalisation*
Only the costs of soil and water can be internalised using the social marginal cost approach. The other costs can be transformed into average cost values and allocated to the infrastructure users as social overhead costs. When it comes to cost–benefit analysis or to social cost comparisons between roads and competing modes, then this type of cost can play an important role. For instance, in urban areas land taking can be a highly relevant criterion while in rural areas the preservation of biodiversity in protected zones can be a prime target of planning.

Urban Separation Effects

a. *Nature of effects and measurement*
In urban areas, a good share of communication and mobility is performed by non-motorised mobility. Building urban expressways or transit lines at grade can lead to serious interruptions of social interaction in agglomerated districts. Therefore, such effects must be considered in the first instance when planning new or upgrading existing infrastructure.

It is possible to measure the impacts by time losses or loss of space for pedestrians or bicyclists.

b. *Evaluation*
The measured time losses can be evaluated by value of time factors, which can be differentiated by type of urban areas.

Upstream and Downstream Effects

a. *Nature of effect and measurement*
Production of energy, vehicles and infrastructure (upstream) causes additional external effect as well as scrapping and disposal of old vehicles (downstream). The most relevant processes are:

- *Energy production (pre-combustion):* This includes the production of fuel (petrol, diesel) for combustion propulsion (most road vehicles, airplanes, waterway shipping) and the production of electrical energy for the railways.
- *Vehicle production, maintenance and disposal*: The production, maintenance and disposal of vehicles and rolling stock causes environmental effects (emission of air, water, soil pollutants, GHGs, etc.) during a long period, considering the life cycles of the vehicle stock.
- *Infrastructure construction, maintenance and disposal*: The construction, maintenance and disposal of infrastructure elements also lead to negative environmental effects (emission of pollutants).

- *Scrapping and disposal of old vehicles*: Scrapping and storing old vehicles as well as transporting scrapped material to the locations of further processing causes energy consumption and negative externalities.

b. *Evaluation*
The methodology of calculating up- and downstream processes is virtually the same in all studies quantifying these costs: The costs are calculated the same way as the direct external cost categories of transport operating, mainly based on additional air pollution and climate change costs. The main difference between the studies is the different kind of cost categories (effects) covered: Some studies only cover climate change costs of up- and downstream processes whereas others also cover air pollution costs and costs due to nuclear power risks (INFRAS and IWW 2000, 2004a; ExternE 1999; NewExt 2004; Friedrich and Bickel 2001; OSD 2006).

c. *Internalisation*
Upstream externalities can be internalised easily through fuel taxes and also through charging systems. Downstream externalities are not directly related to veh.km and therefore not easily internalised through a social marginal cost scheme. A frequently discussed idea is to force the vehicle producer to take back the used vehicle at the end of its economic life and make the producer responsible for proper scrapping and disposal. By this procedure, the producer would add a surcharge for the downstream process to the sales price and thus internalise the costs occurring at the end of the vehicle's life. Internalising upstream external costs of infrastructure provision is a particular problem when it comes to cost comparisons between modes with high infrastructure needs (railways) and others (air transport). The economic opportunity cost principle would suggest calculating the external upstream costs of infrastructure provision as well for existing and for new tracks. In practice, usually the existing infrastructure is ruled out and only new investment is included.

Summary on Internalisation

The discussion of the individual elements of external costs has shown that only some of these can be treated properly by the neoclassical economic instrument, the social marginal cost approach. This does not mean that pricing instruments do not work for the other externalities. But it follows that pricing instruments should be triggered in a different way, either by using Baumol/Oates prices or by integrating prices into a multiple not instrument bundle.

Figure 3.5 shows that there is a host of possible instruments and that in many cases a combination of instruments may achieve the best results and acceptance.

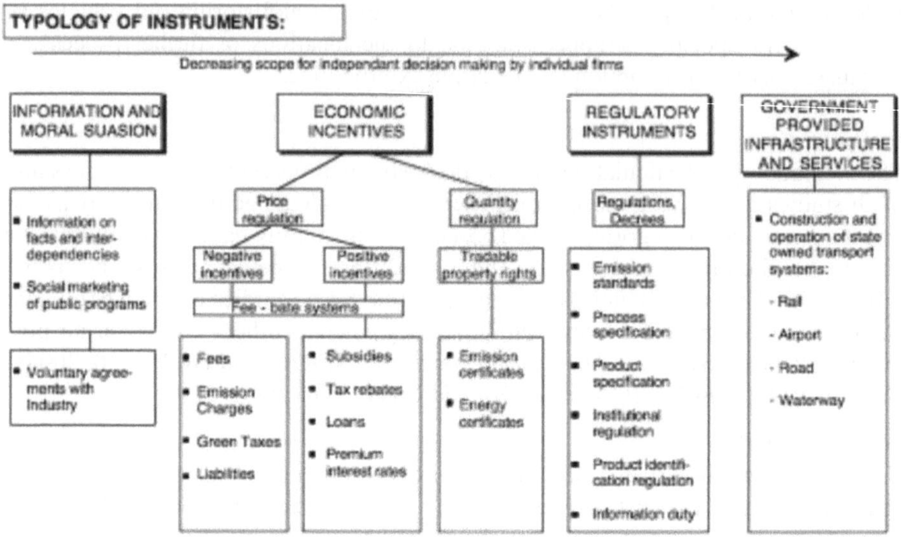

Fig. 3.5 Typology of instruments to internalise external costs of transport. (Source: INFRAS and IWW 1995)

3.2.4 *Special Challenge: Climate Change*

3.2.4.1 Global Problem

The nature of the climate change problem has been presented in Sect. 1.2 based on the IPCC findings and the status of progress of the World Climate Conferences. Figure 3.6 indicates the dimension of necessary reduction of GHG emissions, which sums up to 50 % worldwide in the year 2100 compared to 1990. If the developing and transition countries should be given a chance to develop their economies and to catch up with the industrialised countries, then the latter would have to reduce CO_2 emissions by 80 % (Fig. 3.6).

The Stern Review estimates the cost of reducing CO_2 emissions to a target level of 550 ppm to amount to 1 % of GDP per annum (p.a.). The damage costs, which must be expected in the future in the business-as-usual case, would sum up to a magnitude of 5 % of GDP p.a. or more after the year 2050, or even more. In the meantime, the IPCC has stated that a level of 550 ppm might not be sustainable. A reduction to the level of 500 or even 450 ppm appears to be necessary to stabilise global warming to not more than 2 °C. Nicholas Stern has adjusted his estimations accordingly and has mentioned in his recent speeches that a higher intensity of mitigation measures leading to an increase of the cost of mitigation to about 2 % of GDP might be necessary. Nevertheless, the basic message is left unchanged: That big reduction steps now pay off in the future with high economic benefits in terms of avoided damage costs.

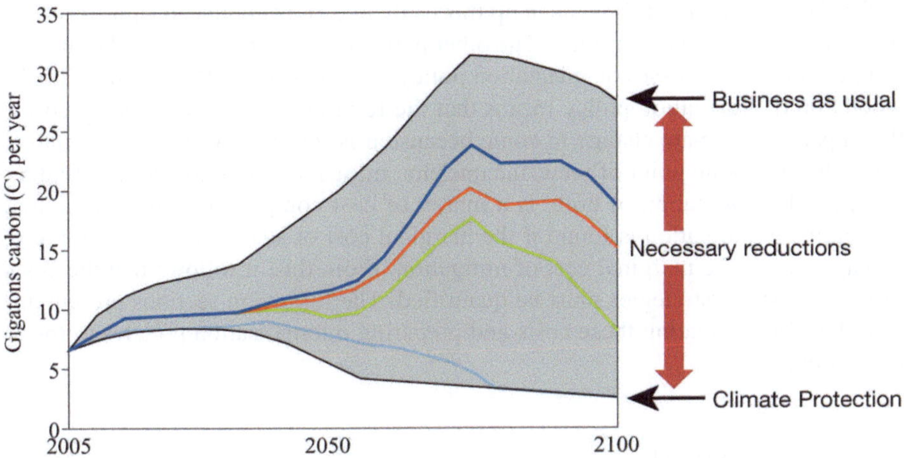

Fig. 3.6 Necessary reduction of CO_2 for climate protection. (Source: Edenhofer 2007)

3.2.4.2 Relevance of the Transport Sector

It has been shown in Sect. 1.2 that the transport sector is a major source of the climate problem and will increase its relative contribution dramatically in the forthcoming decades. While Organisation for Economic Co-operation and Development (OECD) countries currently count for about 70 % of transport induced GHG, this percentage will drop in the next decades and the dynamically developing countries and the countries, which are currently at the threshold to industrialisation (China, India, Latin America, South Africa), will take the lead.

This indicates a basic dilemma of climate policy: The industrialised countries are responsible for the problem of climate change, which the world is currently facing. They have neglected climate effects since the industrial revolution and are responsible for the present threats which reflect in the statement of IPCC at the Valencia meeting in November 2007 that major reduction measures should be started soon to limit the increase of world temperature to 2 °C. In the long-term future, the developing and the threshold countries will be the main accelerators for the problem. Therefore, it is crucially important to integrate them into any form of climate policies from the beginning. There might be a natural interest of these countries to join common policies towards climate change because it can be shown by climate impact analysis that they will be hit substantially by the consequences of climate change.

3.2.4.3 Mitigation, Adaptation and Internalisation Policy

The main IPCC message is that it is still possible to stabilise global warming at a temperature increase of about 2 °C. But this presupposes drastic measures to be taken to reduce GHG emissions, starting already in the next years to come. A part of the World Community has taken up this challenge and announced significant actions to mitigate global warming. The other part is more sceptical about the success of mitigation and stresses on adaptation policy (see Dewar and Wachs 2009; Lockwood 2008). Adaptation policy means that the future generations are prepared for the impacts of climate change to come because it is regarded as almost inevitable. From the economic point of view, the question mitigation or adaptation, or finding an optimal combination of both, is a matter of cost comparison. This means that an optimal combination is found if the marginal cost of adaptation plus remaining damage equals the marginal cost of mitigation. From this, it follows that the costs of the alternative strategies must be quantified. The following sections present approaches for calculating these costs and preparing internalisation policies to foster mitigation.

Damage Cost Approach

The damage cost approach starts from scenarios (here: the IPCC scenarios, see Table 3.2) of physical impacts of global warming in a long-term future, e.g. the year 2100. Starting from this database, the economic impacts of different global warming scenarios can be modelled, as performed in the Stern Review (2006) or previously in a study of Hohmeyer (1996).

Avoidance Cost Approach

The avoidance cost approach is based on three pillars:

- Setting of a reduction target
- Defining a set of instruments and policies to achieve the target
- Calculating the costs for the selected set of instruments

Table 3.2 IPCC scenarios. (Source: IPCC 2007)

	Temperature change °C at 2090–2099 relative to 1980–1999		Sea level rise m at 2090–2099 relative to 1980–1999
Case	Best estimate	Likely range	Model-based range excluding future rapid dynamic changes in ice flow
Constant year 2000 concentrations	0.6	0.3–0.9	NA
B1 scenario	1.8	1.1–2.9	0.18–0.38
A1T scenario	2.4	1.4–3.8	0.20–0.45
B2 scenario	2.4	1.4–3.8	0.20–0.43
A1B scenario	2.8	1.7–4.4	0.21–0.48
A2 scenario	3.4	2.0–5.4	0.23–0.51
A1FI scenario	4.0	2.4–6.4	0.26–0.59

a. *Setting of a reduction target*
When avoidance cost approaches were used as an auxiliary method—applied only if direct measurements of damage costs or if willingness to pay failed—the problem of setting appropriate targets was weakly treated and often led to pragmatic solutions with less ambitious results. Since the IPCC has achieved international acceptance and international organisations as well as national governments have started to use the IPCC long-term reduction targets (50 % worldwide; 80 % for industrialised countries in the year 2050 compared to the 1990 level) as a platform for deriving measures and policies, the theoretical approaches are also converging. IPCC values serve as "safe minimum standards" describing the long-term sustainability of the environmental, economic and social systems.

Technology and human behaviour must be developed and organised in a way that the safe minimum standards for all environmental and safety indicators are guaranteed over time.

b. *Defining a set of instruments and policies to achieve the target*
As the environmentally sustainable transport (EST) project of the OECD (2003) has shown, this is a most crucial step of quantifying avoidance or mitigation costs of climate change. This is because:

- The time horizon of forecast is much longer than the forecasting perspective for technology and behaviour. Therefore, the projections are either based on very conservative assumptions (present knowledge) or rather speculative scenarios for the long-term future.
- The possible future actions must be combined and designed in a way that the costs of achieving the targets are minimised (least-cost approach). This presupposes a series of sophisticated simulation approaches with different technologies and demand reactions.
- Long-term scenarios are necessary for all sectors that interact with the transport sector. This finally leads to a system approach (e.g. using system dynamics) to generate a consistent framework in which the transport sector is embedded.

c. *Calculating the costs for the selected set of instruments*
Cost calculation and selecting the best mix of policy measures is usually carried out in the form of a feedback process. This is because the measures must be triggered according to the least-cost principle. Otherwise, the results might be unreasonable from the economic point of view, which reduces acceptability. Often, the suggestions of environmentalists do not take into account the costs of adjustments of customers and the industry to the behaviours the environmentalists wish them to adopt.

Variance of Possible Valuations and Consistency with Long-term Targets

Both the damage approach and the avoidance/mitigation approach have to cope with the problem of long-time horizon and uncertainty. As a consequence, there is a wide range of cost estimations, spreading from about 10 €/t of CO_2 to about 300 €/t CO_2. The lower bound is taken from the European Trading System for CO_2, which, after the end of the oil price shock, shows very low price figures. Other low figures stem from using the Kyoto targets as benchmarks for avoidance cost calculations (see UNITE 2005). The Kyoto reductions (8 % on the average for the EU between 1990 and 2012) can be achieved through little economic effort and cost.

Recently, the EU Commission and various member states suggested tightened post-Kyoto targets (20–30 % reduction in 2020 compared to 1990 levels). Impact assessment studies for the proposed EU strategy to reduce GHG (e.g. the Handbook impact study) come to the conclusion that carbon prices should rise from 15 €/t in 2010 to at least 65 €/t CO2 in 2030.

Other more recent studies for Germany and Switzerland (e.g. DLR 2006) come out with a central value of 70 €/t of CO_2, with a range of 20 € (short-term EU average, based on Kyoto targets) to 280 € (long-term strategy and risks). The Stern Review arrives at a preliminary value for the damage costs of carbon of US$ 85 or 70 €/t of CO_2. When the EU Commission started to discuss a 120-g limit per pass-carkm for new cars in Europe, the automobile lobby argued that this would mean a cost per tonne of CO_2 of about 150 €. The same holds for the former intention of the EU to add 20 % biofuels to petrol and diesel (meanwhile abolished). These examples show that every target setting and associated regulation policy for reducing CO_2 is associated with a shadow cost of carbon (see Musso and Rothengatter 2013).

It is obvious that ambitious reduction targets, as have been set by the IPCC, can only be met if the carbon prices are comparatively high. This means that the preliminary value of 70 € mentioned in the Stern Review, can be regarded as the lower bound and that values of 100–200 €/t of CO_2 (INFRAS and IWW 2004; they used 140 € in their external cost study) are more consistent with the reduction targets in industrialised countries.

The Problem of Social Discount

A main critique of the Stern scenario has been brought forward by W. Nordhaus (2007), who in particular attacked the very low social rate of discount, which was applied by the Stern team to make all future impacts comparable. Stern's social rate of time preference is of an order of magnitude of 0.1 %, which is obviously far from the market interest rate today. Nordhaus himself had derived a time preference rate of 1.5 % from another modelling approach. Using a higher time preference rate some of the Stern results are no longer valid, so this problem deserves some clarification.

a. *Effect of discounting*

Climate effects happen in a future time horizon of more than 50 years. While the present generation is a main driver of the effects, the future generations are the victims who will suffer from the negative impacts. Economists use discounted future values to make them comparable to a base year. The effect of discounting over a long period, such as 100 years, is dramatic. Taking the Nordhaus rate of 1.5 %, the damage of 1 € paid in 100 years would have a present value of 22 cents today. The higher the discount rate is, the lower is the present value which implies increasing problems for future generations.

b. *Discounting for a society with overlapping generations*
The economic problem of discounting in the case of overlapping generations has been a challenge of resource economics for a long time. Already in the 1970s, basic insights were gained through the economy of exhaustible resources (e.g. Dasgupta and Heal (1976)). Kotz et al. (1987) have constructed a model of overlapping generations that comprises the main factors influencing the choice of social rate of return. These are:

- Technical parameters of the production function and of technical progress, which describe the future production frontier for given resources
- The resource limits or limited ability of nature to absorb human emissions and waste
- Consumer's preferences for material goods
- Ethical parameters of the present society with respect to the well-being of future generations, which are:
  - Rate of impatience (rate of rapacity) as the preference of the present society to enjoy consumption now
  - Preference for intergenerational equity, i.e. a balanced supply of resources to the present and the future generations
  - Preference for the environment (vs. preference for material goods)

c. *Discounting: A matter of empathy for future generations*
This brief summary of intergenerational welfare modelling with overlapping generations underlines that climate policy is a matter of empathy of the present generation for the future generations (who cannot vote today). A high discount rate reflects a high "rate of rapacity" of the present generation and a low "preference for equity"[9] and long-term "environmental sustainability". If future generations should not be compromised with respect to their chances to live in a sustainable environment, then it is necessary to keep the discount rate low.

The treatment of the problem by Nordhaus—which is absolutely correct for neoclassical economic approaches—shows that economists often prefer to neglect the ethical problems behind discounting. Supposing that there is strong empathy of the present for future generations, it is not justified to apply a social rate of discount to climate effects at all. This will favour mitigation compared with adaptation strategies, as underlined in the Stern review.

[9] If future generations will face problems with economic growth.

Internalisation Through Taxes and Charges

There are general and transport-specific instruments to mitigate climate change. Carbon taxation for instance is a general instrument, which could be applied in a consistent way to all emitting sectors. However, there are already many country- and sector-based (partial) solutions such as fuel taxes or vehicle taxes differentiated by specific fuel consumption such that the existing taxation systems would have to be changed in many countries before introduction of a general carbon tax.

Sector-specific instruments apply only for segments of the transport sector, as for instance:

- Fuel taxation (special carbon tax)
- Vehicle taxation based on CO_2 production
- Ticket tax for aviation (introduced in Germany since January 2011)
- User charges for infrastructure use with differentiation or mark-ups for CO_2
- Regulation and standards

While sector-specific instruments can easily be introduced by national governments, it is impossible to control the impacts on GHG emissions accurately.

Internalisation Through Emission Trading

Emission trading is the favourite instrument of the Kyoto protocol and it seems that also the international agreements for the post-Kyoto phase (after 2012) might be based on this instrument. ETSs can be

- Open or closed
- Upstream
- Downstream or
- Midstream

In open schemes, all parties emitting CO_2 are allowed to trade, regardless which sector or region they belong to. Closed schemes are restricted to particular sectors or spatial entities. Upstream trading means that the producers or traders of fossil energy—which is burned in engines and transformed to CO_2—are the trading units. International organisations and states set the targets, which are introduced as yearly caps to reduce emissions over time. The advantage is that the number of traders is limited and the transaction costs are low. For the demand side of the market, this scheme works like a carbon tax.

In downstream schemes, the consumers of fossil energy are the trading parties, such that there will be millions of traders. CO_2 allowances in such a system function like an additional currency and high transaction costs are necessary to make such a market function.

Midstream schemes are usually closed and include the producers of vehicles. For instance, the CO_2 allowances can be auctioned to car manufacturers. Producers of small and fuel-efficient cars would be able to sell allowances while producers of premium cars and sport utility vehicles (SUVs) would have to buy. In this case, every producer would be free to produce an optimal mix of passenger cars in consideration of the purchase/sales of CO_2 allowances.

The price per tonne of CO_2 for the transport sector is expected to be comparatively low if this sector is allowed to enter into an open trading scheme. This is because other sectors like electrical power production or industrial production might achieve CO_2 reduction by lower mitigation costs. The CO_2 price resulting from a closed trading scheme for transport would be much higher (see Jochem 2009).

Internalisation Through Regulation and Standards

a. *European Approach*
In the EU, there has been a heavy debate on the reduction of specific emission values of cars. The vehicle manufacturers had promised by voluntary agreement to reduce CO_2 emissions to 140 g/km by the year 2008. It turned out that this target could not be achieved, as the emission values for 2007 exceeded 160 g/km. The Directorate-General for the Environment (DG Environment) of the EU Commission decided, against this background, to set a reduction target of 120 g/km by the year 2012, as an average value for all newly registered vehicles. The intervention of the German government led to a change insofar as it was decided to allow for adding biofuels in an order of magnitude which leads to a reduction of 10 g/km of CO_2 emissions. Therefore, the target for fuel consumption accordingly was set at 130 g/km. The EU Commission has prepared a directive to reduce this target value to 95 g/km beginning with the year 2020.

Following Directive 2009/33 EC, the EU Commission has introduced soft regulations on fuel consumption of the vehicle fleet of manufacturers in 2012, which are to be followed by rigid rules in the year 2015, limiting CO_2 emissions to 120 g/veh.km (+ 10 g for adding biofuels) in 2015 and reducing this limit to 95 g/veh.km in 2020 (for 95 % of the newly licensed car fleet, in 2010 for 100 %). The penalties foreseen can reach a magnitude of 95 €/g of CO_2 emission exceeding the reference curve. The reference curve allows the producers of bigger cars (e.g. vans) to deviate from the target, but less than proportionally. Very fuel-efficient cars in the fleet can compensate for higher fuel consumption of others. Electrical vehicles receive a bonus multiplier, starting with two (counting as two conventional vehicles) which decreases to one in 2023. The long transition phase of introduction allows the producers of high-powered "premium" passenger cars to adjust their fleet.

b. *US Approach*
In the USA, the Corporate Average Fuel Economy (CAFE) standard was enacted in 1975, as a response to the Organization of the Petroleum Exporting Countries (OPEC) oil embargo 1973. It regulates the fuel consumption of the newly produced car fleet of a company, in terms of the harmonic mean of all passenger car types produced for the domestic market. The limit values were set dynamically and in such a way that producers could adjust without major friction, e.g. 20 miles per

gallon (mpg) 1980, 27.5 mpg 1990–2010. The regulation was successful until the 1990s, but lost its incentive power because it was no longer adjusted dynamically after 1990. Furthermore, particularly trucks and SUVs were exempted or treated more softly, such that vehicle types with high fuel consumption—as for instance the famous Hummer SUV—were not penalised.

In 2007, the CAFE standards received their first overhaul after about three decades. Future fuel efficiency should be increased substantially by setting the performance of 35 mpg until 2020, without exemptions for light trucks and SUVs. Furthermore, a credit trading flexibility was introduced, such that producers of high fuel consumption cars can trade with producers of low fuel consumption cars. Japanese car makers, in the first instance Toyota, are benefiting from this rule. After the change in government, President Obama has pushed forward a plan for a further reduction of fuel consumption of cars, which probably was decided in May 2010. The new requirements mark the first fixed standard for the emission of GHGs by car traffic in the USA and cover the years 2012–2016. The standards set for 2016 are 39 mpg for cars and 25 mpg for trucks. This will cause difficulties for companies that specialise in high-powered premium cars (like Porsche) and lack the possibility of internal compensation by small and energy-efficient cars.

3.2.5 Empirical Measurement of External Costs of Transport

3.2.5.1 Background Studies

There are two streams of analysis in Europe on quantification of external costs of transport. The first stream has been established by INFRAS/IWW in several studies for the International Union of Railways (UIC), beginning with 1995. Their analysis is based on a complete list of externalities, regardless of whether they are directly related to traffic volume or not. Total, average and marginal costs of externalities are calculated, which serve as indicators for necessary internalisation policies. The internalisation strategy then can be based on Baumol/Oates prices, combined with cap-and-trade systems, regulation and other appropriate instruments, combined by a least-cost calculus.

The second stream follows from a series of research projects for the EU Commission led by the ITS, Leeds, such as CAPRI, PETS, UNITE or GRACE. These studies aim at setting prices at marginal costs according to the neoclassical rule of optimal allocation of resources. As transport does not pay the external marginal costs, the latter must be charged through a public charging system. The list of externalities in this case includes external costs only, which are directly related to traffic activity. This means that impacts on biodiversity, landscape, urban separation or upstream/downstream effects are neglected. As a consequence, the resulting figures of marginal external costs are summarised to get the optimal charge. This means that internalisation is to be achieved by charges and taxes only, following this concept.

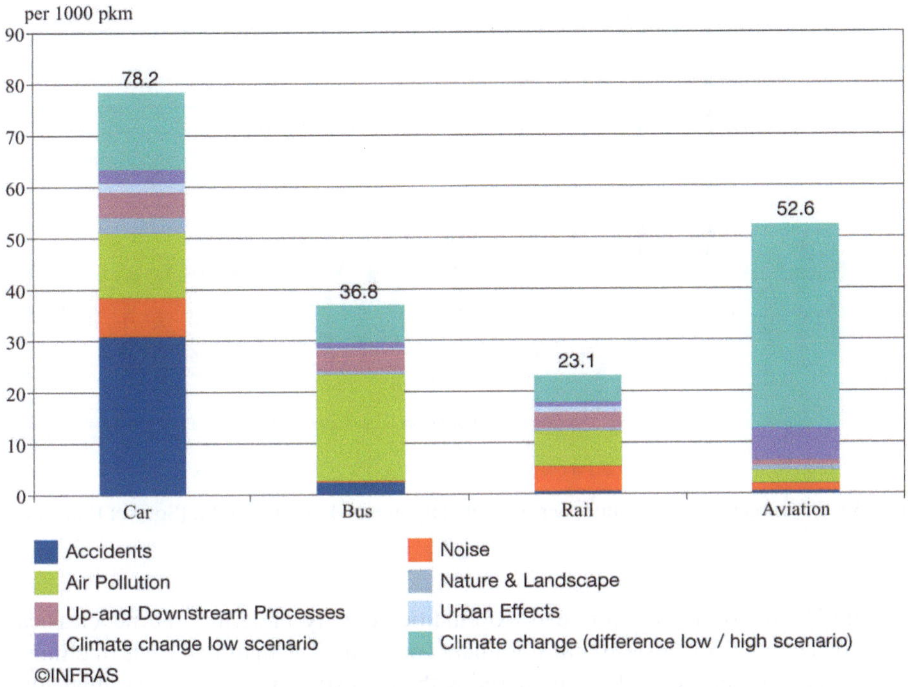

Fig. 3.7 Average external costs of passenger transport. (Source: INFRAS3 and IWW 2004)

INFRAS/IWW Studies for the UIC

In this section, we give some basic results on average external costs of transport, focusing only on passenger transport (Fig. 3.7).

In this study, rail comes out as the mode with lowest external cost per pass-km. Car traffic is producing about 4 times and air traffic about 2.5 times as much external cost. With respect to air, it has to be added that long-distance flights are included in this comparison. When one compares air and rail on distances where competition is possible, the advantage of rail is more significant. The ratio (depending on aircraft type and occupancy rates) in this case is 1:4 in favour of rail or even better.

3.2.5.2 IMPACT Study and Handbook of the EU Commission

Handbook Results

The following figures summarise the essential results of the Handbook of the EU Commission. For some effects, the figures represent marginal cost values, while for others they represent unit costs based on average cost calculations (Fig. 3.8).

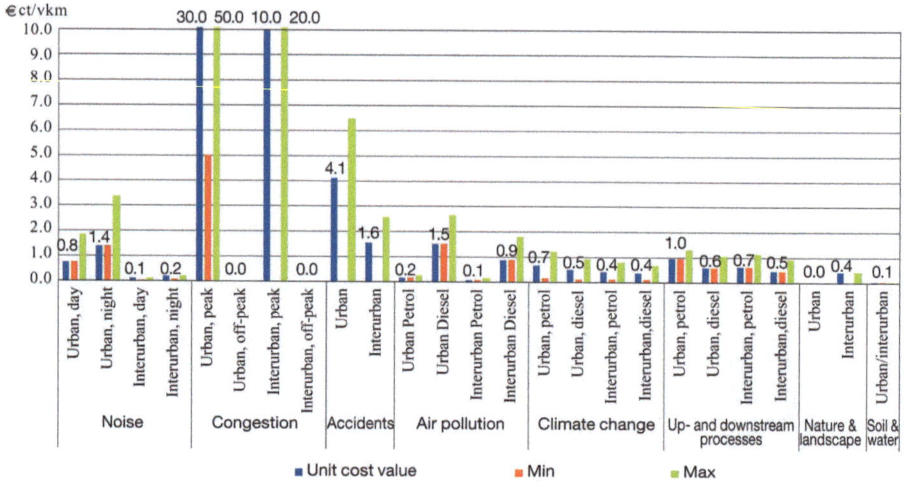

Fig. 3.8 Passenger cars: unit values per cost category in €ct/vkm (in 2000 €). (Source: Handbook 2008)

The Handbook results for road externalities clearly underline the neoclassical view that congestion costs are the dominant cost item. The marginal costs of climate change appear very low when compared with congestion. Of course, marginal congestion costs only occur in peak periods and such peak periods are observed more frequently in urban rather than in interurban transport. Nevertheless, congestion comes out as the dominating externality in the Handbook. This is one of the reasons why the EU Commission decided to suggest capping the unit values in such a way that the relative weight of congestion is reduced (Fig. 3.9).

In the railway sector, the major source of externalities is identified as the air pollution of diesel locomotives. Noise comes second, but only in urban areas this is a relevant factor. Figure 3.10 shows that trains with electric propulsion cause much lower external costs than passenger cars; the ratio is between 1:4 and 1:5 taking the present mix of electricity production.

Marginal External Costs of Intercity Passenger Transport

In the following, we summarise the results on intercity passenger transport given in the Handbook of the EU Commission. As the Handbook is restricted to road and rail, and does not present comparable results for aviation, we add our own estimation of the marginal external costs of aviation based on the sources, which have been used for the Handbook. Congestion costs are left out because it does not make sense to include them in an intermodal comparison. Furthermore, we restrict to day values in the case of noise, because rail passenger transport is very low during night hours and air passenger traffic is prohibited or drastically reduced at many European airports during night-time.

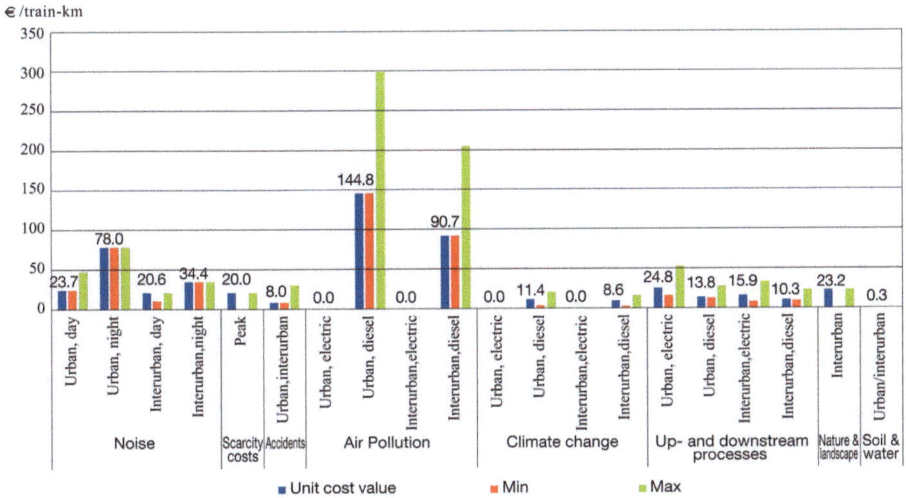

Fig. 3.9 Passenger railways: unit values per cost category in €ct/vkm (in 2000 €). (Source: Handbook 2008)

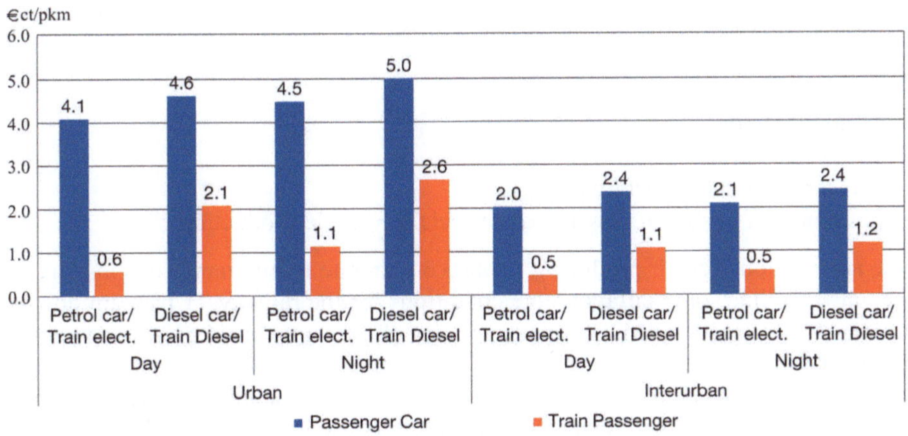

Fig. 3.10 Comparison of road and rail passenger transport: cost per pass-km in €ct/pkm. (Source: Handbook 2008)

According to the results exhibited in Table 3.3, rail intercity passenger transport has high advantages over road transport and aviation. The competing modes produce three to four times more external costs per pass-km, which reconfirms the relationships exhibited in Fig. 3.7.[10]

[10] The advantages of rail compared with car and air transport are lower in Europe compared with Japan. The main reason is the much higher occupancy of Japanese trains (see Sect. 3.3).

Table 3.3 External costs of intercity transport in Europe (Handbook values)

Type of effect	Pass. Car cts/veh.km	Rail centts/train.km	Aviation/flight
Noise	0.12	20.6	228
Accidents	1.57	8.0	118
Air pollution	0.35	9.0	117
Climate change	0.42	0.9	1060
Up-/downstream	0.58	15.3	612
Nature and landsc.	0.4	23.2	–
Soil and water	0.06	0.3	–
Total	3.45	77.30	2135
Ext. cost cents/pass.km	2.06	0.55	1.53

Assumptions
1. Share of diesel cars: 1/3, petrol cars: 2/3
2. Car occupancy rate: 1.7; train and air plane occupancy: 140
3. Share of electrical traction for trains: 90%
4. Air travel distance: 1000 km
5. Mark-up factor for non-CO_2 emissions: 2
Aviation-related assumptions are based on estimations using the model TREMOVE, as quoted in the Handbook Table 50, p. 110 and recommended in the appendix (p. 271).

When it comes to specific comparisons between rail and air with respect to climate impacts, it is most important to set appropriate assumptions with respect to:

- Distance
- Vehicle/aircraft types used
- Loading factors
- CO_2 or all climate gases
- Existence of infrastructure provision
- Existence of upstream/downstream effects of vehicle production.

The assumptions set for producing Table 3.3 generate a fair framework insofar as they also include some unfavourable conditions for the railways. Upstream/downstream effects are included, as well as impacts of infrastructure provision on nature and landscape.

As the electricity production for railway traction is included in the upstream/downstream item and the impacts of infrastructure provision are included in the nature and landscape effects, it makes sense to compare the combined climate change/upstream–downstream and nature-landscape impacts between rail and air. The external costs of climate and nature then result in 0.28 cents/pass-km for rail and 1.19 cents/pass-km for air, which makes a factor of 4 in favour of the railways.

Focusing on CO_2 emissions of operation only, the Institute for Energy and Environmental Research (IFEU Institute 2008) has provided estimations on behalf of Deutsche Bahn AG. According to these estimations, the resulting specific emissions are

- 380 g CO_2/pass-km for air, short distance, and
- 40 g CO_2/pass-km for rail.

High-speed rail is performing in general not worse than the rail average, because the higher energy costs per train.km are compensated by a much better loading of the trains. In Germany, standard intercity (IC) trains have an average occupancy rate of one third while the average occupancy rate of ICEs is about 50 %. IFEU (2008) has prepared a database for an OD-related comparison of CO_2 emissions and energy consumption between car, rail and air. Visiting the webpage of Deutsche Bahn AG, a client will not only get information on usual trip data but also on the environmental consequences of a modal decision. Two examples:

- Frankfurt–London Gatwick: air—103.4 kg CO_2, rail—25.1 kg CO_2
- Frankfurt–Paris CDG: air—82.2 kg CO_2, rail—13.7 kg CO_2

This means that, depending on the essential assumptions, the operational climate advantage of rail on short/medium distances compared with air is between the factors 4 and 8, which means that a high-speed train with high seat occupancy (occupancy rate > 50 %) will have a strong climate advantage compared even to air planes with even higher seat occupancy (assumed occupancy rate 75 %).

Deutsche Bahn AG has introduced the "Green Bahncard" in 2013 which guarantees that every rail trip carried out by a Green Bahncard user is powered by regenerative energy. Objective is to produce 100 % carbon-free electricity until the year 2050. This underlines that a modal shift from car and air to rail will be an efficient climate strategy for a long time, even if a part of future car mobility is performed by electrical vehicles and the airline manufacturers are making progress with fuel efficiency.

Treatment of Aviation

International aviation is regulated under the Convention on International Civil Aviation, signed in Chicago in 1944 by 52 signatory states. This means that only domestic aviation can be regulated by the individual states and only intra-EU aviation can be influenced by EU legislation. For instance, aviation fuel taxes have not been introduced for this reason, because only a minor part of aviation would be included and negative diversion effects would follow.

The major environmental problem of aviation is its contribution to climate change. So, the Kyoto Protocol has addressed this issue and assigned it to International Civil Aviation Organization (ICAO) for further treatment. But as no unanimous decision of the signatory states could be achieved, aviation could not be included in any effective climate protection policy.

Aviation counts for about 3 % of CO_2 emissions worldwide. As CO_2 makes up only about 50 % of the total of GHGs of aviation, one can double the above percentage to result in a very relevant order of magnitude. That means a priori, that the argument according to which aviation is such a small player in the transport markets that their emissions do not matter, does not hold. Furthermore, aviation is growing much faster than other transport sectors, so that emissions are rising drastically.

The increasing pressure of environmental policy encouraged the EU Commission to work out a draft directive in 2006 to integrate aviation into the existing ETS. The European ETS was introduced in 2005 and includes energy producers and some industrial producers as well as transport companies (e.g. railways with their own energy production plants). Emission rights have been allocated by grandfathering at the beginning, while in the second phase of ETS (2008–2012) 10 % of the rights will be auctioned off. This is why the price of the tonne of CO_2 is still very modest. But prices are expected to rise, which is indicated by the rise of prices of the CO_2 allowances (EUA), which were traded at about 16 € at the end of 2008 and the beginning of 2011.[11] In 2013, all allowances for the industries included should be auctioned off according to the original plan.

Following this draft and after a phase of consultations, the EU Commission prepared Directive 2008/101 EC, which was introduced in November 2008. This presents the legal basis for the integration of aviation into the ETS. The following essentials summarise the most important elements of the original plan:

- All flights are included that arrive/depart at EU aerodromes, beginning with January 1, 2012.
- In the year 2012, the number of allowances equals 97 % of historical CO_2 emissions.

From January 1, 2013, the number of allowances will be reduced to 95 % of historical emissions. Further reductions will be decided for each subsequent period.

- The commission will decide on the historical emissions of aviation by August 2, 2009.
- From January 1, 2013, 15 % of allowances will be auctioned. This percentage may be increased as part of the general review of this directive.
- The operators can use certified emission reductions (CER; climate credits according to the clean development mechanism (CDM) of the Kyoto Protocol) and emission reduction units (ERU: climate credits according to the Joint Implementation (JI) option of the Kyoto Protocol) to compensate for allowances, up to a volume of 15 % of the total allowances.
- In each period, a special reserve of 3 % of the total allowances is given to operators entering into business and operators in the start-up phase, showing growth rates of 18 % or more. This is limited to 1 million allowances. At least 15 months before the start of each period, the commission shall calculate and adopt a decision setting out:

  - The total number of allowances
  - The number of allowances to be auctioned
  - The special reserve
  - The number of allowances allocated free of charge
  - The benchmarks to be used to allocate allowances free of charge to aircraft operators (in principle dependent on the fuel consumption of the aircrafts of the fleet)

[11] Due to the drop of oil prices, the CO_2 price went down to US$ 8.00 at the beginning of 2009.

Third-country measures to reduce the climate change impact of aviation can lead to compensation agreements (to avoid double charging)

The European Commission regards this directive as a "top runner" model for aviation, hoping that other states will follow. Meanwhile, 15 other European states have joined this initiative. It is important to remember that other sectors, including the railways, will have to buy 100 % of the allowances beginning with 2013, while aviation has to buy only 15 %. This means that aviation will still have a big advantage through a favourable treatment with respect to CO_2 emissions.

Nevertheless, there have been sharp protests by other countries, in particular, the USA, Russia, China and India against the inclusion of intercontinental air transport into EU emission trading. Facing the tough resistance and the announcement of China to cancel orders for the European aircraft industry, the European Commission has decided to postpone the introduction of the scheme and to start negotiations with ICAO on an internationally harmonised approach. The further process would be decided after the following general assembly of ICAO by the end of 2013.

Although the mechanism suggested by Directive 2008/101 EC seems rather complex, it will certainly not lead to a full internalisation of climate change costs of aviation. Following the discussion in Chap. 4 on climate change, the costs per tonne of CO_2 should be significantly higher than 100 € to generate the desired changes of technology and behaviour. With present CO_2 trading prices of under 10 €/t and expected future values of 20–30 €/t, the ETS in Europe will lead to additional costs of 2–3 €/seat on short air distances (e.g. Frankfurt–London). As emission trading will be an incomplete instrument for a full internalisation of climate costs in the short and medium term, it is obvious that further instruments will be necessary (e.g. subsidisation of rail investment, taxation, charging, regulation of competing modes) to compensate the environmentally friendly modes like railways for their negative treatment in the ETS.

The German government has decided to introduce a ticket tax, which is 8 € on short, 25 € on medium distance and 45 € for international flights. Although this measure was not motivated by environmental but rather by fiscal goals, the effect on aviation is significant and much higher than the expected effect of the eventually forthcoming ETS trading. In particular, some low-cost carriers have cut down their flight programmes at German regional airports.

3.2.6 Conclusions and Recommendations Considering the Case of Japan

The concluding remarks relate to lessons that can be important for the Japanese case. First of all, the magnitude of external effects can be different in Japan, such that a transfer of data and relationships is problematic. For instance, the Japanese insurance system covers more risks than that in European countries, such that the external costs of accidents are probably much lower. Secondly, the environmental standards for the car manufacturers are more advanced, than in Europe. Therefore, the impacts of air pollution, particularly related to NO_x and particulate matter, will be significantly lower. Thirdly, the external costs of noise are expectedly high in Japan, because the major expressways and rail lines are very close to densely populated areas.

With respect to the fourth issue, which is climate change, the Japanese situation for ground transport looks more favourable than the European case. Due to the land use and public transport policy in the past decades, paired with a consequent pricing of road use, the modal split values for public transport are much higher than in Europe, on the average. This implies that the average external costs of travelling will be lower in Japan. This does not mean that there is no possibility for a further reduction of CO_2 emissions by passenger cars and trucks. Comparable with Europe, there is a considerable share of high-fuel-consumption cars such that the average fuel consumption almost equals the European average value (about 30 mpg or 8 l/km). This means that the challenge of reducing fuel consumption of automobiles is almost the same in Europe and in Japan.

Coming to the policy towards short-distance flights, the European policy has been ambivalent in the past. Many regional airports have been established in the past two decades, which has fostered the market expansion of low-cost carriers. At several places, the regional airports have been subsidised with respect to infrastructure costs and in some cases also with respect to operation costs. The EU Commission has announced that a co-funding of airport operation costs will be regarded as a non-allowed subsidisation in the future and that also the co-sponsoring of regional airport investments by regional public authorities will be checked more carefully. From this, one can expect that the public support for short-distance flights of low-cost carriers will go down in the future. In particular, after the opening of new high-speed rail lines, the public support of short-distance air relationships has been cancelled. Many flight schedules have been adjusted and flights were cancelled between OD pairs, which are served well now by high-speed rail. Recent examples are Strasbourg–Paris, Stuttgart–Paris (after opening of the TGV line Paris–Strasbourg–Karlsruhe–Stuttgart in the year 2006) or Madrid–Barcelona (opened in the year 2008). In the latter case, the passenger volume of aviation was cut to 50 % within 1 year. A further improvement of rail service (frequency of service) and the appropriate use of marketing instruments including price differentiation will probably lead to a further drop to about 30 % according to the forecasts. This underlines that the extension of HSR in Europe leads to significant improvements of the rail market share and, as a desired side product, to a significant reduction of overall CO_2 emissions.[12]

[12] In the background paper of Rothengatter and Szimba (2009), the authors give an elaboration on the overall effects of the Madrid–Barcelona high-speed link.

3.3 Comparison of Life-Cycle CO_2 Emissions by Shinkansen and Air Transport Systems

3.3.1 Introduction and Background

In planning roads, bridges and other infrastructure projects, the key considerations for reducing CO_2 emissions are where the new infrastructure is needed and how it should be operated so as to minimise CO_2 emissions. In other words, it is necessary to assess the whole life cycle, from construction through operation. In this study, the LCA method is used to obtain a quantitative grasp of total CO_2 emissions.

LCA is typically used for quantitatively evaluating environmental loadings. The resources input, the resulting environmental load and the impact on ecosystems throughout the life cycle of a product, service or system, from the mining of raw materials up to final disposal, are evaluated scientifically, quantitatively and objectively. LCA was standardised under ISO 14040 in 1997 and is in widespread use in the field of production. Here, the application of LCA to the field of transport aims not only at evaluating the infrastructure or vehicles alone but rather the whole transport system and the ripple effects of the system as well. The environmental load is calculated by summing the raw materials used multiplied by the environmental load caused by their production (the environmental load emission factor). The application of LCA makes it possible to evaluate in advance whether it is appropriate to construct infrastructure given the current passenger demand. In this section, the life-cycle CO_2 emissions of Shinkansen (bullet train) and air transport systems in the domestic passenger transport network are evaluated.

Of total human-generated CO_2 emissions in Japan, the proportion accounted for by passenger transport was 11.9 % in 2008. Air transport made up 6.1 % of the transport sector (with a share of total human-generated emissions of 0.7 %); this was a marked increase of 55 % compared to aviation emissions in 1990. The aviation sector has been quite slow to replace carbon fuels and this is considered to be a major reason for the relatively higher share of CO_2 emissions from air transport.

Japanese railways generally serve long-distance passengers alongside the airlines. In national reports by the Ministry of the Environment (MOE), railways have proven their environmental efficiency by demonstrating significantly less energy consumption and CO_2 emissions than air travel as a national average. However, these reports fail to detail the variations in their assessments among different routes, or origin–destinations pairs, and consider environmental loading only from system operation. However, where new rail lines are being considered, evaluation should begin by considering the construction phase of such huge infrastructure projects. Additionally, operational efficiency (namely capacity) and degree of congestion are also very important factors. Future infrastructure improvements will be different from those of the past because capacities will be smaller and geographical conditions will impose severe design limitations; this will probably lead to a shift from the scope in which environmental concerns have had high priority. In quantitatively assessing any possible shift of interregional demand from air transport to new

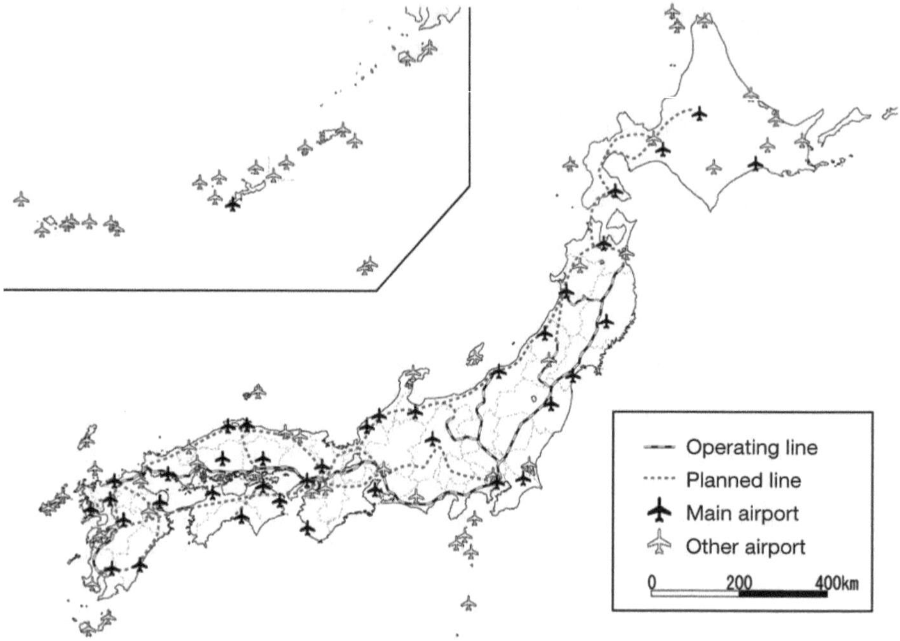

Fig 3.11 Location of airports and Shinkansen lines in Japan (in 2010). (Source: "Railways in Figures 2007", "Civil Aviation in Figures 2007")

Shinkansen lines (either currently being constructed or included in future improvement plans) and the associated reductions in energy use and CO_2 emissions, the construction of this infrastructure should also be taken into account as an important part of the environmental loading.

To date, there has been a big difference in the progress of improvement plans between new airports and Shinkansen lines. Figure 3.11 shows the location of existing (solid) and planned (dashed) airports and Shinkansen lines in Japan (as of November 2010). There are 84 existing airports in 35 prefectures, with two new ones, including one on a remote island, in the future development plans (either being currently constructed or planned). In comparison, the Shinkansen system has less regional spread, with stations in only 23 prefectures. Since the "Act for Country-wide Construction of Shinkansen Lines" came into force and the "Improvement Plan" was developed in 1973, construction of various lines (the so-called new Shinkansen lines) proposed in the plan has been ongoing, while some remain in the planning stage (as indicated in Fig. 3.11). The total length of Shinkansen lines under this plan was 6,853 km, but only 31 % of these have been completed. Today, the Japanese high-speed rail network consists of 2,176 km of Shinkansen lines. In prefectures that are not served by Shinkansen lines, the main modes of transport are aircraft, bus and private car, while the share taken by railways is notably low.

As a practical and useful assessment of the contribution of Shinkansen improvement projects to national CO_2 emissions, this study aims at estimating how CO_2 emissions would change if current air transport passengers were to shift to existing or new Shinkansen lines. The basic approach is to use the LCA methodology. Additionally, inter-prefecture passenger demand is analysed by using actual inter-prefecture trip data and journey distances; any possible changes in demand are examined through sensitivity analysis. Finally, the best high-speed transport mode from an environmental perspective is suggested based on the results obtained from this analysis framework.

However, this study must be treated as a strategic evaluation because a more elaborate analysis within this given framework would require a huge data set and detailed scenario setting, which would be neither practical nor fully available. Still, as a strategic approach, the analysis provides valuable evidence for the main proposition of this study, which is that Shinkansen improvement projects do not always favour reduction in total amount of CO_2 emissions and sometimes the reductions gained by shifts to the Shinkansen system are overwhelmed by the CO_2 emissions generated during the construction phase of the infrastructure.

3.3.2 Framework of LCA for Regional High-Speed Passenger Transport

National average CO_2 emissions generated from energy use in the operation of transport modes provides a useful tool for the comparison of modes, but as noted earlier this approach has some problems and a true comparative analysis requires more special consideration of transport improvements as a whole. Therefore, this study explores modal CO_2 emissions more specifically by taking into account the characteristics of each mode, as listed in Table 3.4.

On this basis, the comparative analysis carried out here first provides insight into the extent of the life-cycle environmental load imposed by each of the above modes. The main components are the infrastructure, vehicles and their operation. Next, possible shifts of air transport passengers to newly constructed Shinkansen lines are estimated. The current state of the respective transport systems provides the starting point and each alternative is analysed according to which of its impacts add more to the environmental load. In this context, increased Shinkansen operations and the relevant additional infrastructure and rolling stock requirements should necessarily

Table 3.4 Characteristics of air and Shinkansen transport systems that are assumed to affect the estimation of CO_2 emissions

Air transport system	Shinkansen transport system
There is no infrastructure between the origin and destination Construction or maintenance of the infrastructure is not necessary	Construction of infrastructure is necessary between the origin and destination Infrastructure-related CO_2 emissions per passenger are higher for sections with low passenger demand
The origin and destination can be connected by a straight line	The origin and destination cannot be connected by a straight line due to geographical and other factors Trains travel longer distances and an equivalent length of infrastructure has to be constructed
The greater the distance travelled, the lower the CO_2 emissions per kilometre	

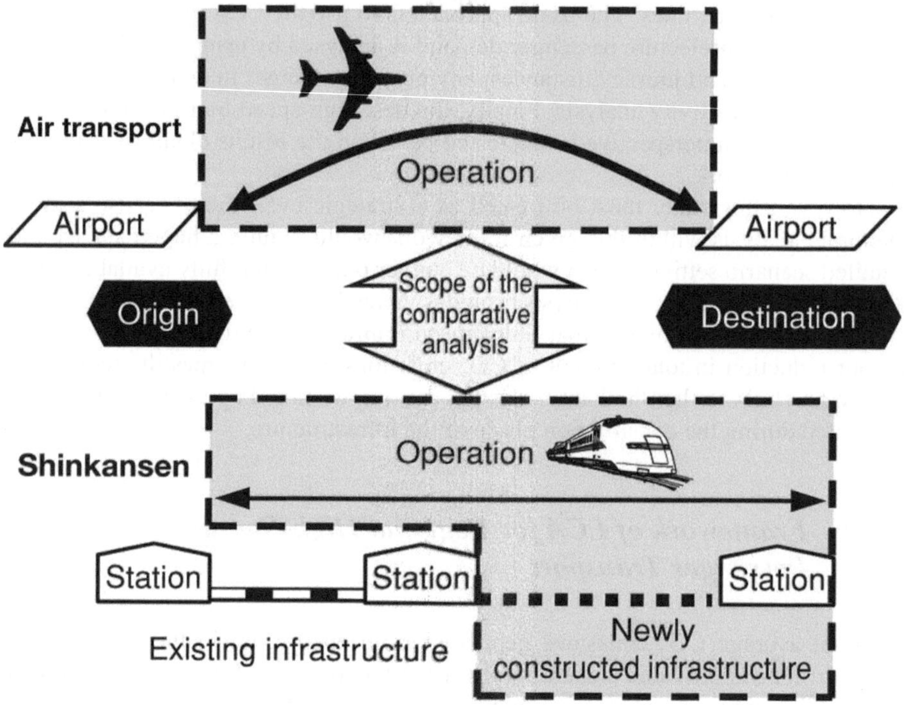

Fig. 3.12 Setting the scope of the comparative analysis (shown by the *dotted lines*)

be considered. On the other hand, if air routes are assumed to be cancelled, it is accepted that such a decrease will not have any effect on the number of aircraft or airports in the short term and so these are not included in the analysis. The scope of the analysis is limited to new Shinkansen lines and rolling stock while it does not necessarily take into account existing infrastructure and aircraft (Fig. 3.12). The reasons for ignoring the emissions generated by airports and the production of aircraft will be explained later in the study.

3.3.3 Method of Analysis

3.3.3.1 Data for Estimation of LC-CO_2

a. Air transport

1. Infrastructure

The emission factors used to substantiate the CO_2 emissions generated by the construction of an airport are extracted from observed data relating to Chubu International Airport. The main parameters are the materials used for the pavement of the runways, taxiways (lifetime: 20 years), and aprons (lifetime: 10 years), as well as their construction (boring and filling). The total number of arrivals and departures recorded in 2006 for Chubu International Airport was 53,450, with an average of 99,313 passenger-km/flight. By using these real figures, CO_2 emissions from the construction phase of an airport are estimated in terms of passenger-kilometres.

2. Aircraft

The assumption is that only a B-777 aircraft with a capacity of 500 passengers serve regional passengers. Obviously, for lower levels of demand, smaller types of aircraft are used and this assumption is thought to overestimate emissions. LCA results obtained by the Japan Carbon Fiber Manufacturing Association[3)] are used (with manufacture of materials for a B-767 aircraft at 65 % capacity set at 0.25 g-CO_2/passenger-km; and assembly set at 1.34 g-CO_2/passenger-km).

3. Operations (Flights)

The CO_2 emissions generated by the aircraft are calculated using Eq. 3.1, where

Q_a total daily CO_2 emissions generated by flights of the aircraft (t-CO_2/day)
$F(x)$ CO_2 emissions generated per aircraft per kilometre (t-CO_2/flight-km)
x distance between origin and the destination (km)
N_a number of flights by the aircraft per day (flight/day)

$$Q_a = F(x) \times x \times N_a \tag{3.1}$$

In this formulation, the CO_2 emissions from an aircraft per kilometre are defined on the basis that the largest energy consumption occurs at the take-off phase and that fuel consumption decreases as the loaded weight of the aircraft falls; that is, the longer the distance travelled, the less CO_2 is emitted per kilometre. In the same way, the IPCC Guide Book (IPCC 1996)[4)] suggests that the landing and take-off (LTO) and cruise phases should be separately evaluated when considering the environmental load generated by aircraft. Bearing in mind this requirement arising from environmental studies, this study suggests computing the amount of CO_2 emitted by aircraft between origin and destination using measured fuel consumption data. The relationship between aircraft CO_2 emissions (B-777) and the distance between origin and destination is shown in Fig. 3.13, where the actual distance travelled by the aircraft is treated as the distance between origin and destination. In addition, Eq. 3.2 is used to convert the fuel consumption of jet aircraft to CO_2 emissions:

y CO_2 emissions (t-CO_2)
J fuel consumption of jet aircraft (kl)
α kl-TOE conversion factor=0.8767 (TOE/kl)
e CO_2 emissions per 1TOE=2.789 (t-CO_2/TOE) (3EID, 2,000 data files)

$$y = J \times \alpha \times e \tag{3.2}$$

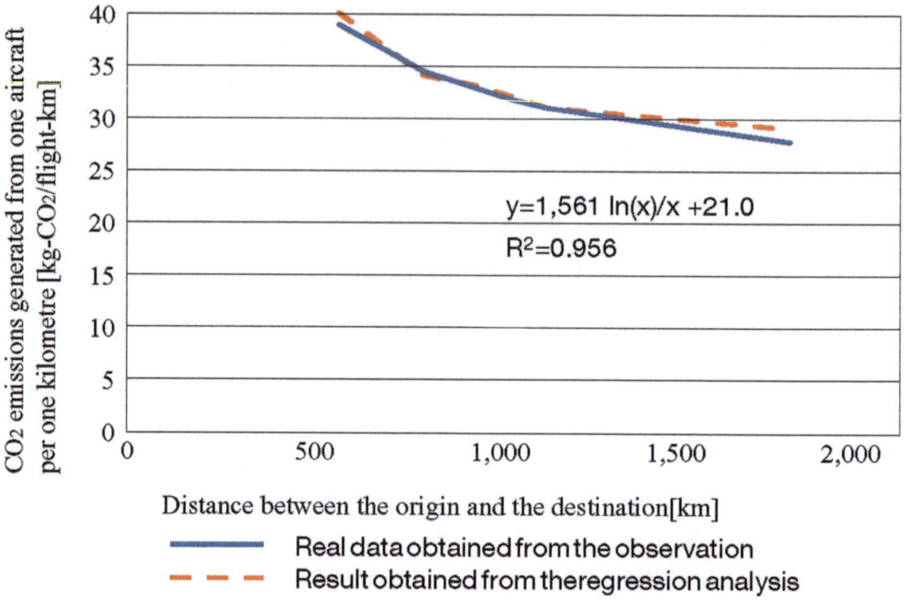

Fig. 3.13 Relationship between aircraft CO_2 emissions (B-777) and distance between origin and destination

From the data given in Fig. 3.13, regression Eq. 3.3 is determined for the amount of CO_2 emitted per kilometre of flight, taking the distance between origin and destination as an explanatory variable. As can be seen from the figure, the longer the distance travelled, the less CO_2 is emitted per kilometre (Fig. 3.13).

$$F(x) = 1561 ln(x) / x + 21.0 \tag{3.3}$$

Here,

F(x) CO_2 emissions generated per aircraft per kilometre (t-CO_2/(flight-km))
x distance between origin and destination (km)

b. Shinkansen

1. Infrastructure

In a similar manner, the necessary CO_2 emission factors are obtained from an earlier work (RTRI 2002)[5)] that provides results separately for elevated sections of track (7,550 t-CO_2/km), tunnel sections (4,160 t-CO_2/km), stations at an average interval of 50 km (1,500 t-CO_2/station) and track work (507 t-CO_2/double line-km).

2. Rolling Stock

One Shinkansen train consisting of N700 type rolling stock has a seating capacity of 1,323. The CO_2 emission factors regarding manufacturing and maintenance of N700 series stock have already been calculated in an earlier study, assuming a lifetime of

20 years per vehicle: 150 t-CO_2/vehicle for manufacturing; and 95 t-CO_2/vehicle for lifetime maintenance (Tsujimura et al. 1998).

3. Operations

The total amount of CO_2 generated by the energy use in operation is formulated using Eq. (3.4) below, where

Q_s total daily CO_2 emissions generated by operation of Shinkansen (t-CO_2/day)
R CO_2 generated per Shinkansen train per kilometre (t-CO_2/(train-km))
x distance between origin and destination (km)
N_a number of Shinkansen trains per day (train/day)

$$Q_s = R \times x \times N_s \tag{3.4}$$

CO_2 emissions per kilometre (R) are calculated by referring to the Central Japan Railway Company database. Unlike flights, CO_2 emissions per kilometre are constant with respect to travelled distance, x. The distance parameter is calculated by taking the lengths of existing lines. For the lengths of planned lines, existing designs are used and in the cases when such designs have not yet been completed, the length of the parallel conventional rail line is used. This is a very practical and correct way of filling such a gap in the data set.

3.3.3.2 Defining the Environmental Efficiency Indicator

For the aims of this comparative analysis, an indicator to measure the unit CO_2 emissions from different types of relevant functions, namely an indicator of environmental efficiency, is necessary. Equation (3.5) below defines a formulation that can be utilized within the framework of this study:

$$\begin{aligned} &\textit{Environmental Efficiency} \\ &\quad = \textit{Performance of product} \,/\, \textit{Environmental load from production} \end{aligned} \tag{3.5}$$

For a transport system, however, the environment is only one aspect of the assessment, and other various measures of performance need to be considered. By using the environmental efficiency indicator, an efficiency evaluation is made possible that takes into account the level of service parameters, such as speed, given by the ratio of distance travelled to trip time, or comfort. The service-based environmental efficiency given by Eq. (3.6) is used in this analysis.

$$\begin{aligned} &\textit{Service-based environmental efficiency of theregional transport} \\ &\quad = ((\textit{Number of passengers} \times \textit{total distance travelled}) \,/\, \textit{total trip time}) \,/ \\ &\qquad \textit{Generated total environmental load} \end{aligned} \tag{3.6}$$

3.3.4 *Analysis Between Tokyo and Osaka*

CO_2 emissions from the Shinkansen and air transport systems between Tokyo and Osaka, which are connected by a 515.4-km-long Shinkansen line and a 450.6-km-long air link, over an assumed lifetime of 60 years, are estimated by the bottom-up method. By reference to the aircraft load factor of 65 %, which is the average of the aviation industry, and actual Shinkansen operations, the number of flights/Shinkansen trains a day is determined.

3.3.4.1 Results of LC-CO_2 Estimations

For the air transport system, CO_2 emissions are estimated based on the observed number of flights in April 2002 and the volume of passenger transport (pass-km divided by length of route) of 22,478 passengers/day (round trip). For the Shinkansen transport system, the estimation is based on actual data for the Tokaido Shinkansen line in 2005 (the number of transported passengers: 144 billion pass-km; 43.8 billion pass-km) and the average passenger volume of 233,000 pass-km/km-day (roundtrip). Figure 3.14 presents the results of LC-CO_2 estimations for Shinkansen and air transport. The amount of CO_2 emissions generated by the whole Shinkansen system through an assumed lifetime of 60 years equals only one ninth of the amount generated by air transport. The weighting of each component in the total estimated LC-CO_2 clearly shows that the contribution of operations is large. In the case of the air transport system, the impact of airport construction and aircraft manufacturing on the environmental load is so small that they can easily be neglected and these contributions are not taken into account in the subsequent analysis. The passenger

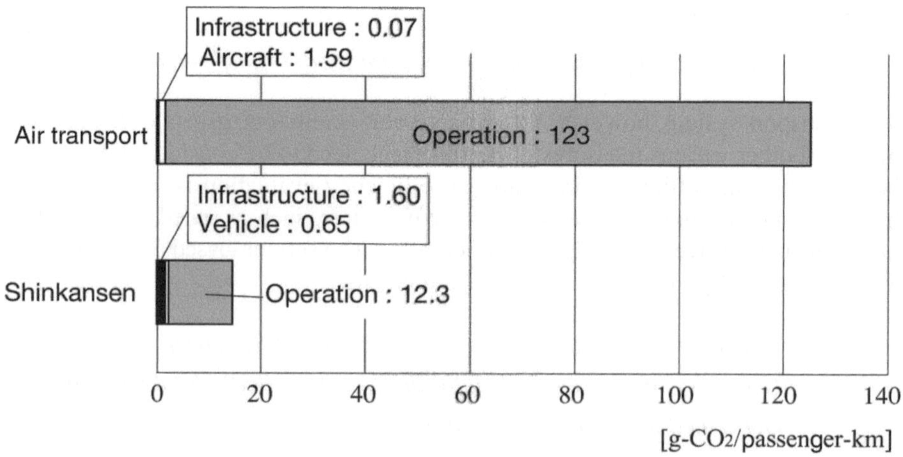

Fig. 3.14 LC-CO_2 by air transport and Shinkansen (between Tokyo and Osaka)

volumes on the Shinkansen transport system is about ten times that of the air transport system between Tokyo and Osaka, so this comparative analysis is not based on equivalent fractional units (Fig. 3.14).

3.3.4.2 Sensitivity Analysis by Passenger Volume

Figure 3.15 shows a plot of the relation between CO_2 emissions per pass-km and the volume of passengers for the two transport systems. In the case of the Shinkansen, there is a strong relation such that as the passenger volume increases, the infrastructure-related CO_2 emissions are distributed among more passengers and therefore LC-CO_2 per pass-km decreases. In the case of air transport, this effect is so small that the environmental load per pass-km proves almost a constant value regardless of the number of passengers because there is no infrastructure between origin and destination. In addition, because there is always an integer number of flights, the graph has a sawtooth shape in the region of low passenger volume. As can be seen from Fig. 3.15, the CO_2 emissions per pass-km of the Shinkansen and air transport systems cross at about 4,000 passengers/day.

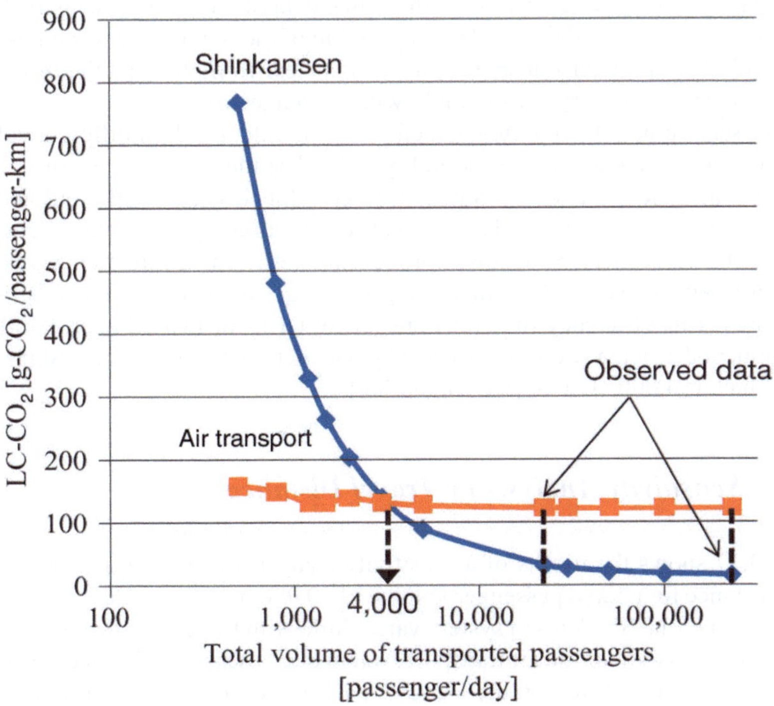

Fig. 3.15 Relationship between LC-CO_2 per pass-km and passenger volume (between Tokyo and Osaka)

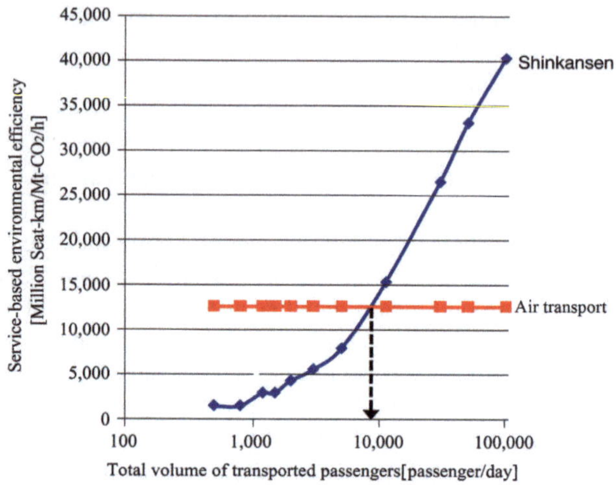

Fig. 3.16 Relationship between service-based environmental efficiency and passenger volume (between Tokyo and Osaka)

Comparing the CO_2 emissions per pass-km based on the actual data of 22,478 passengers/day for the air transport system, the LC-CO_2 per pass-km of the Shinkansen system is about one fourth that of the air transport system. Comparing the CO_2 emissions per pass-km based on the actual data of 233,000 passengers/day for the Shinkansen system, the LC-CO_2 per pass-km of the Shinkansen system is about one eighth that of the air transport system. It is clear from this that the results of a comparative analysis vary considerably with passenger volume.

Next, service-based environmental efficiency is calculated for different volumes of passenger demand and presented in Fig. 3.16. The total trip time by air transport is half of the time travelled by Shinkansen and, for this reason, the efficiency of air transport is initially higher than Shinkansen but after a point of intersection at a daily 10,000 passengers, Shinkansen proves more efficient. Thereafter, Shinkansen shows a steep increase led by increasing passenger demand. As a result, Shinkansen proves its efficiency over air transport in terms of LC-CO_2 per pass-km or environmental efficiency for the projected passenger volume of the new Shinkansen lines, which is 4,000–32,000 [passenger/day].

3.3.5 Sensitivity Analysis by Travel Distance

Figure 3.17 shows the results of a sensitivity analysis of LC-CO_2 per pass-km by travel distance for a fixed passenger volume of 10,000 passengers/day. The LC-CO_2 per pass-km of the Shinkansen system varies little with travel distance, whereas the LC-CO_2 per pass-km of the air transport system decreases with distance for the reasons already given. However, the LC-CO_2 per pass-km of the air transport system

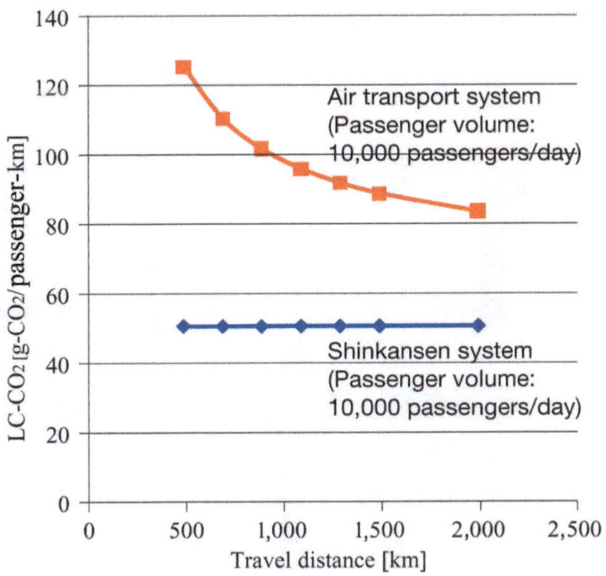

Fig. 3.17 Relationship between LC-CO_2 per pass-km and travel distance. (Passenger volume: 10,000 passengers/day)

is about 1.6 times larger than that of the Shinkansen system even for a distance of 2,000 km and never falls below that of the Shinkansen system.

3.3.6 Analysis for Each Prefecture Origin–Destination Pair

This section discusses the possibility and extent of CO_2 emission reduction resulting from any modal shift from air transport to new Shinkansen lines. For this analysis, origin–destination pairs for which at least one airport exists at each end are included and the total amount of CO_2 emissions generated by air and rail transport are estimated separately using the available distance and passenger demand data as of 2005.

The 2005 Survey of Regional Passenger Flow collected trip volumes on each OD pair across the country based on departures and arrivals in 50 regions (on the basis of prefectures, plus four regions in Hokkaido). Of these regions, the 33 with Shinkansen stations (including planned stations) are subjected to analysis. For comparison, CO_2 emissions are calculated for the origin–destination pairs where air transport is available assuming that passengers shift from air travel to Shinkansen. Where two or more airports exist in the same region, the total passenger demand is assumed to be fulfilled by the airport with a larger number of users (airports marked in solid black in Fig. 3.11). Other airports are coloured white in the figure. Similarly, Shinkansen passenger demand is assumed to be fulfilled by junction stations and those served by express trains.

Table 3.5 Comparison of prefecture origin–destination pairs by LC-CO_2 (*lower left part*) and service-based environmental efficiency (*upper right part*)

	Douhoku	Douou	Dounan	Aomori	Iwate	Miyagi	Akita	Yamagata	Fukushima	Tokyo	Niigata	Toyama	Ishikawa	Nagano	Aichi	Osaka	Hyogo	Tottori	Shimane	Okayama	Hiroshima	Yamaguchi	Tokushima	Kagawa	Ehime	Kouchi	Fukuoka	Saga	Nagasaki	Kumamoto	Ooita	Miyazaki	Kagoshima
Douhoku		**S**	**S**		S					S					A	A					A			A					A				A
Douou	**S**		**S**	**S**	**S**	**S**	**S**	**S**	**S**	**S**	A	A	A	A		**S**	A		A	A	**S**		A	A	**A**		**S**			A		A	A
Dounan	**S**	**S**				**S**				**S**					S	**S**																	
Aomori		**S**								**S**					S	**S**					A				A		A						
Iwate	**S**	**S**													**S**	**S**		A	A								A			A			
Miyagi		**S**	**S**										S		**S**	**S**	S			S	A			S			**S**						A
Akita		**S**								S		S			A	**S**					A			A									
Yamagata		**S**								**S**					S	**S**																	
Fukushima		**S**													S	**S**											S						A
Tokyo	**S**	**S**	**S**	**S**			**S**	**S**				**S**	**S**			**S**	S	A	**S**	**S**	**S**	**S**	S	**S**	**A**	S	**S**	**S**	**S**	**S**	**A**	**A**	A
Niigata		**S**													**S**	**S**											S						
Toyama		S								**S**																	S						
Ishikawa		S				**S**				**S**																	**S**						
Nagano		S														S											A						
Aichi	S	**S**	**S**	**S**	**S**	**S**	S	**S**	**S**		**S**	**S**											**S**		**A**	**S**	**S**		**S**	**S**	**A**	**A**	**S**
Osaka	S	**S**	**S**	**S**	**S**	**S**	**S**	**S**	**S**	**S**	**S**			S			**S**		**S**						**A**	**S**	**S**	**S**	**S**	**S**	**A**	**A**	**S**
Hyogo		**S**				**S**				**S**	**S**					**S**														**S**			**S**
Tottori				**S**						**S**					S																		
Shimane		S		**S**						**S**					**S**	**S**											**A**						S
Okayama		S				**S**				**S**						**S**																	**S**
Hiroshima	S	**S**	**S**	**S**	**S**	**S**	**S**			**S**	**S**								**S**													**A**	**S**
Yamaguchi	S									**S**																							
Tokushima		**S**								**S**					**S**												A						
Kagawa	S	S				**S**	**S**			**S**					**S**	**S**																	S
Ehime		A								**A**					**A**	**A**											**A**						**A**
Kouchi										**S**					**S**	**S**					**S**						S					**A**	
Fukuoka		SS		**S**	**S**	**S**			**S**	**S**	**S**	**S**	**S**	**S**	**S**	**S**		**S**	**A**				A	**S**	**A**	S			**S**	**S**		**A**	**S**
Saga										**S**					**S**	**S**																	
Nagasaki	S									**S**					**S**	**S**											**S**					**A**	**S**
Kumamoto	S	S								**S**					**S**	**S**	**S**								A		**S**						
Ooita										**A**					A	**A**																	
Miyazaki		S								**A**					**A**	**A**				**S**	**A**				**S**	**A**	**A**		**A**				
Kagoshima	S	S				**S**			**S**	S					**S**	**S**	**S**		**S**	**S**	**S**			S	A		**S**		**S**				

The actual number of flights recorded in April 2002 is used for the analysis. The number of Shinkansen trains is estimated from the number of transported passengers and the number of cars per train based on the 2005 survey. In reference to the average number of cars per train in April 2002 on each section of the Shinkansen lines, the region-specific train configuration is taken into account.

Table 3.5 presents the results estimated on the assumption that the new Shinkansen lines will replace air transport. This table is divided into two main parts by the diagonal, where the upper right part shows the results for service-based environmental efficiency and the lower left part gives the CO_2 emissions per passenger kilometre. The letter A represents a significant environmental advantage for air transport (with **A** indicating a difference of two times or more) and, similarly, S represents a significant advantage for Shinkansen in that origin–destination pair (with **S** indicating a difference of two times or more). The cells for origin–destination pairs where there is no available air transport data are left empty. This empirical study provides a basis for policy discussions based on the results given by a model that incorporates certain assumptions. Therefore, numerical results that do not have a suitable level of significance are ignored and not presented here.

In the case of a transport corridor through which more than one origin–destination pair is linked, the infrastructure-generated CO_2 emissions are distributed according to trip volumes on each OD pair so as to obtain a better representation of environmental load in terms of passenger-kilometres (Eq. 3.7). The total trip volume used for the distribution of CO_2 emissions is set by assuming that all origin–destination pairs receive Shinkansen lines.

$$\begin{aligned}&\textit{Distribution of infrastructure generated } CO_2\ (t-CO_2) \textit{ to one origin}\\&\quad -\textit{destination pair in section } A = (\textit{Total infrastructure}\\&\quad -\textit{generated } CO_2 \textit{ emissions } (t-CO_2)\times \textit{Volume of trips on the origin}\\&\quad -\textit{destination pair in section } A(\textit{passenger}))\ /\ \textit{volume in section } A(\textit{passenger})\end{aligned} \tag{3.7}$$

The infrastructure-generated emissions are not considered in corridors through which the given means of transport is currently providing services and where no new infrastructure development is required; if development was required, it would definitely add more to the environmental load. However, the infrastructure-generated emissions between Hachinohe and Shin Aomori on the Tohoku Shinkansen line (scheduled to open in December 2010) and between Hakata and Shin Yatsushiro on the Kyusyu Shinkansen line (scheduled to open in March 2011) are taken into account.

The results for CO_2 emissions per pass-km indicate that origin–destination pairs terminating in Tokyo have high passenger demand and for this reason the Shinkansen is advantageous in nearly all cases (the lower left part of Table 3.5). Similarly, for some of the other large metropolitan areas (for example, Aichi or Fukuoka), Shinkansen generates less CO_2 emissions per pass-km than air. In contrast, for regions that have not yet been connected to the Shinkansen system (such as Shikoku or Kyushu), the advantageous position of the railways is lost to air transport due to the large environmental load imposed by the infrastructure construction phase. However, among origin–destination pairs linked by Shinkansen, those prefectures that are connected by short air routes show a smaller difference between Shinkansen and air transport (for example, the Osaka–Iwate pair).

Service-based environmental efficiency results show a similar trend to those for CO_2 emissions (the upper right part of Table 3.5), but the number of origin–destination pairs where air transport efficiency exceeds that of Shinkansen is higher, and this trend is particularly noticeable in origin–destinations pairs linking Aichi and Osaka to Shikoku and Kyushu. Air transport between Tottori and Tokyo also turns out to be more environmentally advantageous.

The assumption that the current air transport passenger volumes prefer the planned Shinkansen improvements proves the environmentally advantageous position of Shinkansen in terms of both CO_2 emissions per pass-km and service-based environmental efficiency particularly for the region of Douo, where construction of Shinkansen lines is planned, and the origin–destination pairs from Hokuriku/northern Kyushu to Tokyo.

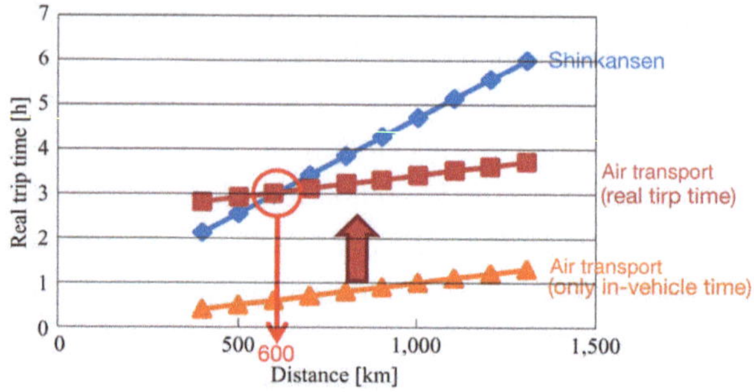

Fig. 3.18 Computation of real trip time

3.3.7 Service-based Environmental Efficiency Defined in Terms of Real Trip Time

3.3.7.1 The Concept of Real Trip Time

The modal split for regional travel indicates that the share taken by air transport generally increases for longer journeys. However, there is a problem in comparing the modal shares of air and Shinkansen transport using only the time spent in the vehicle; even if the actual journey time on an origin–destination pair is shorter for air transport, rail transport sometimes gains a greater share because of the far longer out-of-vehicle time (comprising waiting, boarding and de-boarding times) for air transport (Fig. 3.18).

Bearing in mind the significance of this out-of-vehicle time in mode choices, this study defines a "real trip time" which attempts to include out-of-vehicle time for air transport but does not seek to substantiate such out-of-vehicle time in the common way of estimating each of its three components for different modes. Rather, the real trip time approach assumes that at distances where the modal shares of air and rail are equal, the real trip times are also equal.

In Fig. 3.18, the relationship between observed trip time and distance for Shinkansen along the Tokaido and Sanyo lines is plotted. Similarly for airlines, the in-flight time is plotted for the same corridor starting from Tokyo International Airport (Haneda Airport). Then, the trip distance at which rail and air transport have nearly equal modal shares is determined from observations and a line drawn on the figure (600 km). Then, in order to equate the real trip time (3 h) for air and Shinkansen at 600 km, the air plot is shifted upwards until it intersects with the Shinkansen line at this distance. The shifted line then represents the relation between distance and real trip time for air transport, while the difference between this line and the one for in-vehicle line gives an estimate of the out-of-vehicle time. Based on this shift, the out-of-flight time for air transport is estimated at approximately 2.5 h.

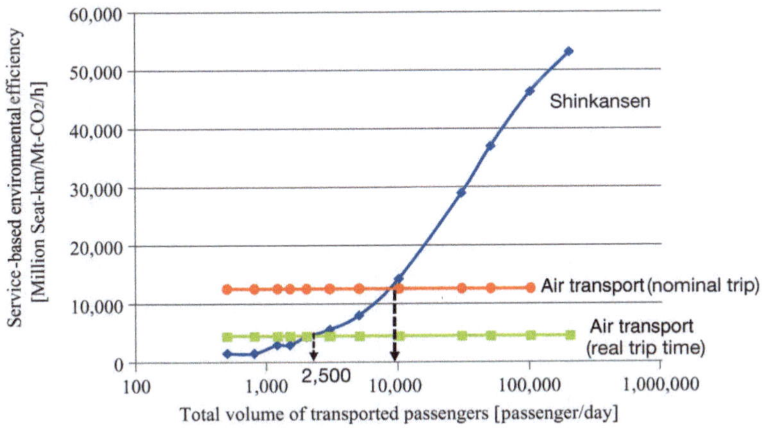

Fig. 3.19 Relationship between service-based environmental efficiency and passenger volume (between Tokyo and Osaka)

3.3.7.2 Analysis Between Tokyo and Osaka

Figure 3.19 shows the results of a sensitivity analysis conducted for the Tokyo–Osaka pair to represent the relationship between passenger volume and service-based environmental efficiency in terms of real trip time. This type of environmental efficiency based on real trip time reduces the efficiency of air transport compared with the previous nominal value given in terms of only the in-flight time (Fig. 3.16). By applying this service-based efficiency, the point at which Shinkansen exceeds air transport in environmental efficiency drops to 2,500 passengers a day.

3.3.7.3 Sensitivity Analysis by Travel Distance

Figure 3.20 shows the results of a sensitivity analysis of the LC-CO_2 per pass-km to represent the relationship between travel distance and service-based environmental efficiency (both the nominal environmental efficiency and the environmental efficiency by real trip time for the air transport system) when the passenger volume is 10,000 passengers/day. The service-based environmental efficiency of the air transport system exceeds the Shinkansen transport system at about 700 km for the nominal trip time and about 1,600 km for the real trip time, indicating the environmental advantage of air transport over Shinkansen for longer distances.

3.3.7.4 Analysis for Each Prefecture Origin–Destination Pair

Table 3.6 shows the results of a comparison of prefecture origin–destination pairs by the LC-CO_2 per pass-km and service-based environmental efficiency of the air transport system (using real trip time) and the Shinkansen transport system. The left

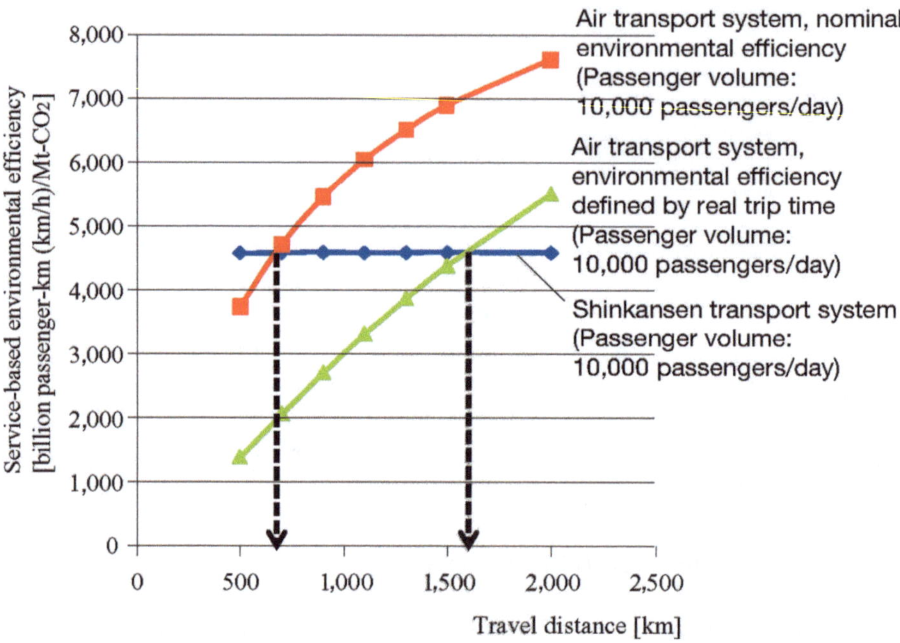

Fig. 3.20 Relationship between service-based environmental efficiency and travel distance

Table 3.6 Comparison of prefecture origin–destination pairs by LC-CO_2 and service-based environmental efficiency (using real trip time for air transport)

	Douhoku	Douou	Dounan	Aomori	Iwate	Miyagi	Akita	Yamagata	Fukushima	Tokyo	Niigata	Toyama	Ishikawa	Nagano	Aichi	Osaka	Hyogo	Tottori	Shimane	Okayama	Hiroshima	Yamaguchi	Tokushima	Kagawa	Ehime	Kouchi	Fukuoka	Saga	Nagasaki	Kumamoto	Ooita	Miyazaki	Kagoshima
Douhoku		S	S		S					S					S	S					S			S					S				A
Douou	S		S	S	S	S	S	S	S	S	S	S	S	S	S	S	S		S	S	S		S	S	A		S			S		A	S
Dounan	S	S				S				S					S	S																	
Aomori		S								S					S	S					S				S		S						
Iwate	S	S													S	S		S	S								S			S			
Miyagi		S	S										S		S	S	S			S	S			S			S						S
Akita		S								S		S			S	S					S			S									
Yamagata		S								S					S	S																	
Fukushima		S													S	S											S						S
Tokyo	S	S	S	S			S	S				S	S			S	S	S	S	S	S	S	S	S	A	S	S	S	S	S	A	A	S
Niigata		S													S	S											S						
Toyama		S								S																	S						
Ishikawa		S				S				S																	S						
Nagano		S														S											S						
Aichi	S	S	S	S	S	S	S	S	S		S	S											S		A	S	S		S	S	A	A	S
Osaka	S	S	S	S	S	S	S	S	S	S	S			S			S		S						A	S	S	S	S	S	A	A	S
Hyogo		S				S				S	S					S														S			S
Tottori				S						S					S																		
Shimane		S		S						S					S	S											S						S
Okayama		S				S				S						S																	S
Hiroshima	S	S	S	S	S	S	S			S	S								S													A	S
Yamaguchi	S									S																							
Tokushima		S								S					S												S						
Kagawa	S	S				S	S			S					S	S																	S
Ehime		A								A					A	A											A						S
Kouchi										S					S	S					S						S					A	
Fukuoka		S		S	S	S			S	S	S	S	S	S	S	S		S	A				A	S	A	S			S	S		A	S
Saga										S					S	S																	
Nagasaki	S									S					S	S											S					A	S
Kumamoto	S	S								S					S	S	A								A		S						
Ooita										A					A	A																	
Miyazaki		S								A					A	A				S	A				S	A	A		A				
Kagoshima	S	S				S			S	S					S	S	S		S	S	S			S	A		S		S				

lower part of the table is exactly the same as in Table 3.6. Comparing this with the service-based environmental efficiency (in the upper right part of the table) indicates that both the efficiency and CO_2 analysis yield the same result as to which of the modes is advantageous for the same origin–destination pair in most cases. This is clearly because there are more origin–destination pairs in which trip times for Shinkansen and air transport are equal or closer when real trip time is considered.

3.3.8 Conclusions

This study uses the LCA method to estimate CO_2 emissions and thereby verify the effectiveness of reducing airline routes and shifting passenger demand to the Shinkansen transport system as a measure to reduce CO_2 emissions in the transport sector in Japan. The main results obtained from this study are as follows:

1. Considering trips between Tokyo and Osaka, the LC-CO_2 emissions per pass-km and the service-based environmental efficiency of the air and Shinkansen transport systems for a given passenger volume are estimated. The Shinkansen system is proven to be environmentally advantageous for assumed passenger-kilometre figures on existing and planned Shinkansen lines.
2. CO_2 emissions resulting from the operation of the air transport system are estimated using actual air transport data between the prefectures of Japan, and the change in CO_2 emissions caused by a shift from air transport to Shinkansen lines is estimated. Air transport is shown to be environmentally advantageous over the Shinkansen system between origin–destination pairs with low passenger demand where infrastructure would have to be constructed for new Shinkansen lines.
3. In some of the origin–destination pairs where Shinkansen proves its environmental advantage in terms of LC-CO_2, air transport turns out to be advantageous in terms of service-based environmental efficiency assessment for the same pairs.
4. In the assessment of service-based environmental efficiency, which uses real trip times including out-of-vehicle time (waiting, boarding and de-boarding times) for air transport, Shinkansen is advantageous for most sections, and similar results in favour of Shinkansen are also obtained from the comparison of LC-CO_2. However, cautious use of this real trip time as defined in this study is necessary because its validity has not yet been sufficiently studied.

3.4 Policy Implication

It is accepted as common sense that the Shinkansen transport system is less CO_2 intensive than the air transport system since it consumes less energy in operation. This study of CO_2 emissions based on LCA, taking into account the construction of infrastructure and the production of aircraft/vehicles as well as variations in passenger demand, reveals that this is not always true for every given rail line or rail section. A comparative analysis of Shinkansen infrastructure projects for prefecture origin–destination pairs show that CO_2 emissions will be reduced for new Shinkansen lines that have a passenger volume of at least 2,500 passengers/day. Further, when service-based environmental efficiency (including trip time) is considered, a similar trend is seen. That is, the results indicate that it is effective to abandon airline routes and shift demand to Shinkansen for shorter travel distances and higher passenger demand. In this case, however, the shift will have a large effect on demand on other routes/lines and the quantitative trend will also vary. This confirms the importance of constructing domestic transport networks with CO_2 emissions in view while making the most of the characteristics of each system.

3.5 Economic Incentives to Achieve Low-Carbon Transport Systems: CO_2 Reduction Effect of Environmental Taxation

3.5.1 Background and Aim

The contribution of intercity passenger transport to domestic CO_2 emissions is steadily increasing in Japan. Economic growth is obviously a major factor in this increase, but the principle of market mechanism introduced into the inefficient transport sector through deregulation including liberalisation of airfares and privatisation of national rail lines is recognised as another important factor. In the competitive market, mechanism that has been shaped by deregulation, improvements in the inefficient transport sector such as reduced fares and increased service levels have led to a substantial improvement in travel convenience. As a result, demand for transport services has grown, further adding to CO_2 emissions. Since deregulation, changes in the modal shares of air, rail and passenger car transport have been particularly conspicuous, with passenger cars taking a greater share of shorter trips and air transport increasing for long-distance trips. Better prices and shorter journey times have also brought changes, resulting in more passengers being attracted to the rail transport system and generating larger amounts of CO_2 (Fig. 3.21).

However, in the future, the issue of CO_2 emissions generated by intercity passenger transport will require more attention at the time of policy design and implementation, and this will require a two-sided approach. The first need is to develop policies regarding the supply dimension; for example, the private sector can achieve technological improvements for the production of more energy-efficient vehicles, or a policy framework can be applied to accelerate improvements in the high-speed rail network in corridors where Shinkansen improvements can be expected to reduce life-cycle CO_2 emissions (discussed in the previous section). The second dimension is to foster policies to manage the demand side of transport in a manner that prompts travellers to change their travel behaviour; for example, passenger cars

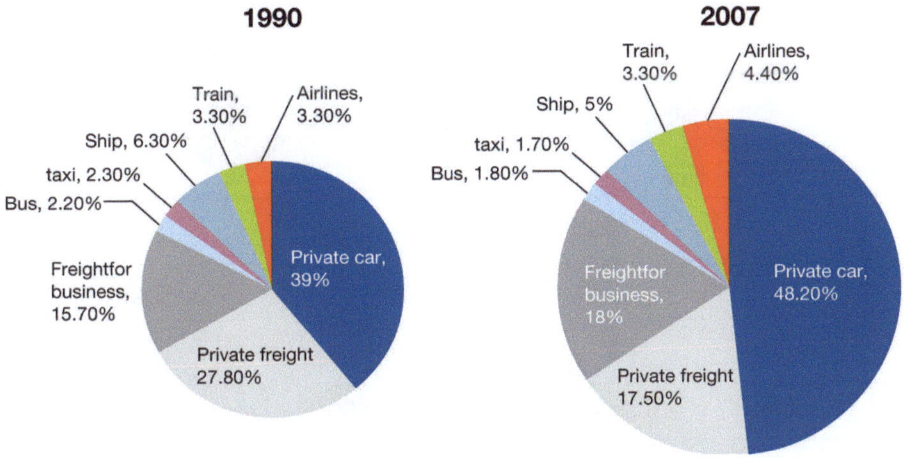

Fig. 3.21 Share change in CO_2 emissions in transport sector (1990–2007): Ministry of Land, Infrastructure, and Transport and Tourism

with very high-unit CO_2 emissions can be controlled through adopting a revised tax system or other supplementary programs. It is now essential to design policies that employ this two-sided approach in order to maximise the merits of a comprehensive approach and develop frameworks that will guide Japan to a low carbon society.

In Britain, the target for cutting domestic CO_2 emissions was set at 12.5 %, and policies and programs developed in the economic approach were launched early on. For example, a petrol tax was introduced as a demand control policy appealing to the economic motive; as a result, rising petrol use has been mitigated and the increasing modal share of passenger cars has been controlled[1)]. It is highly important to set out clear targets, design policies and develop suitable programs aimed at achieving CO_2 reduction targets.

Japanese efforts to cut domestic CO_2 emissions have been more successful in applying supply-side policies. These were led by the Top Runner System and other schemes established under the revised “Energy Saving Act (1999)”. On the other hand, demand-side policies such as influencing individual travel behaviour in the environmental context have been relatively retarded. The main reason for the slow progress in such implementations is the failure of the “Comprehensive National Development Plan” to establish a uniform transport network across the whole country, mostly for cost effectiveness and financial reasons. This means that tax-levying policies could cause discrepancies between the outlying regions and large metropolitan areas; it would be difficult to apply such economic policy measures unless they were applied in a context that would mitigate and control such regional disparities.

Bearing in mind these issues in Japan, this section sets a target for reducing CO_2 emissions generated from intercity passenger transport and discuss the introduction of a “Baumol model” taxation system (as proposed by W. Oates and Baumol (1971)) as an alternative demand-side policy that would ensure attainment of the target and promote changes in travel behaviour. Further, a policy design will be

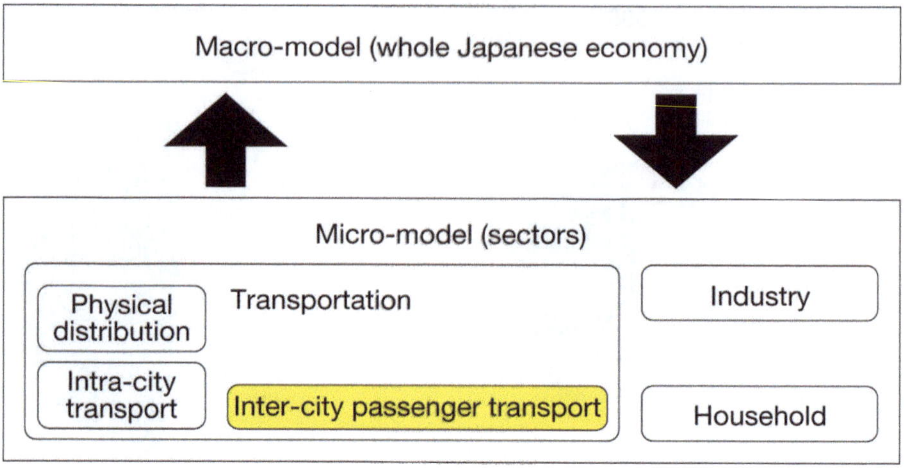

Fig. 3.22 Scope of this study

sought to ensure equity through the reallocation of tax resources. In undertaking these discussions, a systematic method of evaluation is proposed that allows for quantitative examination of these two policies as to their ability to influence travel behaviour, their contribution to CO_2 reduction and their effectiveness in controlling regional disparities. The purpose is not to track secular changes brought about by the introduction of a tax levy, but rather to identify its temporary effects. Therefore, since the purpose is to measure only the effectiveness of environmental tax regulations over a certain time, any technological innovations that may take place in this given period are disregarded. This analysis also deals only with intercity passenger transport and not the entire domestic economy (Fig. 3.22).

3.5.2 Case Study Analysis of Effects of a Petrol Tax

This part introduces the outline of the empirical analysis by Fujisaki et al. (2011) on the effects of fluctuation in petrol prices on transport behaviours in Japan, such as usage of private passenger cars as well as that of public transport. The analysis focuses on verifying whether and to what extent the change not only of petrol prices but also of income affects the transport volume (passenger-kilometres) per capita travelled by public transport, as well as that by private passenger cars. The approach taken by Fujisaki et al. estimated the elasticities of petrol price and income both in real and nominal pair on per capita transport volume of public transport and private passenger cars, respectively, as well as on modal share of public transport and private passenger cars, respectively, by means of econometric models. The data examined were both yearly and quarterly datasets and were adapted from Annual Statistical Report on Railway Transport and Annual Statistical Report on Motor

Table 3.7 Short-run and long-run elasticities on transport volume per capita. (Source: Adapted from Fujisaki et al. (2011, Table 5))

	Short-run petrol price elasticity		Short-run income elasticity		Long-run petrol price elasticity		Long-run income elasticity	
Pass-km per capita	Nominal	Real	Nominal	Real	Nominal	Real	Nominal	Real
Public transport	0.16 (4.096)	0.08 (2.914)	0.22 (4.489)	0.46 (5.320)	0.25	0.14	0.34	0.78
Private passenger cars	−0.15 (−4.086)	−0.19 (−4.869)	0.40 (5.940)	0.36 (5.432)	−0.22	−0.26	0.58	0.50

Table 3.8 Example of estimated effect of the hypothetical abolition of the provisional additional petrol tax rate on surface transport. (Source: Adapted from Fujisaki et al. (2011, Table 7))

		Effect (%)
Surface transport	Passenger-km per capita	+ 2.6 – 3.5
Public transport	Passenger-km per capita	− 2.3 – 2.5
	Modal share	− 5.0 – 5.8
Private passenger cars	Passenger-km per capita	+ 6.2 – 8.5
	Modal share	+ 3.7 – 5.8

Vehicle Transport by MLIT (n.d. b,c), Monthly Gasoline Retail Price by the Oil Information Center (n.d.), National Accounts by the Cabinet Office (2010), as well as Population Estimates and Consumer Price Index by the Ministry of Internal Affairs and Communications (MIAC) (n.d., 2010). As a result, regarding the effect of petrol prices and a level of income on traffic volumes and the shares in the transport behaviours of private passenger cars and public transport after fiscal year (FY) 1987, the analysis found that "petrol prices have a negative effect on private passenger cars, while they have a positive effect on public transport". One of the calculated examples is shown in Table 3.7. As for the other values of elasticity and analysis results, please refer to Supplementary note 1.

In addition, by using the values of elasticity (see, Supplementary Note 1) obtained from the analysis, Fujisaki et al. (2011) performed a simulation of the hypothetical case of abolishing a provisional additional petrol tax rate fixed at ¥ 25.1 per litre, which would be equivalent to a 20.4 % drop in price in the case of petrol price being ¥ 123 per litre (average nominal price between FY1987 and FY2008). Rough estimation about the effect on traffic volume is shown in Table 3.8.

Under the assumption of such a tax reduction on a traffic volume of FY2008, the CO_2 emissions in this sector would have increased about 5.4–7.6 % from a macro perspective. In turn, the result of the simulation might suggest that the existing provisional additional petrol tax rate has been actually preventing such an extent of increase in CO_2 emissions.

Apart from this, in order to identify the effect of discounting of expressway tolls which was introduced in March 2009 in Japan, Fujisaki et al. (2011) calculated the disparity by comparing the theoretical values in FY2009 of per capita transport volume in public transport and private passenger cars, which were estimated by applying the same econometric models that represent the impact of petrol price and income in a way to exclude the impact of the discounting, with the observed respective actual values in the same fiscal year. The disparity, which means the impact of the discounting, was −1.6 to −1.8 % and +2.3 to +4.4 % in per capita transport volume in public transport and private passenger cars, respectively. In other words, the discounting of expressway tolls was found to reduce usage of public transport and increase usage of private passenger cars. Fujisaki et al. further made a rough estimation of adverse effect of the hypothetical abolition of expressway tolls which the Japanese premier once proposed to implement: Public transport would be used much less and private passenger cars would be used much more.

3.5.3 Proposed Baumol Taxation Model and Reallocation of Tax Revenues

In this section, the methods of applying environmental taxes using the Baumol model and of allocating tax revenues will be explained. Various concerns related to this framework will also be clearly identified.

3.5.3.1 Merits and Demerits of Baumol Taxation

1. Backcasting
Backcasting is a method proposed by Kari-Henrik Robert, founder of the Swedish "The Natural Step" (NPO; Holmberg 1998), to develop the necessary process for achieving a new goal that may require a series of actions. The distinctiveness of the method arises from its approach based not on currently possible issues but on seeking how to achieve aims in the future. In applying the backcasting method to transport evaluations, the main policies of concerns are: (1) economic approaches such as emissions trading or environmental taxes; and (2) regulations under the Top Runner System. The particular focus of this study is the Baumol method of environmental taxation, which is an important economic instrument to mitigating the increasing environmental load from the transport sector.

2. Backcasting and imposition of Baumol tax
The imposition of environmental taxes is suggested as a way of internalizing social costs generated by external diseconomies in the market pricing mechanism. In other words, public welfare is not maximised by applying the Pareto optimisation, so environmental levies can be applied in an effort toward such maximisation. The theoretical merits of an environmental tax are: (1) targets for reducing environmental load can be realised at minimum social cost, (2) there is a price effect, (3) technological innovation is promoted and (4) the quality of the double dividend can be increased.

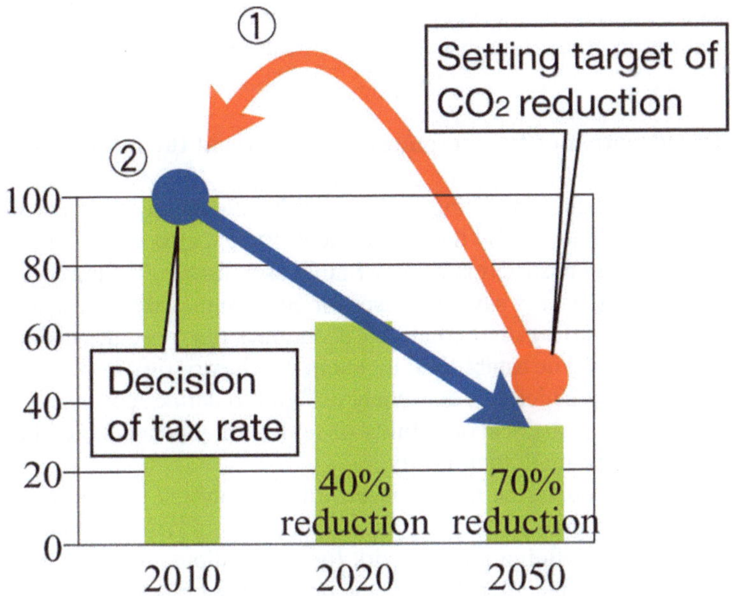

Fig. 3.23 Baumol model of environmental tax

In addition to these theoretical merits of environmental taxes in general, Baumol levies can be set at the level required to meet set targets by utilising the backcasting technique (Fig. 3.23).

In a Baumol taxation system, a unit tax rate is allocated per unit amount of generated pollution for a uniform increase in the environmental levy. In this study, it is assumed that a tax is imposed on intercity passenger transport by allocating a tax rate per tonne of CO_2. The merit of the method lies in minimising the costs for cutting CO_2 by the amount set in policy through a uniformly increasing taxation system in which the marginal costs for emission reduction are equalised. Using this type of environmental levy, a taxation policy for cutting CO_2 emissions can be very efficiently implemented. It should be noted, however, that although the applicable tax rate is even among all transport modes, tax revenues vary according to the tax rate multiplied by the unit emissions of CO_2 by air, rail and passenger cars.

However, besides the merits of such a tax regime, there are also some demerits in applying the Baumol model to environmental tax systems. In countries like Japan, where there is a transport network with some regional disparities in level of services, a uniform environmental levy on a non-uniform transport system may be expected to have different effects in different regions. Such an environmental levy may be more influential in regions where the transport system generates higher unit values of CO_2 emissions per passenger-kilometre, whereas its influence may be more moderate in the large metropolitan areas that have well-developed railway networks with relatively low unit values of CO_2 emissions per passenger-kilometre.

Thus, a Baumol-type environmental tax is efficient in general but has not proved its efficiency in the context of equity. Therefore, it is essential to develop an environmental tax framework that also includes policies to ensure more balanced implementations by mitigating disparity in application.

3.5.3.2 Re-allocation of Tax Revenues to Supplement the Baumol Model

One appropriate way of minimising inequity in the allocation of collected taxes is the re-allocation of revenues by the government. Revenues could be used in many ways, such as for the construction of new Shinkansen lines or the purchase of low-energy equipment for energy saving. If uses for tax revenues are to be selected from among such a wide variety of alternatives, economics theory suggests that a social welfare function should be established to determine the priority in social utility that results. However, in practice it is extremely difficult to define such a social welfare function, so there is no practical possibility of setting an implementation priority for the alternatives according to their utility.

Given this background, the main concern of this study in the context of social welfare and taxation does not lie in determining certain desired levels of utility change, but rather in defining a framework for the re-allocation of tax revenues in such a way that, before and after the levy, the rate of utility change in each region remains constant. The standard utility is set by consumption activity subject to budget constraint before the imposition of the tax. The utility varies substantially among the regions but the standard utility prior to imposing the levy is set to unity so as to measure the rate of change in each region. Here, the re-allocated amount Eq. 3.8) is defined on the basis of the principle of equivalent variation (EV) in economics, which substantiates the amount of compensation required to bring regional utility, which is diminished as a result of the levy, back to its level prior to the levy. The necessary condition for the equivalence of compensation amount per person at the national level is set in the following formulation:

$$EV(Re - allocated\,amount.) = \left(\frac{u_i - \overline{u_i}}{\overline{u_i}} \right) \times I_i, \tag{3.8}$$

where

u_i Utility after imposition of tax
$\overline{u}_i$ Utility before imposition of tax
I_i Income in region i

It is worth underlining the point that the utilisation of re-allocated tax revenues should be such that the re-allocated amount computed by this equation mitigates regional disparities as much as possible, without regard to the utility of the re-allocated tax.

3.5.3.3 Perspectives on Policy Evaluations

The scope of this policy assessment is wide in that it considers a policy package including a tax re-allocation to mitigate regional disparities arising from a new environmental taxation program. In this context, the aim is to achieve more effective policy implementations and to design a better framework for environmental measures. Assessment of the impact of the two combined policies—the levy of an environmental tax and re-allocation of the tax revenues—requires a quantitative analysis. The three main issues in this quantitative analysis are as follows:

1. After levying a tax in the form of the Baumol model, what sorts of trip pattern changes are observed in each region?
2. After levying a tax in the form of the Baumol model, how much are CO_2 emissions reduced in each region?
3. After levying a tax in the form of the Baumol model, how much re-allocated revenue should be provided to each region?

By finding answers to these three questions, the impact of the levy policy on transport users is analysed, then the associated CO_2 reductions in each region are evaluated, and finally the cost of mitigating undesired impacts in each region is determined.

3.5.4 *Development of the Analytical Model*

3.5.4.1 Properties of the Analytical Model

The previous section defined three parameters that must necessarily be quantified for each region in the policy assessment. This section describes the development of an analytical model for the quantification of these parameters and clearly explains the three main properties on which the model is based:

1. Changing the scope of the trip demand estimation model
2. Changing the scope of the tax revenue re-allocation model
3. Integrating the tax model and the tax revenue re-allocation model

In changing the scope of the trip demand estimation model, the main approach is to use the backcasting principle to propose an alternative modelling framework. Figure 3.24 compares the conventional transport demand estimation method with

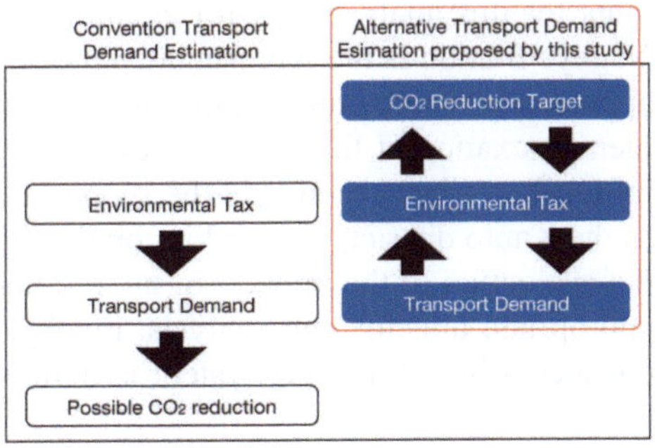

Fig. 3.24 Scope of trip demand estimation in conventional models and in the proposed analytical model

this alternative model. The conventional model estimates transport demand changes regarding trip time or trip cost changes, which represent policy implementations in the model. Thus, the total amount of future CO_2 emissions is estimated at future transport volumes already determined in the previous modelling stage. However, this study proposes an alternative model in which future targets for CO_2 emissions are first set and then, in the next step, the Baumol environmental tax needed to achieve that level of emissions is obtained. Finally, the results of these two steps are used to estimate future transport volumes that will lead to the determined CO_2 reductions (Fig. 3.24).

In changing the scope of a tax revenue re-allocation model, the amount to be re-allocated is defined based on assumptions relating to the impact of re-allocation. The common approach to the allocation of tax revenues is to spend the tax resources when there is a new need for financial resources. However, this study suggests an alternative method, seeking a model that re-allocates revenues to each region in such a way that the rate of change in utility before and after the environmental levy remains constant.

By integrating the above two policies, the proposed analytical method provides an integrated approach to defining individual trip behaviour changes due to the levy of the tax and discussing better ways of allocating the collected tax revenues by government in the context of interregional equity. In other words, this analytical model incorporates individual trip choices and government public policy as well as the integration of both. This enables the design and implementation of an effective environmental tax policy whereby the aimed-for CO_2 reductions are achieved and tax revenues are allocated among the regions based on constant variation in the rate of utility, or transport demand and tax rate, for each prefecture.

3.5.4.2 The Model's Prerequisites

Although this analytical method is developed for the evaluation of intercity transport policy in Japan, it is defined in such a way that application is not limited to Japan and can be readily applied in many other regions. In general, after setting the area subject to the evaluation, it is divided into prefectures in order to explain trip attributes. All arrivals and departures are represented by a single selected main node within the intercity. In the case of Japan, the prefectures are considered appropriate units for this analysis; so, for example, Aichi prefecture is one regional unit and Nagoya Station is the representative node. The main reason for such a macro-level approach is that the model aims at grasping trip behaviour changes due to environmental taxation at the intercity level across the large area of Japan and therefore intra-urban trips are outside the scope of the analysis. Larger regional units, such as the Kanto district, are too big; on the other hand, the urban level is too small for the evaluation of the impacts of tax re-allocation. Hence, the prefectural level is an appropriate unit for this analysis. Finally, the trip demand pattern among regions connected by a transport system is defined with the help of a trip demand utility function. This is developed through the application of the principle of consumer behaviour in economics.

3.5.4.3 Main Variables in the Model

The model consists of five main variables:

1. Utility: u_i
2. Transport volume: x_{ij}^k
3. Transport cost: c_{ij}^k
4. Baumol tax rate: t
5. CO_2 emissions: e_{ij}^k

A transport user in region i acts to maximise the utility (u_i) resulting from a trip and this utility may either increase or decrease depending on the transport cost. Transport cost is defined in the form of a generalised cost (c_{ij}^k), which uses transport volumes (x_{ij}^k) representing the mode-specific high demand between regions i and j, with transport mode k (x_{ij}^k). Finally, the amount of CO_2 emissions (e_{ij}^k) generated after the adoption of the Baumol levy is obtained by a defined tax rate (t).

Details of the model are provided in supplementary note 2.

3.5.5 Results of the Analysis

3.5.5.1 Rationale for Reduction Target Setting

This analysis assumes that the CO_2 reduction target for intercity passenger transport is 5 %. It should be noted that the study covers a single period of time and involves no time-series analysis.

According to the National Institute for Environmental Studies (AIM), Keio University (KEO) and others, a cut of approximately 20 % will be required for the entire transport sector to achieve a 20 % reduction by 2020. The transport sector includes not only intercity passenger transport but also intra-urban passenger transport, physical distribution and others, and no specific reduction target for intercity passenger transport has been clearly defined. Further, using a backcasting approach, it is necessary to establish more specific reduction levels for the period preceding 2020 if a 20 % reduction is to be achieved by 2020. Against this background, this analysis sets the reduction level for the transport sector in 2015 at 10 % over the 1990 level, as shown in Table 3.9. For intercity passenger transport, which has lower CO_2 emissions, a 5 % target is assumed, less than the target for the transport sector as a whole. It should be noted that since this study employs a static model with no time axis, the 5 % curtailment setting in intercity passenger transport is for the analysis of its effects in the scenario where a reduction of about 10 % is required by 2015 for Japan as a whole.

Table 3.9 CO_2 reduction target setting for regional passenger transport (set by the author based on the AIM and KEO calculations)

Year	1990	2015	2020	…	2050
National target	0%	10%	20%	…	60%
Transportation target	0%	10%	20%	…	–
Regional passenger transport target	0%	5%	10%	…	–

3.5.5.2 Trip Demand and Mode Variations

The national target for cutting the total emissions is set at a rate of 5% (2.42 million t per year over the whole Japan). The Tokyo metropolitan area is selected as the case study for explaining trip demand and mode variation after applying the Baumol levy and then re-allocating this tax revenue among the regions on an original area basis (that is, where tax is levied and re-allocated in the same area).

Figure 3.25 is a map showing the estimated decrease in total demand for all trips originating in Tokyo by the three main modes of transport as a result of levying an environmental tax. The largest decrease is observed in trips that terminate not far

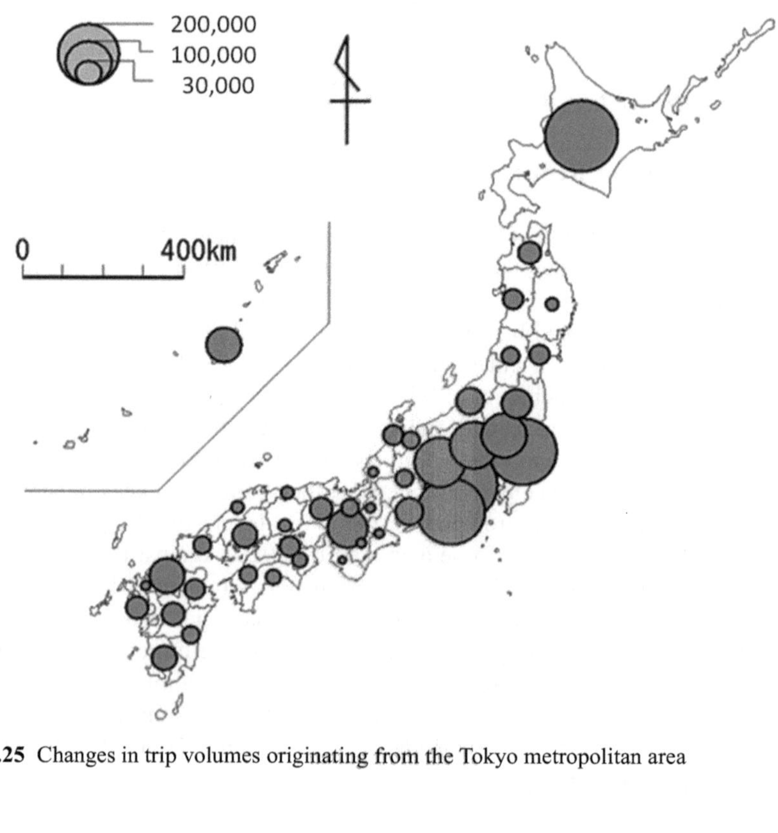

Fig. 3.25 Changes in trip volumes originating from the Tokyo metropolitan area

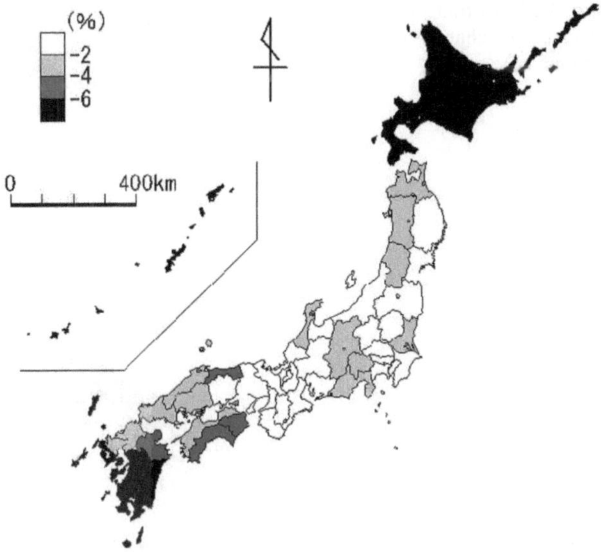

Fig. 3.26 Ratio of trip changes originating from the Tokyo metropolitan area

from the Tokyo area; annual reductions in passengers are 280,000 to Yamanashi prefecture, 240,000 to Ibaraki prefecture and 240,000 to Shizuoka prefecture. The reason for these reductions is that the tax levy influences the predominantly passenger car trips to these prefectures, where the modal share attributed to passenger cars is 92 % to Yamanashi prefecture, 75 % to Ibaraki prefecture and 56 % to Shizuoka prefecture. Another notable decline is seen in trips between Tokyo and Hokkaido, a destination far from Tokyo and indeed one of the longest domestic routes from Tokyo. On this route, demand decreases by 270,000 passengers, meaning a decrease of 8 % in trips between Hokkaido and Tokyo. This specific result for Hokkaido is mainly attributable to the increase in transport costs between Hokkaido and Tokyo and the lack of good alternatives to the air transport mode between these regions. It suggests that much travel activity between these regions is abandoned (Fig. 3.26).

Figure 3.26 represents the distribution of the ratio of decline in the number of passengers after levying the tax. In comparison with the previous map, a different pattern is observed for Hokkaido, Kyushu, Shikoku and Okinawa prefectures, where the decline is relatively high at around 6–8 %. Because the share of air transport along these routes is high, at approximately 78 %, the levied tax has more potential to add to the generalised cost of transport connecting these regions to Tokyo. This forces a considerable number of passengers to give up making trips over these routes, which are poorly provided with alternative modes.

The three maps in Figs. 3.27–3.29 show the mode share changes for trips originating from the Tokyo metropolitan area. Figure 3.17, for the aviation sector, shows marked falls for the Japan Sea coastal areas of Tohoku, Chugoku, Shikoku and Northern Kyushu. Similarly, Fig. 3.28 shows the changes for passenger car trips originating from Tokyo, in which the loss of car share is particularly evident for trips of less than 200–300 km. In these shorter trips from Tokyo to Yamanashi,

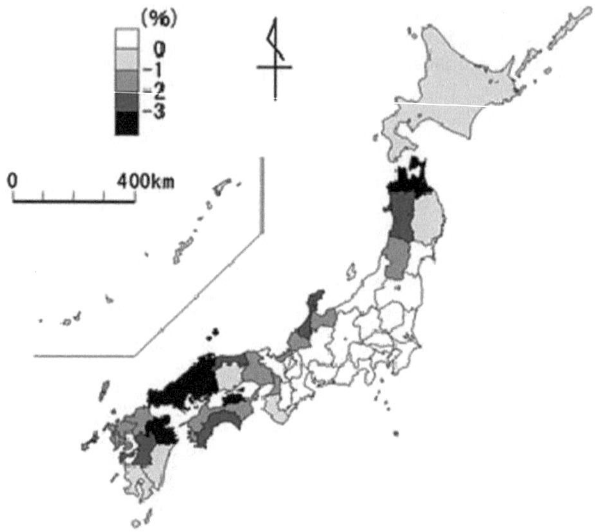

Fig. 3.27 Air transport modal share changes (Tokyo originated)

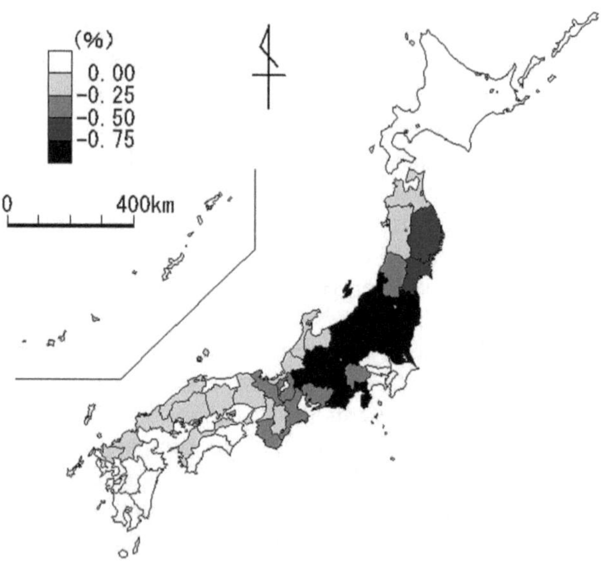

Fig. 3.28 Passenger car modal share changes (Tokyo originated)

Ibaraki and Shizuoka prefectures, trip demand decreases by 250,000 car passengers, mainly because of the high dependency on cars in these areas as in the interpretation of Fig. 3.25. However the impacts of environmental tax on car users are limited and the modal share decline for car trips does not exceed 1 %.

Figure 3.29 shows how the modal share of railway trips originating from Tokyo increases. In regions where the air transport share decline is higher in Fig. 3.27 (the Japan Sea coastal area of Tohoku District, Chugoku, Shikoku and Northern

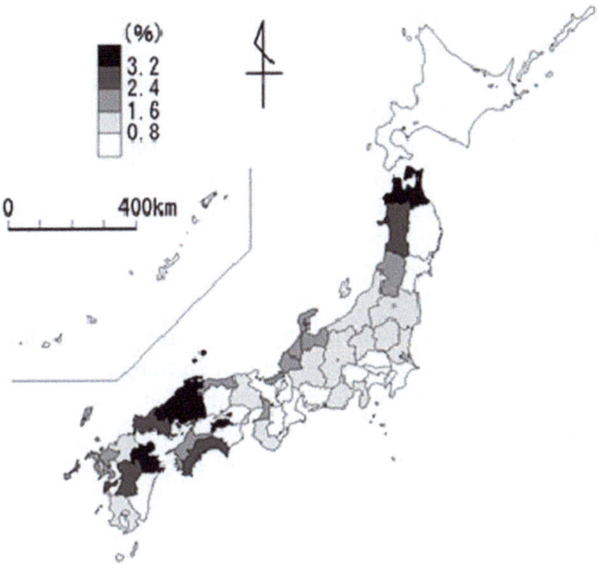

Fig. 3.29 Rail transport modal share changes (Tokyo originated)

Kyushu), the railways gain higher modal shares after the environmental tax. In other regions showing smaller aviation decline rates of approximately 3 %, such as Aomori, Hiroshima or Yamaguchi prefectures, the railway modal share increases by a similar amount of 3.3 % on average. In comparison with the reduction in total transport volume (Fig. 3.25), the modal shift to railways is considerably less; for instance, for travel to Aomori prefecture, 34,000 passengers cease to travel by air and only 25 % of them shift to railway (7800 passengers). For Yamaguchi prefecture, air transport users decrease by 12,000 passengers, of which approximately 68 % shift to railway. This ratio is even smaller for Hiroshima prefecture, where only 36 % of former air users move to railway.

Figure 3.30 reveals the proportional relationship between increase in railway modal share and decrease in air modal share for all the plotted results over the whole of Japan. A similar graph was plotted to seek a relationship between railway and car

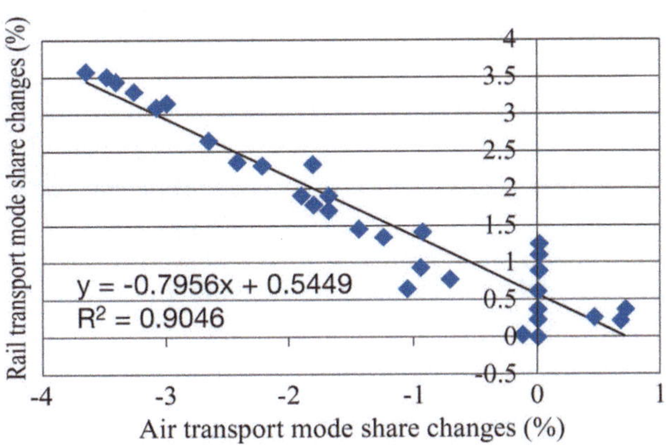

Fig. 3.30 The relation between air and rail transport modal share changes

modal share; however, no such relation was observed. That is, the degree of substitutability of rail and air transport is high, producing higher values of elasticity of substitution; whereas the degree of passenger car substitution by rail is notably low.

To sum up the findings above, for journeys of medium distance starting from the Tokyo metropolitan area, the number of passengers giving up car trips is extremely high; however, the fall in passenger car modal share (Fig. 3.26) does not even reach 1 %. This leads to the conclusion that the impact of this type of environmental tax on overall transport demand is somehow limited. In the case of longer journeys and for prefectures where a modal shift from air transport to rail is possible, because of the substitutability of air with rail transport (Fig. 3.30), the effectiveness of tax policy in reducing overall trip volumes is diminished. In this sense, for regions that are not connected by a well-developed transport network with a mix of alternative modes of transport, a significant reduction in transport volumes is evident because of the constraints of the transport system, which provides no better alternatives than giving up trips (such as between the Tokyo metropolitan area and Kyushu, Shikoku or Okinawa).

3.5.5.3 Changes in Regional CO_2 Emissions

The national target for cutting the CO_2 emissions is set at 5 %. The regional variations in CO_2 reductions resulting from the levy of the Baumol tax and the re-allocation of the collected tax resources are estimated using the analytical model. Figure 3.31 shows the reduction in total CO_2 emissions in each region. The five regions with

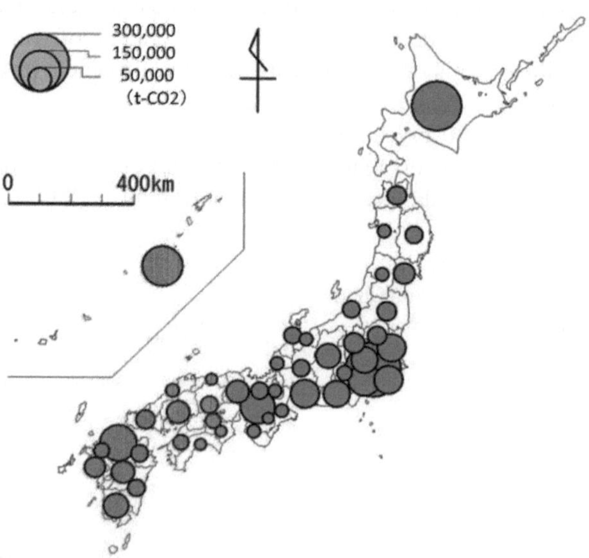

Fig. 3.31 Amount of reduced CO_2 emissions based on original area

Fig. 3.32 Ratios of CO_2 emissions reductions based on original area

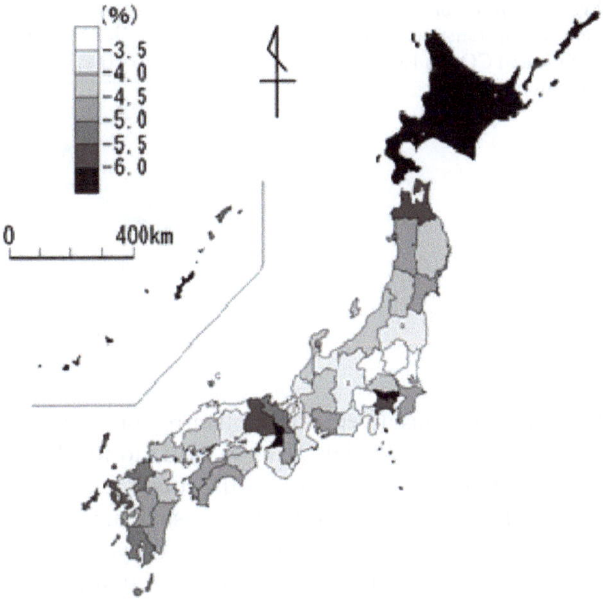

Fig. 3.33 Relationship between total trip demand decline and CO_2 emissions reductions

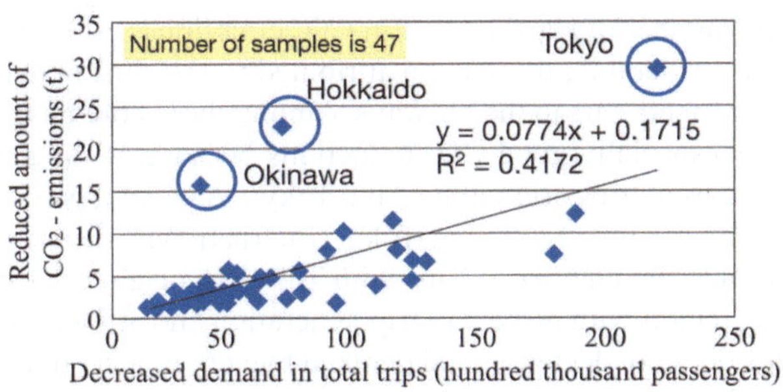

the most substantial reductions in CO_2 emissions are distributed across the country: 230,000 t in Hokkaido; 290,000 t in the Tokyo metropolitan area; 1.1 million t in Osaka; 120,000 t in Fukuoka; and 160,000 t in Okinawa. Figure 3.32 gives the relative CO_2 reduction ratios calculated from the target emission amounts given in Fig. 3.11 for comparison with the absolute values. The regions with greatest ratios of reduction are Hokkaido (9 %), Tokyo (6 %), Osaka (6 %) and Okinawa (10 %), all of which exceed the national Figure (5 %) (Figs. 3.31 and 3.32).

Trips either starting or ending at Hokkaido and Okinawa rely on air transport because there is poor provision of alternatives (the air transport share is 81 % for Hokkaido and 100 % for Okinawa). This leads to the remarkable decrease in CO_2 emissions noted above for these regions, which is around 8–9 %. Moreover, the ratio of CO_2 emissions reductions to decline in trip demand is notably high for Hokkaido and Okinawa, and even for Tokyo. Figure 3.33 clearly illustrates how far

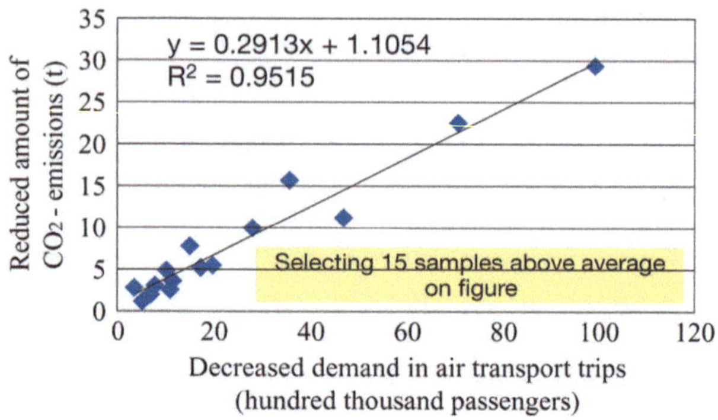

Fig. 3.34 Relationship between air transport demand decline and CO_2 emissions reductions

these three regions diverge from other regions in this respect; the ratio is 4–5 times greater for Hokkaido and Okinawa, and 1.6 for the Tokyo metropolitan area. The specific reason for these results is the strong relation between CO_2 emissions reductions and trip demand decline in the case of air travel (Fig. 3.34).

On a simple interpretation regarding Tokyo, Osaka and Fukuoka, which are central metropolitan areas with good intercity transport, transport demand to and from these regions is particularly high, but the decline in the number of trips since the tax has been levied is limited and does not exceed 3 %. In spite of this small fall in the number of trips, the cut in CO_2 emissions resulting from the environmental tax is considerable; this can be attributed to the shift in demand from air transport and passengers cars to the railways. Among these core metropolitan areas, Tokyo generates especially good CO_2 reductions because of the greater number of destination regions in trips originating from Tokyo. A general pattern in regional trips is the reliance on Tokyo Station or Tokyo International Airport for transfers and this further adds to the number of trips with Tokyo as origin or destination, making it the leading region with respect to trip generation and attraction. Overall, the environmental tax does not decrease total demand by very much, but it does lead to a notable modal shift onto rail and, as a result, delivers the desired emissions reductions.

3.5.5.4 Relationship Between Amounts of Tax Collected and Re-allocated

Figure 3.35 shows the regional distribution of environmental taxes substantiated by setting the national level CO_2 emissions reductions at 5 %. Of the total annual revenue brought in by the new tax, which is 260 billion JPY, the largest single beneficiary is the Tokyo metropolitan area (24.2 billion JPY), followed by Hokkaido, Osaka, Fukuoka and Okinawa (with 10 billion JPY on average). But the amount of tax paid differs between urban areas and rural areas: the tax revenue per head is \6501 for Okinawa and about \4000 for Ibaraki, Tottori, Yamanashi, Gunma and other prefectures, while for urban areas it is in the range \1000 to \2000. This illuminates the greater burden of tax on the provinces. The given regional tax levy

Fig. 3.35 Collected tax amount by regions

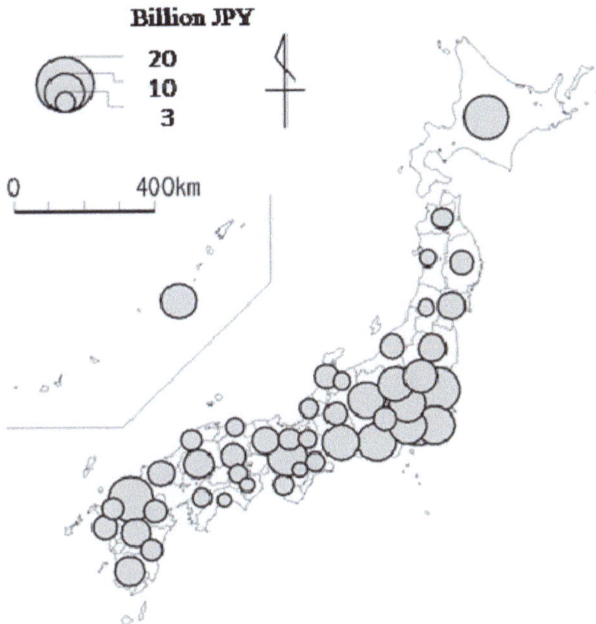

Fig. 3.36 Relationship between total trip demand and collected tax amount

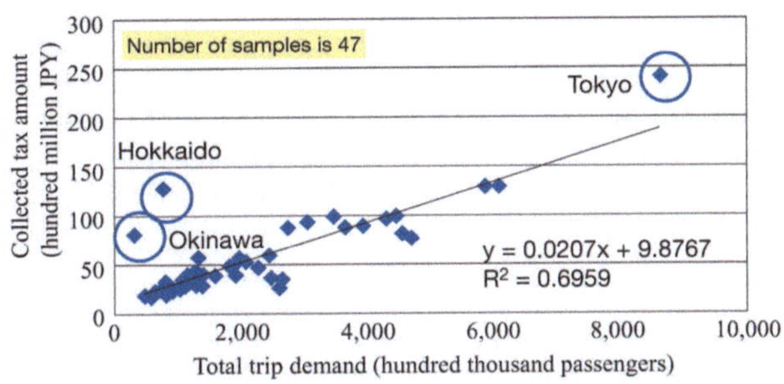

amounts are defined by multiplying the CO_2 emissions per person and the total passenger volume in the area of movement. There is a correlation between total trip demand and collected taxes, with only Hokkaido, Tokyo and Okinawa failing to fit the general relation shown by the curve in Fig. 3.36. These three regions have particularly good correlations for air transport demand (correlation coefficient = 0.91).

The regional distribution of the tax resources, re-allocated to the prefectures by utilising the equivalent utility principle, is shown in Fig. 3.37, where Tokyo receives 24.2 billion JPY and each of Hokkaido, Osaka, Fukuoka and Okinawa receives 10 billion JPY at average. In the prefectures gaining the highest benefit from the re-allocation of revenues, the levied tax amounts and the returned re-allocations are almost equal. The main reason for this is that the aim of reducing regional tax levy

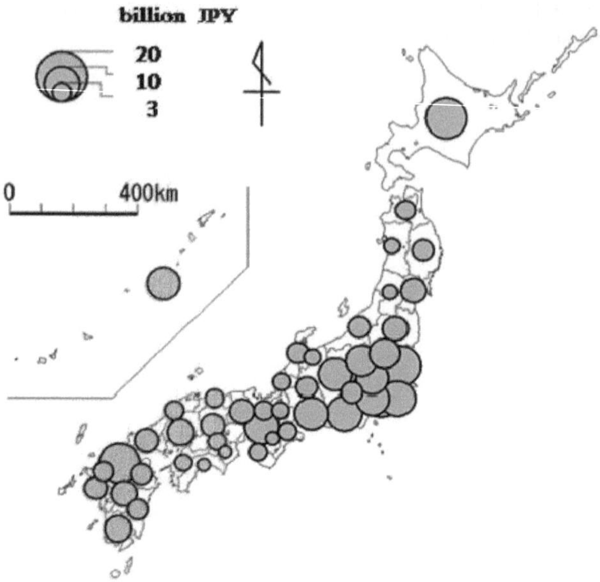

Fig. 3.37 Amount of taxes re-allocated to regions

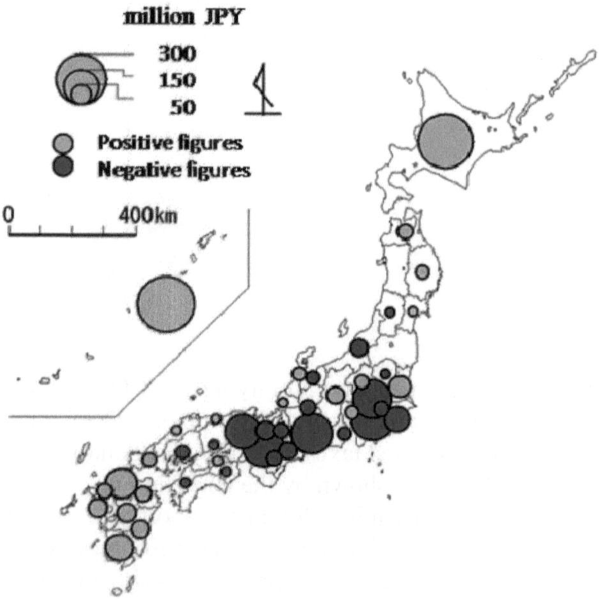

Fig. 3.38 Difference between levied and returned tax amounts (obtained by subtracting the re-allocated amount from the collected amount)

disparities through a re-allocation mechanism in which the rate of utility change in each prefecture is set equal has been successful; that is, utility is maintained in prefectures with high rates of tax collection. In this context, Fig. 3.38 presents the remaining degree of unbalance among the prefectures by subtracting the re-allocated amounts from the collected amounts.

Fig. 3.39 Relationship between change in total trip demand and excess amount of re-allocated tax

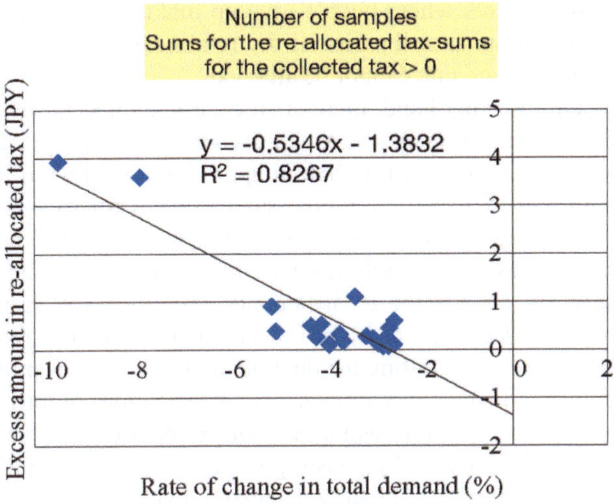

Fig. 3.40 Relationship between reduced air transport demand and excess amount of re-allocated tax

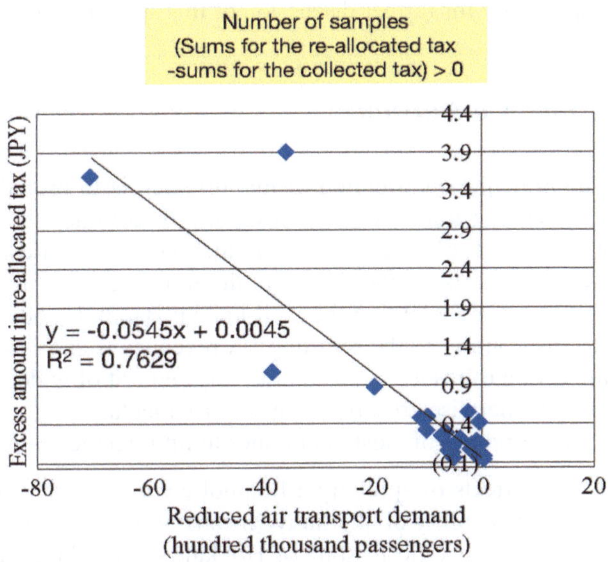

The highest excess is for Okinawa at about\320/head, followed by Hokkaido and Kagoshima at about\60/head. It is necessary to re-allocate more funds to these regions. Figure 3.39 reveals that the change in total transport demand is proportional to the excess amount of re-allocated revenues. Similarly, in regions with higher excess amounts, there is a proportional relation between reduced air transport demand and re-allocated revenues (Fig. 3.40). These regions are fifth or higher in automobile share, except Hokkaido and Okinawa. From the calculations carried out above, one notable interpretation is that unless more tax resources are re-allocated to prefectures where people give up making trips, that is, where air transport demand diminishes, the aim of the re-allocation mechanism to maintain a constant utility change rate cannot be attained.

On the other hand, there is an excess amount in the sums of the collected taxes for Saitama, Kanagawa, Aichi, Osaka and Hyogo prefectures, for which the average excess amount for each is approximately 180 million JPY. After levying the environmental tax, the modal shares of rail for four of the above five prefectures (Kanagawa, Aichi, Osaka and Hyogo prefectures) increase such that they are rated in the national top eight railway use shares. In regions with high levels of railway services, even if the environmental levy increases transport costs, the possibility of choosing to travel by rail means that the overall reduction in travel demand is small, thereby limiting the decline in utility. This eventually results in the excess in collected taxes. Tokyo ranks 21st in ascending order of per head tax revenue and re-allocated amount, while the four prefectures above are within the top six places. This is because, Tokyo has high transport demand and total cost and, moreover, a larger proportion of air-travel demand. In the case of Saitama prefecture, no features of this prefecture that account for the excess amounts can be grasped, so there is no opportunity for further discussion of this specific case.

3.5.6 Conclusions

The principal contribution of this research is to provide a basis for discussion of a systematic framework for designing future intercity transport policies aimed at certain levels of CO_2 emissions reduction. An analytical model is proposed that has two main parts. In the first step, the model defines the application of an environmental tax by adopting the well-known Baumol model. Then, in the second step, the distribution of the resulting tax revenues is assessed in such a way that the re-allocation of resources overcomes the demerit of widening regional disparities that would otherwise arise from any environmental levy.

The findings of most importance to future policy implementations are as follows:

1. The effects of applying a Baumol environmental tax, which vary according to the availability of alternative means of transport, are quantitatively analysed. In the two specific regions of Hokkaido and Okinawa, where intercity transport relies heavily on the air mode, a substantial decrease of 8.5 % in total demand is observed because the poor transport system does not provide modes that compete with air. However, where railways are present, demand shifts to rail, thereby limiting the decrease in transport demand. In proving such modal shifts, the statistical significance of the relationship between the decrease in air modal share and the increase in rail share is confirmed. However, a similar relationship for passenger car trips and the rail mode cannot be proven. This is clearly because of the high elasticity of substitution between air and rail transport, in contrast with the low elasticity of substitution between car and any mode of public transport.

2. In Hokkaido and Okinawa prefectures, the ratio of reduction in CO_2 emissions is 4–5 times higher than the associated fall in transport demand after the levy is imposed. The main mode in regional trips to and from these two regions is air, a CO_2 emissions intensive mode, and the considerable decline in trips resulting from the lack of alternative modes leads to the large cut in CO_2 emissions. On the contrary, in the core regional areas of Tokyo, Osaka and Fukuoka, the decline in total demand does not exceed 3 %, but still the fall in CO_2 emissions is substantial. This is attributed to a shift by passenger car and air users to rail. The estimated modal shift is around 3 %, but its impact in reducing the environmental load is much higher.
3. In returning environmental tax revenues to the prefectures so as to maintain the level of utility before and after the levy, Hokkaido, Kyushu and Okinawa require as much as 400 million JPY, more than the amount collected in tax in those regions. This great difference provides a fair trade-off for the large decline in trip demand in these prefectures. On the other hand, in the three metropolitan areas where collected tax sums exceed the re-allocated sums, the railway network provides a high level of service as an alternative transport mode, limiting the decline in trips and utility. The result is the high level of tax collection.

3.6 Specific Policy Implications

On the basis of the findings of this analysis of the effects on regional passenger transport of stricter environmental regulations, two policy proposals are offered as follows:

1. Improvements to the railway system are significant. The curtailment of regional travel that would otherwise result from stricter environmental controls can be eased by developing the railway network, allowing journeys to shift from air to rail. Thus, the railways play an important role in maintaining a certain level of transport services when an environmental levy is imposed.
2. An effective application of tax revenues gained from an environmental levy will be their use in developing new railways (so as to protect certain levels of transport services) in addition to their application to other environmental measures. Given that environmental restrictions will be further tightened in the future, an urgent task is to establish a new railway system befitting the stricter regulations.

Two points regarding this analysis should be noted:

1. This study only deals with taxation effects within intercity passenger transport. That is, there is no consideration of propagation effects, kickback effects, or others and the analysis results are evaluated only with respect to the direct effects of the levy. Accordingly, the tax ratio settings might be overestimated and tax revenues might be less than the estimated\260 billion. Hence, it is essential to estimate tax revenues in consideration of their propagation effects.

2. Seating capacities in railway, air and other transport modes are not considered, so there might be a possibility of railway users being inconvenienced as a result of an excessive shift to railways. To prevent this, analysis that factors in transport capacities should be carried out to make it possible to develop a more specific railway improvement plan including service schedules and other details.

Supplementary note 1: Case study analysis by Fujisaki et al. etc. of effects of fluctuation in gasoline prices

The approach taken by Fujisaki et al. (2011) estimated the elasticities of petrol price and income both in real and nominal pair on per capita transport volume of public transport and private passenger cars, respectively, as well as on modal share of public transport and private passenger cars, respectively, by means of econometric models The data examined were both yearly and quarterly datasets for some 20 years between the beginning of FY1987 and the end of FY2008.

The result of analysis 1: Effect on transport volume of public transport versus private passenger cars

Firstly, Table 3.10 shows the basic analysis results for a yearly dataset and Table 3.11 for a quarterly dataset:

- If petrol prices rise, they are shown to be related to higher usage of public transport and lower usage of private passenger cars. Then, the transport volume of private passenger cars declines more than the public transport increases and thus the volume of surface transport decreases as a sum of both. Meanwhile, the modal share of public transport increases since the volume of public transport as a numerator increases, whereas the volume of surface transport as a denominator decreases. The absolute value of petrol price elasticities turned out to be less than 1.0.
- Income elasticities have a positive sign on transport volume of private passenger cars and surface transport as well as on modal share of private passenger cars, whereas they have a positive sign on transport volume of public transport and a negative sign on modal share of public transport. Because income affects the

Table 3.10 Estimated elasticities in the case of a yearly dataset. (Source: Adapted from Fujisaki et al. (2011, Table 3))

		Petrol price elasticity		Income elasticity	
		Nominal	Real	Nominal	Real
Surface transport	Passenger-km per capita	Minus		Plus	
		−0.13	−0.17	0.60	0.66
Public transport	Passenger-km per capita	Plus		Plus	
		0.12	0.12	0.23	0.27
	Modal share	Plus		Minus	
		0.25	0.28	−0.33	−0.42
Private passenger cars	Passenger-km per capita	Minus		Plus	
		−0.30	−0.41	0.86	1.02
	Modal share	Minus		Plus	
		−0.18	−0.22	0.27	0.36

Private light freight cars are not included in "Private passenger cars"
The result of the first-order auto-regressive models of error term

transport volume of private passenger cars more than that of public transport, which constitutes the smaller part of surface transport, the sign of elasticity on transport volume is different from the sign on modal share with regard to public transport.
- It was confirmed that the significance is higher overall when the length of time-lag is 1 year for a yearly dataset and four quarters for a quarterly dataset than with no or shorter lag, in the case of both a nominal and real pair of datasets.

In comparing both datasets, the following can be seen:

- Table 3.11 shows that the result for a quarterly dataset has a similar characteristic to that for a yearly dataset regarding the signs and the value of the elasticities. The value of petrol elasticities and income elasticities on private passenger cars and surface transport are basically larger for a yearly dataset than for a quarterly dataset. A similar tendency can be seen from the next section, namely the comparison between short-run and long-run elasticities.

Secondly, short-run and long-run elasticities were estimated under distinction between them. Table 3.12 shows the results of their estimation in the case of a quarterly dataset. It can be seen that the values of long-run elasticities are larger than those of short-run elasticities. It may be in line with another hypothesis that

Table 3.11 Estimated elasticities in the case of a quarterly dataset. (Source: Adapted from Fujisaki et al. (2011, Table 4))

		Petrol price elasticity		Income elasticity	
		Nominal	Real	Nominal	Real
Surface transport	Passenger-km per capita	Minus		Plus	
		−0.09	**−0.12**	**0.42**	**0.43**
Public transport	Passenger-km per capita	Plus		Plus	
		0.10	**0.08**	**0.23**	**0.27**
	Modal share	Plus		Minus	
		0.27	**0.21**	−0.09	**−0.23**
Private passenger cars	Passenger-km per capita	Minus		Plus	
		−0.21	**−0.26**	**0.57**	**0.57**
	Modal share	Minus		Plus	
		−0.16	**−0.13**	0.07	**0.15**

Private light freight cars are included in "Private passenger cars"
Bold figures are the result of the seasonal first-order auto-regressive models of error terms

Table 3.12 Short-run and long-run elasticities on transport volume per capita. (Source: Adapted from Fujisaki et al. (2011, Table 5))

	Short-run petrol price elasticity		Short-run income elasticity		Long-run petrol price elasticity		Long-run income elasticity	
Passenger-km per capita	Nominal	Real	Nominal	Real	Nominal	Real	Nominal	Real
Public transport	0.16 (4.096)	0.08 (2.914)	0.22 (4.489)	0.46 (5.320)	0.25	0.14	0.34	0.78
Private passenger cars	−0.15 (−4.086)	−0.19 (−4.869)	0.40 (5.940)	0.36 (5.432)	−0.22	−0.26	0.58	0.50

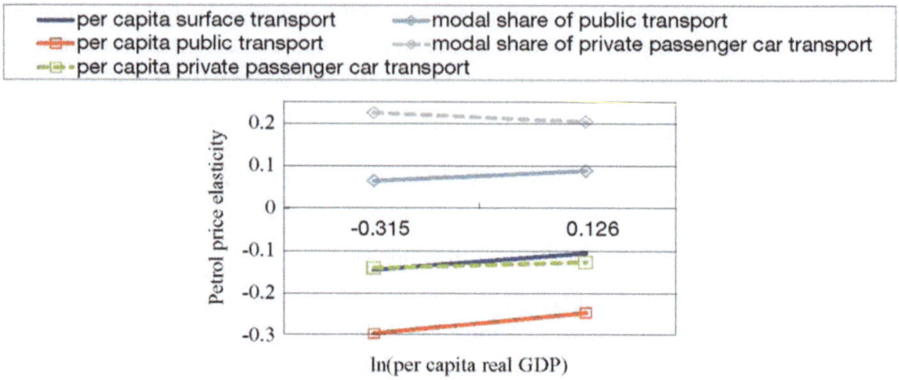

Fig. 3.41 Petrol price elasticities versus income level for a real dataset. (Source: Adapted from Fujisaki et al. (2011, Figure 2))

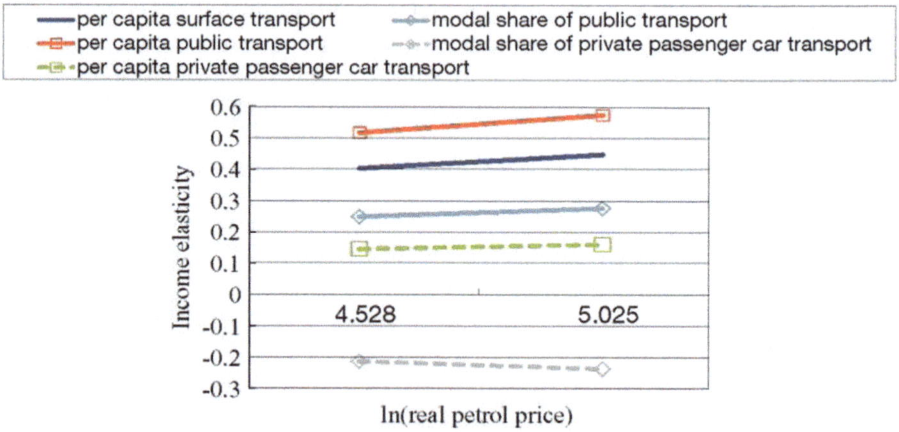

Fig. 3.42 Income elasticities versus petrol price level for a real dataset. (Source: Adapted from Fujisaki et al. (2011, Figure 3))

petrol price or income affects transport volume not only directly but also indirectly through the effect on car ownership over time, as seen in the later paragraph.

Thirdly, elasticities were estimated incorporating the assumption that those elasticities change according to the level of other independent variables, by means of models with an interaction term. Figures 3.41 and 3.42 show the relation between these elasticities and the level of other independent variables, calculated according to the results of the estimation in the case of a quarterly real dataset. The result of the estimation in the case of a quarterly nominal dataset is similar. From these two figures, the following general policy implications can be deduced.

- The higher the level of real income, the smaller the absolute value of petrol price elasticities, excepting petrol price elasticity on the per capita volume of

Table 3.13 Elasticities on per capita petrol consumption on private cars. (Data Source: Adapted from Fujisaki et al. (2011, p 362))

	Petrol price elasticity		Income elasticity		Elasticity of average number of loaded passengers per car	
	Nominal	Real	Nominal	Real	Nominal	Real
Petrol consumption on private passenger cars per capita	Minus		Plus		Minus	
	−0.35	−0.50	0.88	1.0	−1.1	−1.45

Private light freight cars are excluded in "Private passenger cars"
The result of the first-order auto-regressive models of error terms

public transport (Fig. 3.41). Therefore, in pricing measures such as a petrol tax to reduce the usage of private passenger cars, the higher the level of income, the more the petrol price has to be raised, to deter usage of private passenger cars to a certain extent.

- The higher the level of real petrol price, the larger the absolute value of income elasticities (Fig. 3.42). Therefore, in economic measures to reduce the usage of private passenger cars, the lower the level of petrol price, the greater the economic load has to be put on income to deter usage of private passenger cars to a certain extent.

The result of analysis 2: Effect on petrol consumption on private passenger cars

Table 3.13 shows the estimated elasticities on per capita petrol consumption on private passenger cars for a yearly dataset. It turned out that petrol price affects petrol consumption negatively as with usage of private passenger cars.

The result of analysis 3: Effect on ownership and purchase of private passenger cars

The data for monthly sales (in vehicles) of normal passenger cars, small passenger cars and light passenger cars with four wheels from April 2000 to March 2009 were obtained from the statistics of the Japan Automobile Dealers Association (n.d.) and the Japan Light Motor Vehicle and Motorcycle Association (n.d.). The quarterly data for the share of light passenger cars with four wheels in the total sales of these three kinds of car were calculated. The direct impact of petrol price and income on sales share of the light passenger cars was examined by means of a method similar to that of the analysis on transport volume. Lag dependent variable models with seasonal dummies turned out to be the best and estimated the elasticity of petrol price with one quarter lag on sales share of light passenger cars: In the nominal, it is 0.13 and in the real it is 0.14. It was reasoned that people were aware that light passenger cars are more fuel-efficient.

Following the analysis at the national level by Fujisaki et al. and overcoming its limited result, Fujisaki (2013) made a pooled analysis using a cross-sectional time-series data from different regions in Japan, which were prepared being adapted from the regional version of the data source Fujisaki et al. used, including Motor Vehicles Owned by MLIT (n.d. d) as well as Annual Report on Prefectural Accounts compiled by the Cabinet Office (n.d. b). Fujisaki estimated the direct impact of petrol price and income on ownership per 1000-population of private passenger cars, as well as their components: registered private passenger cars and light cars with more than 2 wheels; Petrol price elasticity tuned out −0.13 in the nominal and −0.20 in the real for total personal automobiles; −0.24 in the nominal and −0.35 in the real for registered private passenger cars; 0.10 in the nominal and 0.15 for light cars. That confirms the reasoning behind that the long-run elasticities on transport volume of private passenger cars are larger than the short-run elasticities on that.

The result of the analysis by Fujisaki et al. (2011) etc. of the effect of petrol prices and income on transport behaviour quantified in the transport volume and the modal share of public transport and private passenger cars since FY1987 confirmed the following phenomena:

- Petrol prices affect public transport positively and usage of private passenger cars negatively.
- Through the examination of distinctions between short-run and long-run elasticities, it was revealed that petrol prices and income affect the usage of private passenger cars more in the longer term. It is in harmony with the assumption that the indirect effect on the usage of private passenger cars through the verified change of car ownership caused by petrol prices and income adds to the direct effect on the transport volume.
- Petrol price elasticities are variable responding to the level of income: The higher the level of income, the more the petrol price has to be raised, to deter usage of private passenger cars to a certain extent. Considering income elasticities responding to the level of petrol prices, the lower the level of petrol price, the greater economic load has to be put on income for the same purpose.
- Higher petrol prices are related to lower consumption of petrol on private passenger cars.
- Higher petrol prices are related to lower ownership (stock) of registered private passenger cars and total of private passenger cars. Such prices are related to higher share in purchase and higher ownership of light cars, which suggests that consumers are aware of fuel efficiency in purchasing cars.

Supplementary note 2: Model Details

1. Defining the formula

1. Constant elasticity of substitution (CES) utility function

Travel behaviour is defined in terms of the CES utility function, the mechanism of which is depicted in Fig. 3.43.

Since the concern in this study is intercity passenger transport, the individual's total utility is first classified into two components, travel and consumption of other goods, in such a way that other goods defines all types of consumption other than transport. Therefore, in substantiating individual utility, intercity passenger transport is the main component and the other markets are outside the scope of this analysis. Three forms of the CES type utility function, its general form and the two

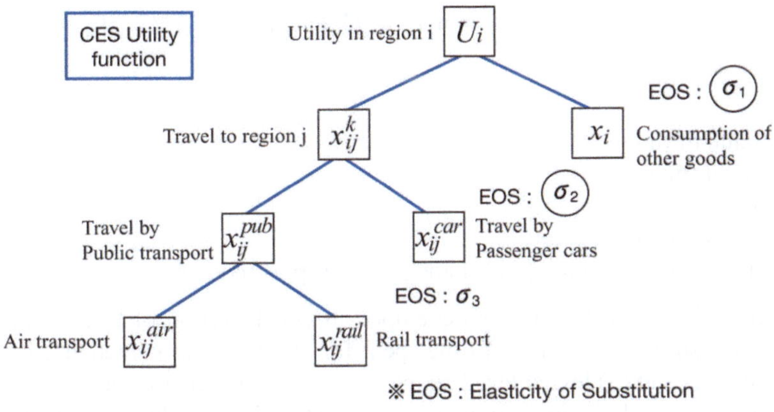

Fig. 3.43 CES utility function

applications used here to define the utility of inter-city passenger transport and of consumption of the other goods are given by the following formulas:

$$u_i = \left\{ \sum\nolimits_j \alpha_{ij}^{\frac{1}{\sigma_1}} x_{ij}^{\frac{\sigma_1-1}{\sigma_1}} + \alpha_i^{\frac{1}{\sigma}} x_i^{\frac{\sigma_1-1}{\sigma_1}} \right\}^{\frac{\sigma_1}{\sigma_1-1}} \tag{3.9}$$

$$x_{ij} = \left\{ \alpha_{ij}^{pub\,\frac{1}{\sigma_2}} x_{ij}^{pub\,\frac{\sigma_2-1}{\sigma_2}} + \alpha_{ij}^{car\,\frac{1}{\sigma_2}} x_{ij}^{car\,\frac{\sigma_2-1}{\sigma_2}} \right\}^{\frac{\sigma_2}{\sigma_2-1}} \tag{3.10}$$

$$x_{ij}^{pub} = \left\{ \alpha_{ij}^{air\,\frac{1}{\sigma_3}} x_{ij}^{air\,\frac{\sigma_3-1}{\sigma_3}} + \alpha_{ij}^{rail\,\frac{1}{\sigma_3}} x_{ij}^{rail\,\frac{\sigma_3-1}{\sigma_3}} \right\}^{\frac{\sigma_3}{\sigma_3-1}} \tag{3.11}$$

where $\alpha_{ij}, \alpha_i, \alpha_{ij}^{pub}, \alpha_{ij}^{car}, \alpha_{ij}^{air}, \alpha_{ij}^{rail}$ represent the scale parameters of the CES utility function.

The CES utility function incorporates all travel spending to determine the mode choice for a journey as indicated in the figure above. First, the choice between passenger car and public transport is determined, and then whether the public transport choice is air or rail. In this mechanism, the elasticity of substitution, given by σ, is the most important factor as it determines the extent of each mode share. Elasticity of substitution values are estimated from existing trip data.

2. Generalised cost of transport

Generalised cost is a commonly used concept for incorporating travel time into the total cost of making a trip. In its general form, including the environmental levy, it is given by Eq. 3.12 below:

$$c_{ij}^{k} = w_{ij}t_{ij}^{K} + f_{ij}^{k} + te_{ij}^{k}, \tag{3.12}$$

where

$w_{i,j}$ value of time
t_{ij}^{k} travel time
f_{ij}^{k} other costs
t environmental tax rate
e_{ij}^{k} emissions between regions *i* and *j* by transport mode *k*

This generalised cost formula is developed in consideration of three main components: (1) sum of the value of time spent for the trip, (2) maintenance costs for passenger cars and (3) levied tax. The main reason for taking the value of time into consideration in calculating the generalised cost is the possibility of determining travel costs by assigning a certain value to time. This is particularly important in the case of air transport users who place more value on time and therefore choose the air transport mode even at comparatively high prices. Therefore, Eq. 3.12 calculates the generalised cost for each mode at different values of time, specific to that mode. Modal shifts can then be determined in terms of this generalised cost.

The final component of generalised cost in the definition above is specific to this study and represents the introduction of the Baumol model to specify the environmental tax levy. This tax rate, *t,* is in the unit of tonne-CO^2/JPY and is multiplied by the unit emissions resulting from the transport of one passenger over 1 km separately for each mode. This substantiates the total levy on each trip. Ranked by their unit emissions, the three modes are passenger car first, air transport second and railway third (with the least burden on the environment). In the Baumol model of taxation, the amount of tax is higher for the higher unit emissions; where the levy is smaller, trip patterns are little affected. That is, to set a strict levy against emissions, a higher rate of *t* is applied to passenger car and air transport trips (which have high unit emissions) and a significant shift can be expected from these two modes to the railways.

3. Budget constraints

Many trip behaviour models assume budget constraints and, similarly, this study applies the same principle and defines the budget constraint by the below formula:

$$\sum_{j}\sum_{k} c_{ij}^{k}x_{ij}^{k} + p_{i}x_{i} \leq I_{i} + Z_{i}, \tag{3.13}$$

where

p_i price for other goods
x_i demand for other goods
I_i consumer income in region i
Z_i re-allocated amount of collected tax in region *i*

On the right side of the formula, I_i is the total income in region i. Since travel patterns are defined in a region by one representative node for all arrivals and departures, total income should be obtained by multiplying regional average income by the number of inhabitants in the region. The second component on the right side, Z_i, represents the re-allocated amount of tax.

The first component on the left side, which is the summary of the product of generalised cost and transport demand originating from the representative node of region i for all modes, gives the total cost associated with transport in that region.

The second part, p_i and x_i, represents all costs and demand other than intercity passenger transport, combined in one component and calculated by multiplying the unit price by the demand for such goods. For this, the basic price provides an appropriate unit price for a compatible analysis with transport costs. Then, by the above additive formula, transport users either choose to spend on travel or on other consumption subject to the budget constraint.

In this way, the choice of consuming a transport service or other goods is defined by the associated costs and incomes within a given budget.

2. Utility maximisation problem

The travel demand function, which is the main scope of modelling in this study, is developed using the above CES utility function and the principle of budget constraint. In this context, individual travel behaviour is defined by adopting a principle of consumer behaviour: That a consumer behaves in such a way as to maximise the utility (u) gained from consuming goods and services. By bringing in the consumer behaviour principle, a travel demand function is described that maximises the CES type utility function subject to the budget constraint. The first-order Lagrange theorem is used to derive the travel demand function, shown by the below set of Eq. 3.14:

$$x_{ij} = \alpha_{ij}\left(\frac{c_{ij}}{c_i}\right)^{-\alpha_1} \frac{I_i + t(1-r)\left(\sum_i \sum_j \sum_k e_{ij}^k \overline{x_{ij}^k}\right)}{c_i} \tag{3.14}$$

$$x_{ij}^{pub} = \alpha_{ij}^{pub}\left(\frac{c_{ij}^{pub}}{c_{ij}}\right)^{-\sigma_2} x_{ij} \tag{3.15}$$

$$x_{ij}^{car} = \alpha_{ij}^{car}\left(\frac{c_{ij}^{car}}{c_{ij}}\right)^{-\sigma_2} x_{ij} \tag{3.16}$$

$$x_{ij}^{air} = \alpha_{ij}^{air}\left(\frac{c_{ij}^{air}}{c_{ij}}\right)^{-\sigma_3} x_{ij}^{pub} \tag{3.17}$$

$$x_{ij}^{rail} = \alpha_{ij}^{rail}\left(\frac{c_{ij}^{rail}}{c_{ij}}\right)^{-\sigma_3} x_{ij}^{pub}. \tag{3.18}$$

Here, the generalised cost of transport and the indirect utility functions are as follows:

$$c_{ij}^{pub} = \left\{\alpha_{ij}^{air}\left(c_{ij}^{air}\right)^{1-\sigma_3} + \alpha_{ij}^{rail}\left(c_{ij}^{rail}\right)^{1-\sigma_3}\right\}^{\frac{1}{1-\sigma_3}} \tag{3.19}$$

$$c_{ij} = \left\{\alpha_{ij}^{pub} c_{ij}^{pub^{1-\sigma_2}} + \alpha_{ij}^{car}\left(c_{ij}^{car}\right)^{1-\sigma_2}\right\}^{\frac{1}{1-\sigma_2}} \tag{3.20}$$

$$c_{ij} = \left\{\alpha_{ij} c_{ij}^{1-\sigma_1} + \alpha_i p^{1-\sigma_1}\right\}^{\frac{1}{1-\sigma_1}} \tag{3.21}$$

$$u_i = \frac{I_i + Z_i}{c_i}. \tag{3.22}$$

3. Conditional expressions in setting model variables

1. Determining the tax rate

Using the concept of backcasting, Japan's target for CO_2 emissions reduction from intercity passenger transport is first defined. Then further analysis is carried out to achieve this amount of reduction. The total amount of CO_2 emissions after levying the environmental tax is calculated by multiplying the travel demand between regions *i* and *j* for transport mode *k* (x_{ij}^k), which changes when the levy is imposed, by the associated unit emissions (e_{ij}^k) at the given tax rate (*t*). This calculation is summed over all origin–destination pairs and the three main modes of transport. The emissions before the levy are calculated in the same way but taking the demand before the taxation ($\overline{x}_{ij}^k$). The ratio of emissions before and after the levy is set equal to the target rate for CO_2 emissions reduction, given by *r* in the below formula (Eq. 3.23).

$$\sum_i \sum_j \sum_k e_{ij}^k x_{ij}^k = (1-r)\left(\sum_i \sum_j \sum_k e_{ij}^k \overline{xx_{ij}^k}\right), \tag{3.23}$$

where

$\overline{x}_{ij}^k$ traffic demand before the levy
x_{ij}^k traffic demand after the levy
r CO_2 reduction target

2. Balance of tax collection and re-allocation

As discussed in the main text, re-allocating the tax revenues equally among all regions would have an undesirably unbalanced effect across the regions. To overcome this problem, tax re-allocation is substantiated by adopting the principle of equivalent variation, as given by Eq. 3.24. The main idea here is to re-allocate the collected tax in such a way that the ratio of utility change before imposition of the levy (U_i) and after imposition of the levy ($\overline{u}_i$) is equal in all regions (C).

$$\frac{\Delta u_i}{\overline{u}_i} = \frac{u_i - \overline{u}_i}{\overline{u}_i} = const(C), \tag{3.24}$$

where

u_i utility after tax reallocation
$\overline{u}_i$ utility of region i before the levy

Ultimately, the government levies the tax according to the tax rate (t) that satisfies the relation in Eq. 3.23, the travel demand (x_{ij}^k) and associated unit emissions (e_{ij}^k). Then this levied total amount is returned by allocating an amount to each region such that the utility change caused by imposition of the tax is the same in each region. The position of the government here is to ensure that the total national tax amount collected equals the total re-allocated amount given by Eq. 3.25 below:

$$t\left(\sum_i \sum_j \sum_k e_{ij}^k x_{ij}^k\right) = \sum_i Z_i. \tag{3.25}$$

In this formula, the left side is the levied tax as calculated by considering the change in travel demand caused by the Baumol environmental tax, unit CO_2 emissions and the tax rate per unit emissions. The right side is the total tax re-allocated to the regions.

The necessary parameters in this analytical model t, C, and Z are then estimated by using the above three conditional expressions.

4. Analytical Model Developed by Three Conditional Expressions

The use of the three conditional expressions defined by Eqs. 3.23, 3.24 and 3.25 makes possible a single analytical model that can both explain the effects of the levy and the effects of tax re-allocation. The tax rate in Eq. 3.25 is defined by including Eq. 3.23, while the re-allocation equality condition of Eq. 3.24 is provided by Eq. 3.25. The three expressions can then be solved by the equilibrium method.

The analytical approach described here can be summarised as follows. A consumer chooses a travel pattern that maximises his or her utility subject to budget constraints, which will change as a result of environmental tax regulations and the re-allocation of tax revenues to the region. The tax rate and re-allocated tax, which obviously influence the budget constraints, are calculated by solving the system of equations containing the three conditional expressions given above. In determining the tax rate and the amount of tax re-allocation, parameters t, Z and C are determined by solving for equilibrium of the three set conditions, which are: the CO_2 emissions reduction target; the equality of utility change; and balanced tax collection and re-allocation on a national scale. By numerically solving for these three parameters, the increase in generalised cost of transport caused by levying an environmental tax (t) and tax re-allocation in each region are calculated. This further identifies the budget constraints. Then the consumer makes a travel pattern decision that maximises the gained utility from travelling from one place to another by the selected means of transport, subject to the identified budget constraints.

Once the parameters are estimated, each is tested as to whether it can satisfy each of the given three conditional expressions. If one of the parameters fails to meet one of the conditions, then the whole process of estimation is repeated. In this estimation and control process, first a tax rate is identified and then the total emission reduction percentage is calculated; if the reduction far exceeds the target, then the tax rate is decreased slightly and the total emissions are re-calculated. In a case where none of the conditions are met, by establishing a linkage between the initially set value and the result and using it to assign a new value to the parameter, the process is repeated until the sought result is achieved. In this iterative manner, three parameters that satisfy the three conditions are determined. The results can be used to carry on detailed policy deliberations as regards the extent of declining trip demand, CO_2 emissions reductions and re-allocation of tax revenues. Importantly, these deliberations can be on a regional level.

5. Estimation

1. Parameter estimation

The developed analytical model is applied to intercity passenger transport in Japan and the results are shown here. In estimating the parameters for travel demand functions, the following formulas are used:

$$ln\frac{x_{ij}}{x_i} = ln\frac{\alpha_{ij}}{\alpha_i} - \sigma_1 ln\frac{c_{ij}}{c_i} \tag{3.26}$$

$$ln\frac{x_{ij}^{car}}{x_{ij}^{pub}} = ln\frac{\alpha_{ij}^{car}}{\alpha_{ij}^{pub}} - \sigma_2 ln\frac{c_{ij}^{car}}{c_{ij}^{pub}} \tag{3.27}$$

$$ln\frac{x_{ij}^{rail}}{x_{ij}^{air}} = ln\frac{\alpha_{ij}^{rail}}{\alpha_{ij}^{air}} - \sigma_3 ln\frac{c_{ij}^{rail}}{c_{ij}^{air}}. \tag{3.28}$$

Here, the parameters are elasticity of substitution (σ), defined in (1)-1) between travel and consumption of other goods Eq. (3.26), travel by passenger car or public transport Eq. (3.27), and finally the choice of air or rail public transport Eq. (3.28). A regression model estimates the elasticity parameters (σ_1, σ_2, σ_3) by using available trip data.

Table 3.14 Unit CO_2 emissions

	Unit CO_2 emissions ($g\text{-}CO_2$/capit-km)	Index
Rail transport	18.7	100
Air transport	1107	592
Passenger car	172.9	925

Table 3.15 Results for parameter estimation

Coefficient (n) (A and B)	σ_1 Travel and other goods	σ_2 Passenger car and public transport	σ_3 Rail and air
Elasticity of substitution	2.42 (55.8)	3.42 (41.9)	6.58 (5907)
Constant	16.9 (35.1)	2.48 (60.7)	2.92 (49.1)
Goodness of fit	0.7	0.521	0.702
Correlation constant	0.836	0.716	0.838
t value	–	41.9 (60.7)	5.97 (49.1)

2. Estimation results

By applying the described estimation method and using the unit CO_2 emissions given in Table 3.1, the three parameters are calculated and presented in Table 3.2 together with test results for statistical significance (Table 3.14).

In this estimation process, travel demand and the generalised cost of transport are two required sets of data. Such intercity passenger data is extracted from the "Inter-regional Travel Survey" conducted by the Ministry of Transport in 1990.

The generalised costs between the prefectures are calculated by using the travel time and cost data sourced from the Japan Travel Bureau (JTB) for 1990, the road network in 1990, and applying the formula for generalised cost given by Eq. (3.12). The "Inter-regional Travel Survey" does not include transport circulating within metropolitan areas and prefectures. The total sample is 2132. Table 3.15 summarises the estimation results. The value for σ_2 indicates that, if the travel cost is the same for public transport and passenger cars, then demand for car travel will increase. Comparing σ_3 to σ_2, the elasticity of substitution between air and rail transport is twice as high as that between passenger car and public transport. A higher value of elasticity of substitution means a higher potential of substitutability; that is, for the same travel cost, the shift from air transport to railways is higher than the equivalent shift from passenger car to any mode of public transport.

Chapter 4
Spatial Economic Analysis for Intercity Transport Policies

Atsushi Koike, Tomoki Ishikura, Mitsuhiro Miyashita and Kazuyuki Tsuchiya

4.1 Introduction

This chapter analyses the economic impacts of intercity transport policy by measuring the direct and indirect benefits of the policy. The impacts of an intercity transport policy show at first as changes in traffic volume, price and travel time, and then, propagated into non-transport markets such as real estate market (buildings and land). Urban development around a new terminal such as a railway station is a typical impact propagated from the transport market to the real estate market.

We capture the direct benefits of a transport policy which are accounted mainly as saving of travel time and out-of-pocket cost. In contrast, we must measure the indirect benefits as changes in non-transport markets like commodities, services, labour and capital.

The indirect benefits are distributed geographically or spatially. They spill over to the regions related with an intercity transport policy. Any stakeholders located in the regions share those benefits and costs (negative benefits).

Takayuki Ueda is deceased.

A. Koike (✉)
Graduate School of Engineering, Kobe University, Kobe, Japan
e-mail: koike@lion.kobe-u.ac.jp

T. Ishikura
Graduate School of Urban Environmental Sciences, Tokyo Metropolitan University, Tokyo, Japan
e-mail: iskr@tmu.ac.jp

M. Miyashita
Mitsubishi UFJ Research and Consulting Co., Ltd, Tokyo, Japan
e-mail: mitsuhiro.miyashita@murc.jp

K. Tsuchiya
Mitsubishi Research Institute, Inc., Tokyo, Japan
e-mail: kazuyuki@mri.co.jp

Y. Hayashi et al. (eds.), *Intercity Transport and Climate Change*,
Transportation Research, Economics and Policy 15, DOI 10.1007/978-3-319-06523-6_4

The spatial economic analysis of a transport policy provides a policy maker with an informational basis for social fairness or distributive justice. Although it is difficult to build political consensus on benefit/cost sharing at the interregional or intraregional level, the use of information regarding spatial distribution of indirect economic impacts is very important in a policy-making process.

4.2 Economic Models for Impact Analysis of Intercity High-Speed Transport

There are a variety of models or techniques used for economic analysis of transport policy. The conventional cost-benefit analysis (CBA) has provided a theoretical basis for measuring direct benefits. On the other hand, the spatial computable general equilibrium (SCGE) model is applicable for measuring indirect benefits. Impacts of a large transport infrastructure are widely propagated in spatial dimensions. The SCGE model is applicable to wider spatial economies, which ranges from world economy levels to regional economy levels. The key outputs that a regional policy maker is greatly interested in are production, employment, wages and prices in their region, and those in regions competitive or cooperative with them.

A policy maker at the nationwide level is interested in distributions of such key outputs over a nation. The question of whether an intercity transport system may contribute to the well-balanced distribution of economic welfare among all regions in a nation is the most critical issue. Since an intercity transport system impacts a wide spatial economy which includes many urban areas or cities, the SCGE model is to be well employed for such economic analyses.

We will explain the outline of our SCGE model for economic evaluation of intercity transport policies. The details of the model formulation are described in the Supplementary note. Furthermore, readers interested in more theoretical details regarding the SCGE modelling can find them in the literature listed in references.

We divide a spatial economy into discrete regions or zones in the SCGE model. In each region there are corresponding households and industries. Commodities and services are traded not only within each region but also between regions. The SCGE model can then describe the interregional trade. Demand-supply balancing holds in any markets of traded goods so that the equilibrium price for each of the goods is determined simultaneously. That is the Walrasian general equilibrium. Wages are treated as just the price in the labour market and capital rent is treated as the price in capital market. Mathematically, the general equilibrium is a solution of a system of equations, each of which means equality of aggregate demand with aggregate supply in the market of the corresponding goods.

Intercity transport services appear in the model in two ways. One is interregional freight transport for commodity trade. The other is interregional passenger transport. Freight transport service is modelled as intermediate goods in the input–output table, or as an iceberg cost or margin in transaction. Passenger transport service is modelled as an input factor in production of traded goods if a trip is business related and made by an industrial sector. If a trip is a private one made by a household, passenger transport is modelled as final demand or as an input for a household's self-production.

The framework usually employed for a traditional transport behaviour model has a multi-step choice structure or a nested structure choice. The SCGE model can then describe the destination choices, mode choices and route choices in the intercity transport markets.

The benefits and costs of an intercity transport policy can be measured by each category of household, for example, region by region. Since the SCGE model is based on microeconomic theories, we can employ benefit definitions of equivalent variation (EV), compensating variation (CV) and Marshall-Dupuit consumer surplus (MD) which are popular in applied economics and also in the stylised CBA.

4.3 Economic Effects of Railway Investment Measured by SCGE Models

This section collects spatial economic models that have been developed in the framework of the SCGE model and applied to the impact analysis of high-speed rail (HSR) policies.

Economic impacts of railway investment projects are widely spread to various economic fields as shown in Fig. 4.1. The economic impacts are generally distinguished into direct effects and indirect effects. Typical direct effects are the effects on reduction of transport time and transport cost. Decrease in transport time due to high-speed systems, higher frequency and improvement of punctuality in scheduled transport service contributes to travel time reduction for users. Increased capacity of transport equipment reduces unit operation cost per passenger or per weight. Direct effects are defined as the impacts on transport markets such as freight transport, private passenger travel and business travel. CBA by means of consumers' surplus measures the direct effects as benefit of a project.

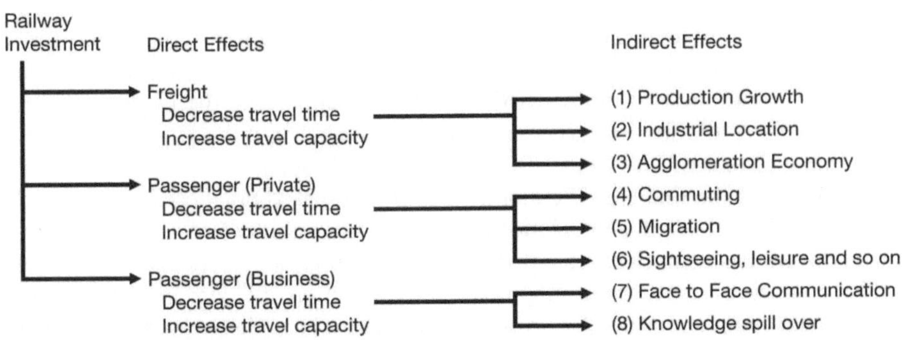

Fig. 4.1 Overview of economic impacts by the railway investment

Direct effects generated in the rail transport market will spread to other markets in the economy as indirect effects through various economic activities. However, indirect effects are not "additional effects" over direct effects. The difference between direct and indirect effects is just the viewpoint of where the effects are observed. In other words, their relationship is the effects measured upstream and effects measured downstream of the propagation. The two effects are essentially equivalent. If not, a double-counting problem occurs and the result should be corrected.

The indirect effects are observed in, for instance, production growth, industrial location, agglomeration of economy, commuting, migration, sightseeing, leisure, face-to-face communication and knowledge spillover. Direct effects of rail freight transport improve the productivity of industries (1). Then, impacts on productivity influence industrial locations (2). Change of location pattern influences the conditions of industrial agglomeration and agglomeration of economy may accelerate more improvements of productivity (3). Direct effects to private passenger travel will widen the range of commuting and influence commuting travel behaviour (4). Change of commuting conditions causes migration as a result of exploring better residential locations (5). Furthermore, in the scene of private activity, sightseeing and leisure travel are influenced by direct effects (6). Direct effects to business passenger travel will widen the range of spatial business opportunity then increase the opportunity of face-to-face communication (7). This effect contributes to knowledge spillover and it will eventually improve the productivity of the affected industries (8).

4.4 International Comparison of Impact Analysis of HSR

As seen in Fig. 4.1, the various indirect effects are interactive rather than independent. SCGE models, which are able to illustrate the direct effects explicitly, have an advantage over traditional CBA. However, SCGE models do not always capture all indirect effects. Practical SCGE models are usually modified, or simplified, to emphasise important impacts of the target project.

In this section, we compare SCGE models that have been applied in economic analysis of HSR policies and the results of the analyses. A similar review study is implemented in Gunn (2004). Our comparison focuses on the treatment of rail transport in the models. We collect spatial economic models which have been developed in the framework of the SCGE model and applied to the impact analysis of HSR policies (Table 4.1). The case studies covered in this section are for the Netherlands, Norway, EU, China, Taiwan, Korea and Japan.

CGEurope is the first New-Economic-Geography-based practical SCGE model, for which the original conceptual model structure was built by Bröcker (1998). The model covers EU countries, separated into over 1000 regions at the NUTS 3 level. Impact analysis of transport developments and evaluation of environmental policies are the main applications of CGEurope. Transport cost is modelled as transport margin of transport services produced by road transport and railway transport service

Table 4.1 General aspects of the SCGE models

Model	Country	The number of regions	Size of region	The number of sectors
CGEurope[a]	EU	1341	NUTS3	6 activities (including services)
		1083 (EU member)		
		162 (candidate)		
		91 (rest of Europe)		
		5 (other)		
PINGO[b]	Norway	19+1 (ROW)	Country	32 commodity groups
				6 services sector
RAEM[c]	The Netherlands	40	NUTS 3	14 industrial sectors
Sino-TERM[d]	China	31	Province (27)	137 commodities
			Municipality (4)	
J-SCGE	Japan	47	Prefecture	7 industrial sectors
K-SCGE	Korea	6	Province	7 industrial sectors
T-SCGE	Taiwan	15	Province	13 industrial sectors

[a] Bröcker et al. 2004
[b] Vold and Jean-Hansen 2007
[c] Knaap and Oosterhaven 2002
[d] Horridge and Wittwer 2007

firms in the TEN-T project evaluation. The impact by HSR network development and expressway network development is assessed by CGEurope. The model can calculate changes in regional gross domestic product (GDP), changes in regional output, changes in employment and so on. However, welfare per regional GDP is the most emphasised index for policy evaluation by CGEurope.

PINGO is an SCGE model that illustrates the interaction of 19 regions in Norway. PINGO was built as a module for a project that predicts regional and interregional freight transport. Regional and interregional trade in the future is predicted by the PINGO model, and other logistics modules estimate freight transport flow at the detailed network level by importing predetermined OD (Origin-Destination) transport demand given by PINGO. Commodity production functions in PINGO are assumed to be two-level Cobb–Douglas technologies. The lower nest contains technologies for substitution between labour and capital, while the upper nest contains technologies for substitution between pooled commodities and factors of production. PINGO represents additional trade agent sectors per region per commodity/service sector. The trade agents purchase commodities from production sectors in any region and from the import sectors and use transport services provided by the corresponding transport sectors to gather the goods into a pool. Transport margin in PINGO is treated as transport service input to commodity production sectors. Explicit transport service sector per commodity per region, which input principal factors and intermediate demands with nested (constant elasticity of substitution) CES technologies, is represented in PINGO. Railway transport is implicitly included in the transport sector. Social accounting matrix (SAM) for base year is used for calibration of PINGO. Parameters of the technologies of the transport sector are estimated using other statistical data including account statistics of transport operators. Essential exogenous variables of PINGO are consumer endowments and export demands. Policy evaluations and forecasts are implemented by changing of the exogenous variables and technological parameters. Vold and Jean Hansen (2007) compute forecasts of production outputs and commodity trades in 2040 by inputting exogenously given future consumption demand by the distributed bottom-up approach.

RAEM is an SCGE model that separates the Netherlands economy into 40 regions, and was also applied to the evaluation of HSR infrastructure projects on which the Dutch government is about to decide. Firms maximise profits by having a production function that is a function of labour and intermediate goods, and operate under conditions of monopolistic competition. Consumers are assumed to maximise a CES utility function that is the product of sub-utilities from consumptions of each sector product. The sub-utilities are based on mixtures of sector products from each region, reflecting a stated assumption of monopolistic competition. RAEM describes commuting travel by importing the theoretical foundation of New Economic Geography, and treats interregional freight transport as well as general SCGE models. Transport cost is modelled as the transport margin of transport services provided by transport firms. An HSR expansion project between Amsterdam and Groningen is assessed by RAEM. The impact on freight transport and the impact on nationwide land use structure by expanding commuting range are analysed in the project evaluation. Input–output tables used in this model are taken from bi-regional input–output tables compiled by RUG and CBS (RUG/CBS (1999)). Some of the model's parameters come directly from these tables while others are estimated by fitting trade flows predicted by the model to flows observed in the tables. The model for a base scenario for the year 2020 was calibrated and five scenarios were also evaluated using this SCGE model. The only differences between these scenarios are in the matrix with travel times, change in the number of jobs and change in the price index of consumption.

Sino TERM is a Chinese SCGE model for assessment. Horridge and Wittwer (2007) applied the model to the impact analysis of the Chongqing Railway Project. The enormous regional model (TERM) of Australia is the original model framework of Sino TERM. Calibration of Sino TERM uses a multi-regional input–output table of China, which separates the Chinese economy into 31 regions and 137 industrial sectors. Interregional freight transport is described by the amount of interregional trade in the model. Although intraregional passenger travel is treated as demand of the passenger transport sector of the region, interregional passenger travel does not appear explicitly. Production technology is basically represented as nested CES production functions taking into account the service margin input. Freight transport cost is treated as transport margin over f.o.b. commodity price. The cost of margin of pipelines, warehousing, trade (retail and wholesale) and insurance are covered as well as transport cost margins. Sino TERM distinguishes transport modes explicitly: freight cost of rail, road, water and air. Assuming Leontief technology in service margins input, commodity production requires a transport cost proportional to the output. If the transport cost of an origin–destination pair decreases, the commodity price at the destination will also decrease. Interregional freight transport is described by the amount of interregional trade in the model. Although intraregional passenger travel is treated as the demand of the passenger transport sector of the region, interregional passenger travel does not appear explicitly. Horridge and Wittwer (2007) gives exogenous change in capital endowment in the rail passenger and rail freight sectors in Chongqing and Hubei where railway construction investment will take place, reduction of rail freight margin in some OD pairs expected to enjoy transport condition development from the project, and productivity improvement in coal mining, gas drilling, rail passenger services and tourism sectors in the regions. The project assessment is implemented by observing macroeconomic impacts, such as changes in GDP, regional income and final demand in each sector, and impacts in each sector, such as changes in output and employment by region. The results represent positive impacts on output and income in the regions receiving investment, and negative impacts on the employment of rail transport sectors in the region.

Various indirect effects are interactive rather than independent. SCGE models, which are able to illustrate the direct effects explicitly, have advantages over traditional CBA. However, no SCGE model can always capture all indirect effects. Practical SCGE models are usually modified, or simplified, to emphasise important impacts of the target project. We select several existing SCGE models for impact analysis of railway investment projects in this section. The following Table 4.2 shows which effects among those in Fig. 4.1 are captured in each model.

Since passenger transport and freight transport are not separated, PINGO and CGEurope mention the effects with regard to trade or freight transport. RAEM treats both passenger travel and freight transport explicitly. Sino TERM also includes passenger travel and freight transport in the model. RAEM contains both

Table 4.2 Economic effects captured in each model

	Production growth	Industrial location	Agglomeration economy	Commuting	Migration	Sightseeing leisure, etc.	Face-to-face communication	Knowledge spillover
RAEM	O	O	O	O	O			
PINGO	O	O						
CGEurope	O	O	O					
Sino-TERM	O	O				O		
T-SCGE						O	O	O
K-SCGE						O	O	O
J-SCGE						O	O	O

passenger travel and commuting travel. Therefore, the spatial scope of RAEM is relatively small. Although interregional trade is treated in Sino TERM, interregional passenger transport is not. Passenger travel is regarded as regional demand of the tourism sector, which is to say, the spatial scope of Sino TERM regarding passenger travel is also relatively small.

On the other hand, J-SCGE, T-SCGE and K-SCGE emphasise interregional passenger transport by high-speed railway. Details of J-SCGE, K-SCGE and T-SCGE are given in the following sections. The range of railway passenger travel in these three models is longer than that of the above models. The situation of interregional HSR transport of East Asian countries is quite different from European countries. HSR transport services mainly compete with air transport in Asian countries in spite of domestic transport. Few passengers use HSR for commuting. The interests of J, T, K-SCGE models are business passenger travels and its economic impacts. Transport in the model, aspects of applications and outputs indicators are shown in Tables 4.3, 4.4 and 4.5, respectively.

Table 4.3 Transport in the SCGE models

Model	Freight and/or passenger	Modelling of transport	Description of intercity transport
CGEurope[a]	Freight	Transport margin	Freight flow as commodity trade between regions
PINGO[b]	Freight	Transport margin, including explicit transport sector	Freight flow as commodity trade between countries
RAEM[c]	Freight passenger	Transport margin	Freight flow as commodity trade, commuting passenger flow as labour migration
Sino-TERM[d]	Freight passenger	Freight transport margin, demand of passenger transport sector	Freight flow as commodity trade, passenger flow as regional output
J-SCGE	Passenger	Input factor, transport margin	Business passenger flow as transport input by firm, private passenger flow as consumption of transport service
K-SCGE	Passenger	Input factor, transport margin	Business passenger flow as transport input by firm, private passenger flow as consumption of transport service
T-SCGE	Passenger	Input factor, transport margin	Business passenger flow as transport input by firm, private passenger flow as consumption of transport service

[a] Bröcker et al. 2004
[b] Vold and Jean-Hansen 2007
[c] Knaap and Oosterhaven 2002
[d] Horridge and Wittwer 2007

Table 4.4 Aspects of applications

Model	Policy scenarios	Exogenous shock
CGEurope[a]	TEN-T	Reduction of interregional transport cost margin
PINGO[b]	Future freight flow forecast	Change in population, change in export demand, changes in investment and public consumption, future consumption demand
RAEM[c]	6 scenarios for HSR, e.g. Amsterdam to Groningen	Change in travel time between municipalities
Sino-TERM[d]	Chongqing Railway Project	Increase in capital endowment in rail passenger and rail, freight sectors, reduction of rail freight margins, improvement of productivity in specific sectors in specific regions
J-SCGE	Maglev Railway	Reduction of travel cost and time
K-SCGE	KTX Railway Project	Reduction of travel cost and time
T-SCGE	Taiwan HSR Project	Reduction of travel cost and time

[a] Bröcker et al. 2004
[b] Vold and Jean-Hansen 2007
[c] Knaap and Oosterhaven 2002
[d] Horridge and Wittwer 2007

Table 4.5 Output indicators of the SCGE models

Model	Output indicators
CGEurope[a]	Welfare/regional GDP
PINGO[b]	Domestic production outputs, import, demand of freight transport service sector
RAEM[c]	Change in job per region
	Change in sectoral outputs
	Change in prices
	Change in consumer welfare
Sino-TERM[d]	Change in national and regional GDP
	Change in sectoral outputs and employment
J-SCGE	Benefit, B/C
	Changes in regional GDP
	CO_2 emission
	Change in passenger transport
K-SCGE	Benefit, B/C
	Changes in regional GDP
	CO_2 emission
	Change in passenger transport
T-SCGE	Benefit, B/C,
	Changes in regional GDP
	CO_2 emission
	Change in passenger transport

B/C benefit–cost ratio
[a] Bröcker et al. 2004
[b] Vold and Jean-Hansen 2007
[c] Knaap and Oosterhaven 2002
[d] Horridge and Wittwer 2007

4.5 Outline of the Economic Impact Analysis of HSR Projects (Japan, South Korea and Taiwan)

4.5.1 Commonality of Frameworks of the Theoretical Models and Consistency of Analytical Results

In South Korea, Taiwan and Japan, high-speed railway projects are planned or implemented, and indirect effects on the economy are caused in each country. These economic effects do not differ only in the difference between national land-use structure and industrial structure but also in differences between the framework of the special economic model and prior condition. This framework of economic model must differ according to the economic conditions in each country, for instance, whether the target of transport project is freight transport or passenger transport.

However, it is comparatively easy to use the same framework for the analysis of these transport projects because the transport systems for Maglev of Japan, KTX of South Korea and HSR of Taiwan have a common function in that these transport systems are all inter-city transport facilities. Therefore, the employed SCGE models for the high-speed railway investment in Japan, South Korea and Taiwan have the same theoretical framework, and it will be easy to compare the results of each. In this chapter, we analyse the economic impacts of HSR project in these three countries by the spatial computable equilibrium model. However, there are some limitations in each model, for instance, definition of the industrial sector classification and the regional segmentation, according to the data feasibilities. So, even if the theoretical structures of the models are the same, it is necessary to keep in mind the limitations in comparing results among the countries.

This chapter shows details of the model assumption and the data, and then, reports individual analysis results in each model (J-SCGE, K-SCGE and T-SCGE). Finally, this chapter compares the results on the economic effects of HSR in the three countries.

The three models have similar structure in terms of theoretical model formulation. Basically, the difference of models appears in area classification and industrial sector classification as shown in Table 4.6.

Table 4.6 Comparison of each country model framework

Country	Area classification		Industrial sectors
Japan	47	Prefectures level	7 sectors
Korea	6	Provinces level	7 sectors
Taiwan	15	Provinces level (except for small islands)	13 sectors

Table 4.7 Concept of transportation mode choice

	Maglev	KTX	T-HSR
Transport mode covered	Air and railway	Railway and car	Air, railway and car
Method of estimation of modal share	Estimated by the logit model, calculation from relation "generalized cost" between each OD and "modal split share" in "without Maglev situation"	Estimated by the logit model, calculation from relation "generalized cost" between each OD and "modal split share" in "with KTX situation"	The quantity demand of T-HSR at with T-HSR situation is a predicted value estimated by Taiwanese HSR authority; based on this result, the all air passengers are assumed to shift to T-HSR and 52.7% of long-haul car traveller shift to T-HSR traveller

4.5.2 *Methodology and Data*

1. Modal Choice in Transportation

The modal split share of the passenger transport in each region pair OD is exogenous value of the SCGE model. Existing travel demand forecast models based on discrete choice (nested logit type) models can give the modal split share after HSR operation begins in the three countries. Although theoretical concept of the demand forecast models is common, the choice alternatives of intercity travel and method of parameter estimation differ among the countries as follows (Table 4.7).

2. Basic Assumptions of the SCGE Models

Our SCGE models assume perfectly competitive economy similar to standard general equilibrium models. Representative households classified into region maximise their utility and representative industrial sectors classified into region maximise their profit. Transport sector is explicitly distinguished from other industrial sector in our models.

Households get income from production factor (capital and labour) supply and spend on consumption at hometown and leisure travel with consumption at destination. Industrial sectors produce goods and services by inputting intermediate input goods, capital, labour and the business trip. Transport sector produces transport service inputting only capital and labour. All markets including goods, services and production factor are closed in each region except for capital market of T-SCGE model which is closed in the country, not in the region.

The parameters of the production functions and the utility functions are given by the calibration method. Interregional input–output table and several relating data source shown in Table 4.8 are the data set of the benchmark equilibrium to be calibrated. However, elasticities of substitution remain as unknown exogenous parameters of SCGE models. We assume the elasticitiy of substitution between regions is 0.5 and the other elasticity (between goods) is 0.8 based on existing research (see Supplementary note: the detail of model formulation).

Table 4.8 Parameters and data in SCGE models

Parameter	Source
(1) Industrial sector	
The 1st stage (production function)	
Input coefficient	Interregional input–output table
Production capacity rate	Interregional input–output table
The 2nd stage (value-added function)	
Scale parameter	Interregional input–output table
Share parameter	Interregional input–output table
Elasticity of substitution (between production elements)	0.8
The 3rd stage (regional selection)	
Scale parameter	Interregional input–output table, traveller data, transportation
Share parameter	Interregional input–output table, traveller data, transportation
Elasticity of substitution (between regions)	0.8
(2) Household	
The 1st stage (utility function)	
Share parameter in household	Interregional input–output table, traveller data, transportation
Share parameter of destination	Interregional input–output table, traveller data, transportation
Elasticity of substitution (between consumer goods)	0.5
The 2nd stage (regional selection)	
Share parameter	Interregional input–output table, traveller data, transportation
Elasticity of substitution (between regions)	0.8
The 3rd stage (composition of recreation and transport demand)	
Share parameter	Interregional input–output table, the family income and expenditure survey, etc.
The 4th stage (household production function)	
Scale parameter of leisure	Interregional input–output table, the family income and expenditure survey, etc.
Share parameter in household	Interregional input–output table, the family income and expenditure survey, etc.
Elasticity of substitution (between consumer goods)	0.5
(3) Transportation sector	
Input coefficient	Interregional input–output table
(4) Exogenous variable	
Total labour	Interregional input–output table
Total capital	Interregional input–output table

Table 4.9 The CO_2 emission for various transport modes in each country. (Reference: Japan: Ministry of Land, Infrastructure, Transport and Tourism (2007), Central Japan Railway Company (JR Tokai; 2007))

Mode	Maglev	KTX	T-HSR
	g-CO_2/passenger-km	g-CO_2/passenger-km	g-CO_2/passenger-km
Car	–	173	151
Air	139	–	274
Conventional rail	19	19	13
HSR (Shinkansen)	19	–	–
Maglev, KTX, HSR	47	19	27

Table 4.10 The CO_2 unit cost from industrial sectors in each country

Industrial sector	Maglev	KTX	T-HSR
	t-c/million yen	t-c/million won	t-c/million NT$
Agriculture, forestry, fishery	0.42	1.56	Following table
Manufacturing	0.92	4.27	
Construction	0.09	0.49	
Electric power, gas and water supply	7.95	41.61	
Commerce	0.05	4.99	
Finance and real estate	0.02	0.62	
Service	0.46	0.62	

3. The CO_2 Emission and the Social Cost

The SCGE models give the change in economic indices caused by exogenous impacts but do not calculate the change in CO_2 emission. Therefore, we need a methodology to estimate the change in SCGE results to the change in CO_2 emission. We simply convert the change in transport demand and outputs of industrial sector into the change in CO_2 emission by using unit amount (given by various sources). The monetary value of reduction of CO_2 is also taken into account of measurement of social net benefit.

- The CO_2 unit cost
  CO_2 emission is assumed to be generated through (1) industrial production activity and (2) transport activity.
  The gross CO_2 emission of the whole economy is estimated based on unit cost of CO_2 in each activity (Tables 4.9 and 4.10).

Industrial sector	T-HSR
	(t-c/million NT$)
Agriculture, forestry, fishery and mining	24.25
Manufacturing	42.47
Chemical products	91.11
Metal and machinery	24.46
Electric machinery	19.63
Electric power, gas and water supply	14.24

Industrial sector	T-HSR
	(t-c/million NT$)
Commerce	6.57
Eating, drinking and lodging places	4.82
Transport and communication	61.27
Finance, insurance and real estate	4.82
Medical service, health and social security and nursing	4.82
Other personal services	4.82
Public administration	4.82

4.6 Analysis of the Economic Effect of the Maglev Project and CO_2 Emission (J-SCGE)

4.6.1 Specific Framework of J-SCGE Model Analysis

The region of Japan is classified into following divisions as regions defined in National Spatial Planning (Fig. 4.2).

<Region>	<Prefecture>
Hokkaido	Hokkaido
Tohoku	Aomori, Iwate, Akita, Miyagi, Yamagata, Fukushima, Niigata
Shuto	Ibaraki, Tochigi, Gunma, Saitama, Chiba, Tokyo, Kanagawa, Yamanashi
Hokuriku	Toyama, Ishikawa, Fukui
Chubu	Nagano, Gifu, Shizuoka, Aichi, Mie
Kinki	Shiga, Kyoto, Osaka, Hyogo, Nara, Wakayama
Chugoku	Okayama, Hiroshima, Yamaguchi, Tottori, Shimane
Shikoku	Tokushima, Kagawa, Ehime, Kochi
Kyushu	Fukuoka, Oita, Saga, Nagasaki, Kumamoto, Miyazaki, Kagoshima
Okinawa	Okinawa

The model treats "air" and "railway" as explicit interregional transport modes. They are general interregional transport system and bus and car are very minor. The travel time and fare among all 47 region pairs are based on the shortest length path of the regional central station pair.

The travel time and fare among all each region pairs are based on the shortest length path of each regional central station pair. The travel time and fare of Maglev are based on the publication by JR Central and preconditions for the study by the advisory committee of Ministry of Land, Infrastructure, Transport and Tourism (MLIT).

The (exogenous) transport modal split share in the model is given by the Institute for Transport Studies (ITPS). According to ITPS assumption, two thirds of conventional HSR (Shinkansen) shifts to Maglev in competitive OD pairs after

Fig. 4.2 Regional classification and Maglev plan

Maglev operation. Then we can assume the modal share in railway both "with Maglev" scenario and "without Maglev" scenario. As an example of the transport modal shares of the railway (Shinkansen and Maglev) for both scenarios, Table 4.11 shows the share in major OD pairs.

Table 4.11 Transportation modal share of railway in major OD

	Without (%)	With (%)
Tokyo–Osaka	73	88
Tokyo–Hiroshima	54	62
Tokyo–Fukuoka	9	12

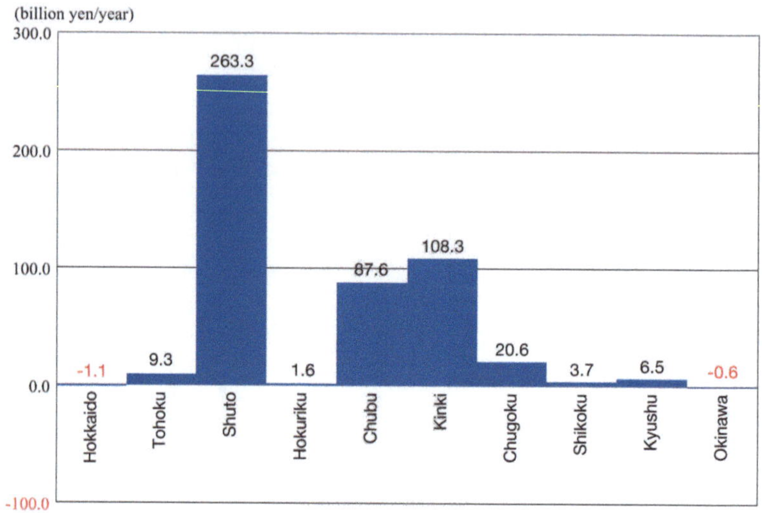

Fig. 4.3 Net benefit in each region

4.6.2 The Results

Regarding total benefit of Maglev project, our analysis estimates that annual benefit is 499.1 billion yen/year and net present value of the benefits is 11.2 trillion yen (project life: 50 years, social discount rate: 4%). From the point of view of benefit incidence, the effects on the Kanto region (around Tokyo), the Kansai region (around Osaka) and the Chubu region (around Nagoya) are relatively larger than other regions. However, the benefit incidence distribution shows the effects widely spread throughout Japan. Moreover, the region where Maglev station will be located, such as Mie, Yamanashi, Nagano and Nara, can enjoy a relatively larger benefit. It implies that marginal productivity in these regions is still higher than others (Figs 4.3, 4.4 and 4.5).

The sum of outputs of entire industrial sector increases about 0.1%. Increase rates in finance, commerce and service, which input larger amount of business travel, are relatively higher.

The number of trips increases 5.8%, which may be a kind of indirect effects, the so called induced demand. That is because passenger traffic can be substituted for goods or services (non-transport) demand or production factor when price system is changed. On the other hand, increase of the sum of whole trip length is 2.0%, which is less than the increase ratio of the number of trip. It means that the average travel length per trip decreases by Maglev opening (Table 4.12).

Our main target of the analysis is how much Maglev project can decrease (increase) CO_2 emissions from transport sector and industrial sector. CO_2 emissions from the Maglev system are large, however, majority of the passengers of Maglev are the shifted users who have used air transport whose CO_2 emissions are quite large. As a result, overall CO_2 emissions from the transport sector will decrease 0.6%. Although

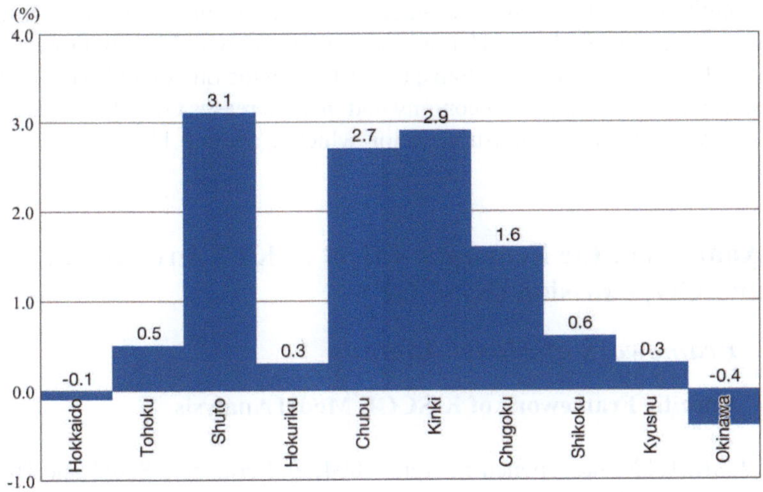

Fig. 4.4 Ratio of net benefit to GDP in each region

Fig. 4.5 Change in industrial outputs

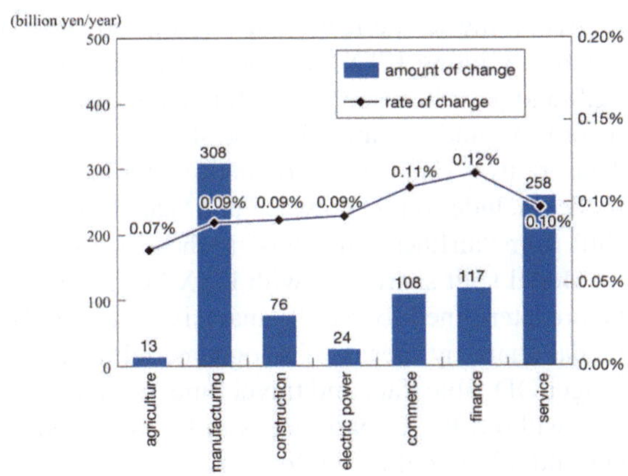

Table 4.12 Summary of J-SCGE analysis (trip and CO_2 emissions)

		Without	With	Change (%)
Number of trips (thousand)		400,876	425,276	6.1
Passenger-kilometre (million)		166,158	161,606	−2.7
CO_2 emissions by transport mode (thousand ton)	Air	9593	8732	−9.0
	Shinkansen	1749	1532	−12.4
	Maglev	0	1011	–
	Total	11,341	11,275	−0.6
CO_2 emissions per passenger-kilometre (without = 100)		100	102	2.0
CO_2 emissions from industrial sector (thousand ton)		148,861	148,999	0.1

unit CO_2 emissions per passenger-kilometre increase, the gross passenger-kilometre decreases as mentioned above. That is also one reason why Maglev project can reduce overall CO_2 emissions from transport sector. On the other hand, Maglev investment project activates Japanese economy and then increases GDP. It causes increase of CO_2 emissions from the industrial sector, which is about 0.1 %.

4.7 Analysis of the Economic Effect of KTX Investment and CO_2 emission (K-SCGE)

4.7.1 Framework of Model Analysis

4.7.1.1 Specific Framework of K-SCGE Model Analysis

Korean HSR, KTX, has opened in April 2004, and connects Seoul and Busan in about 3 h. The new branch line from Daejeon on which a transit station of KTX will be placed to Jeolla region (southwest area of Korea) is also planned. The K-SCGE model classifies Korean economy into six regions in order to analyse the effects of the KTX project.

According to the results of the transport modal share between Seoul and each major city along KTX in 2006, the modal share of car, railway and air are 40.9, 32.8 and 11.7 % respectively for Seoul–Busan (Gyeongsangnam area), the share of railway and car are 41.9 and 41.9 % for Seoul–Taegu (Gyeongsangbuk area). Thus, railway and car are competitive for intercity transport in South Korea. So, our model includes only "railway" and "car" as intercity transport modes and the modal shift from "airlines" and "bus" to these is not considered (Fig. 4.6).

Modal split share for "with KTX"scenario and "without KTX" scenario should be predetermined for SCGE analysis as well as J-SCGE analysis. We estimated the modal share by means of a logit model using a transport data, interregional passenger OD table, fare and travel time, produced by KOTI. Travel time and fare data for "without KTX" case needs to be estimated because KTXis actually operating currently. We applied the level of services of the conventional express train as the level of services of railway of "without KTX" case (Table 4.13).

The estimated modal split share of railway for the main region OD pairs is shown in the Table 4.14. The result shows drastic change of Korean interregional transport condition, especially in Seoul–Busan route, where most of passengers shift to KTX from cars.

4.7.2 The Results

We derive the estimated results of 2029.6 billion won per year as annual benefit and 45.6 trillion won of total benefit over the project life. The benefit spreads out around the country, in particular, to Chungcheong area and Gyeongsangnam area. From a

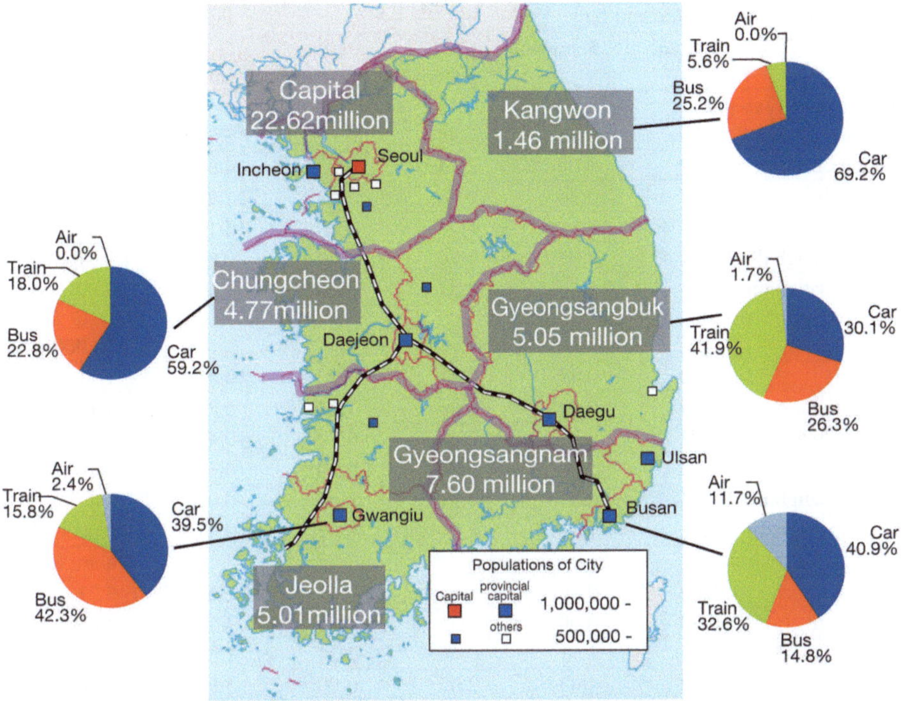

Fig. 4.6 KTX route, regional classification and modal split share. (from/to Seoul)

point of view of relative benefit incidence, namely benefit-regional GDP ratio, the effect on Chungcheong area on which KTXjunction is located is the largest. It implies the locational advantage of centre point in HSR transport network, since travel time to all other major cities is shortened (Figs. 4.7 and 4.8).

Concerning changes in production outputs of industrial sector, 0.24–0.30 % increase are observed. However, the change in outputs of manufacturing sector is remarkable in terms of magnitude of the volume of change. KTX runs through the cities where manufacturing sector concentrate on. It contributes to improve efficiency of production of the sector and, in turn, relatively large increase ratio of outputs (Fig. 4.9).

Total number of trips increases 25.3 %, which is a much larger influence than Japan's Maglev project. The increase of travel demand can be separated substitution effects and income effects in the meaning of economics. Increase of the passenger-kilometre is larger than increase of the number of trip, and it means the average length of travel is extended. It implies that KTX project causes the shift from goods consumption to travel demand. In other words, KTX investment may promote the increase of communication opportunities between the far apart regions (Table 4.15).

Table 4.13 The travel time and fare

	Capital	Kangwon	Chun-gcheong	Jeolla	Gyeong-sangbuk	Gyeong-sangnam
Travel: time (without)						Unit: hour
Capital	0.00	0.90	1.98	4.02	3.96	5.52
Kangwon	0.90	0.00	2.74	4.95	4.05	5.58
Chungcheong	1.98	2.74	0.00	2.32	1.91	3.30
Jeolla	4.02	4.95	2.17	0.00	3.10	3.56
Gyeongsangbuk	3.97	4.05	1.96	3.00	0.00	1.51
Gyeongsangnam	5.51	5.57	3.31	3.56	1.51	0.00
Fare (without)						Unit: 10,000 won
Capital	0.00	0.44	0.71	1.02	0.85	1.11
Kangwon	0.44	0.00	0.96	1.24	0.81	1.12
Chungcheong	0.71	0.96	0.00	0.59	0.49	1.00
Jeolla	1.03	1.24	0.76	0.00	0.64	0.70
Gyeongsangbuk	0.85	0.81	0.51	0.61	0.00	0.47
Gyeongsangnam	1.11	1.11	1.01	0.69	0.47	0.00
Travel: time (with)						Unit: hour
Capital	0.00	0.90	1.61	3.53	2.46	3.21
Kangwon	0.90	0.00	2.52	4.43	3.53	4.45
Chungcheong	1.61	2.51	0.00	2.29	1.05	1.88
Jeolla	3.52	4.43	2.07	0.00	3.08	3.56
Gyeongsangbuk	2.46	3.53	1.04	2.99	0.00	1.24
Gyeongsangnam	3.21	4.45	1.87	3.56	1.24	0.00
Fare (without)						Unit: 10,000 won
Capital	0.00	0.44	0.89	1.48	1.63	2.36
Kangwon	0.44	0.00	1.13	1.74	1.43	2.16
Chungcheong	0.88	1.12	0.00	0.65	0.65	1.37
Jeolla	1.48	1.75	0.84	0.00	0.71	0.73
Gyeongsangnam	1.63	1.43	0.66	0.65	0.00	0.60
Gyeongsangnam	2.36	2.16	1.38	0.72	0.60	0.00

Table 4.14 Transport modal share of railway

	Without (%)	With (%)
Between Seoul and Busan	6	85
Between Seoul and Daejeon	50	62

The model analysis estimates that decrease of CO_2 emissions from the overall transport sector is 3.4 % and it is because the decrease of the emissions from car trips dominates the increase from the railway trips. CO_2 emissions per passenger-kilometre also decrease contrary to the case of Japan because unit CO_2 emission of Maglev is quite large. CO_2 emissions from the industrial sector increase about 0.3 % as well as Japan.

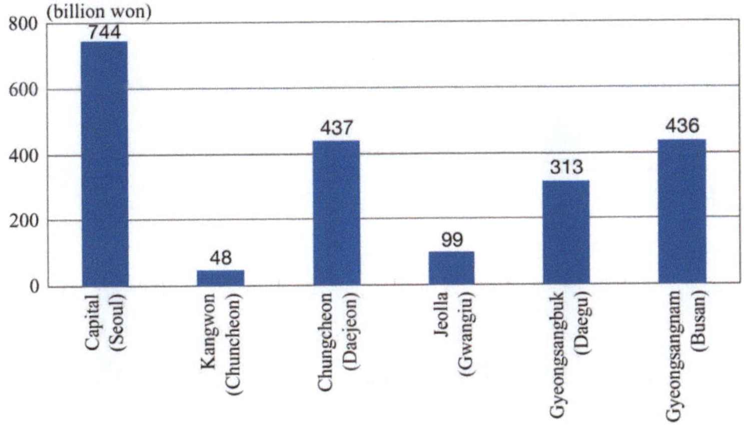

Fig. 4.7 Regional benefit

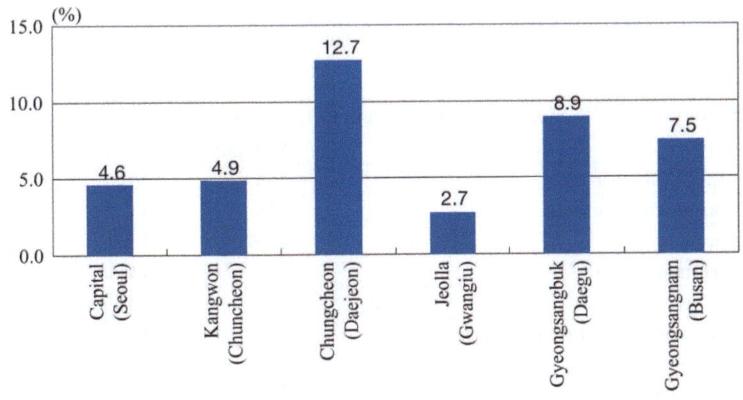

Fig. 4.8 Ratio of regional benefit to regional GDP

4.8 Analysis of the Effects of High-Speed Railway Investment and CO_2 Emission in Taiwan (T-SCGE)

4.8.1 Specific Framework of T-SCGE Model Analysis

Our T-SCGE model classifies Taiwan economy into 15 regions based on economic size of the region and geographical relationship. Each region of the model consists of one or several prefectures as follows. Taiwan's HSR runs west coast of Taiwan connecting between Taipei region and Kaohsiung region (Fig. 4.10).

Our analysis assumes that railway, air and passenger car are competitive transport modes in the interregional transport in Taiwan. Furthermore, the analysis makes

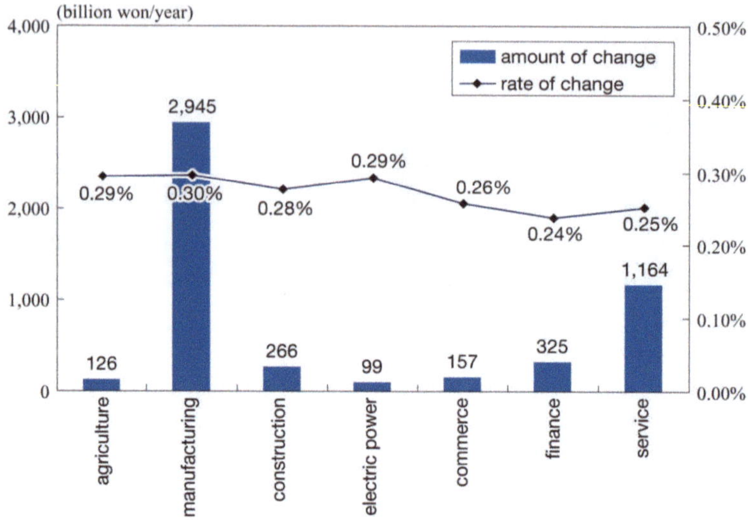

Fig. 4.9 Change in industrial outputs

Table 4.15 Summary of K-SCGE analysis (trip and CO_2 emissions)

		Without	With	Change (%)
Number of trips (thousand)		964	1208	25.3
Passenger-kilometre (million)		139	183	31.4
CO_2 emissions	Car	14,647	13,406	−8.5
By transport mode (ton)	Train	727	1443	98.5
	Total	15,374	14,849	−3.4
CO_2 emissions per passenger-kilometre (without = 100)		100	73	−27.0
CO_2 emissions from industrial sector		23,079	23,146	0.3

several assumptions with regard to transport modal split share conditions based on the existing travel demand forecast in Taiwan. Firstly, all air passengers of the region where HSR is available in the "without HSR case" shift to HSR passengers in case of "with HSR (conventional long-haul train is replaced to HSR)". Secondly, 52.7% of long-haul car travellers in "without HSR case" also shift to HSR users in the "with HSR case".

4.8.2 The Results

Estimated annual benefit is 29.8 billion NT$, and net present benefit is 665 billion NT$. From the point of view of benefit incidence (regional distribution), the effects on Taipei, Gaoxiong and Taichung are larger than other regions. The largest

Fig. 4.10 Regional classification

economic effect appears in Gaoxiong, in terms of benefit GDP ratio. Economic effects of T-HSR investment tend to be concentrated on the high population density regions, which is relatively clear characteristics comparing it with other countries (Figs. 4.11 and 4.12).

Change ratio of outputs in the industrial sector is in the range from 0.04 to 0.09 % except for "public administration" whose value is 0.25 %. In terms of amount of change in level of outputs, relatively large increase is estimated in "metals and machinery", "commerce" and "finance, insurance and real estate" as well as "public administration" (Fig. 4.13).

The total number of trips increases 5.8 %, and the sum of total trips length, passenger-kilometres, also increases 5.0 %. However, CO_2 emission from transport sector decreases 30.4 %, which the emission reduction effect is quite larger than Maglev in Japan and KTX in South Korea. The reduction of CO_2 emissions from cars (and air) dominates the increase of the emissions from HSR as well as Maglev and KTX. Furthermore, CO_2 emissions per passenger-kilometre also decrease similar to KTX project. However, the economic development causes the slight increase (0.07 %) of CO_2 emissions from industrial sector (Table 4.16).

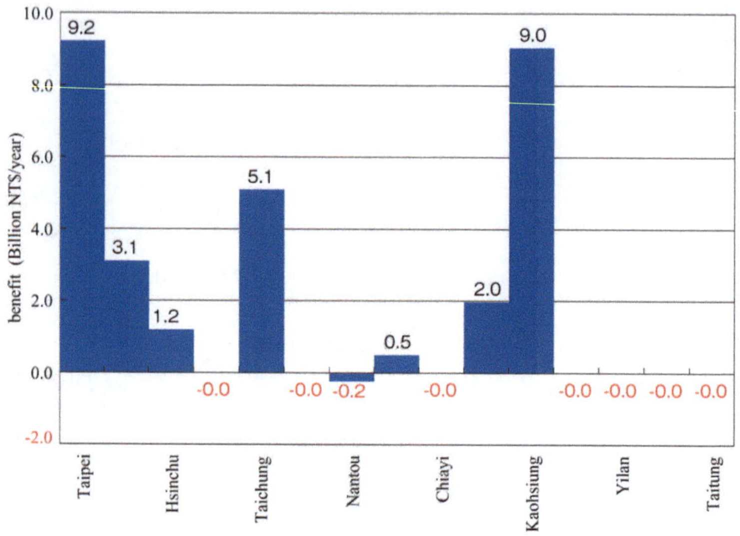

Fig. 4.11 Regional benefit

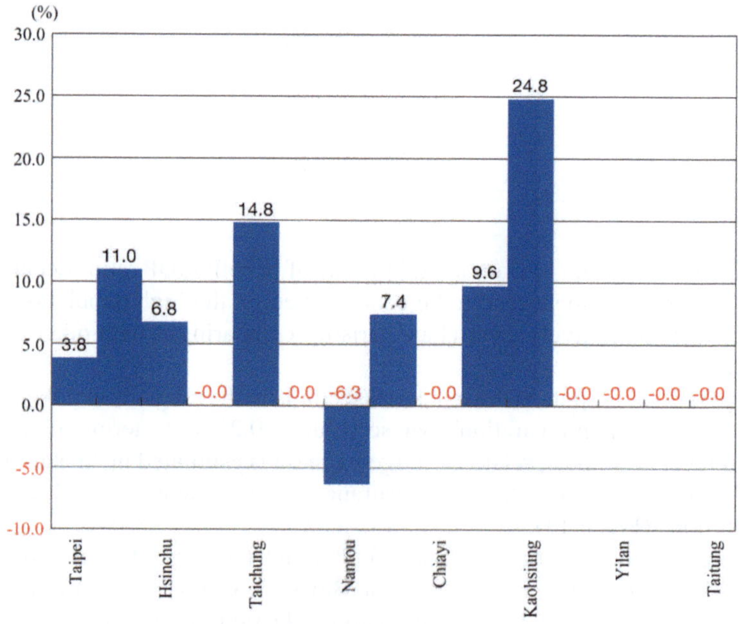

Fig. 4.12 Ratio of regional benefit to regional GDP

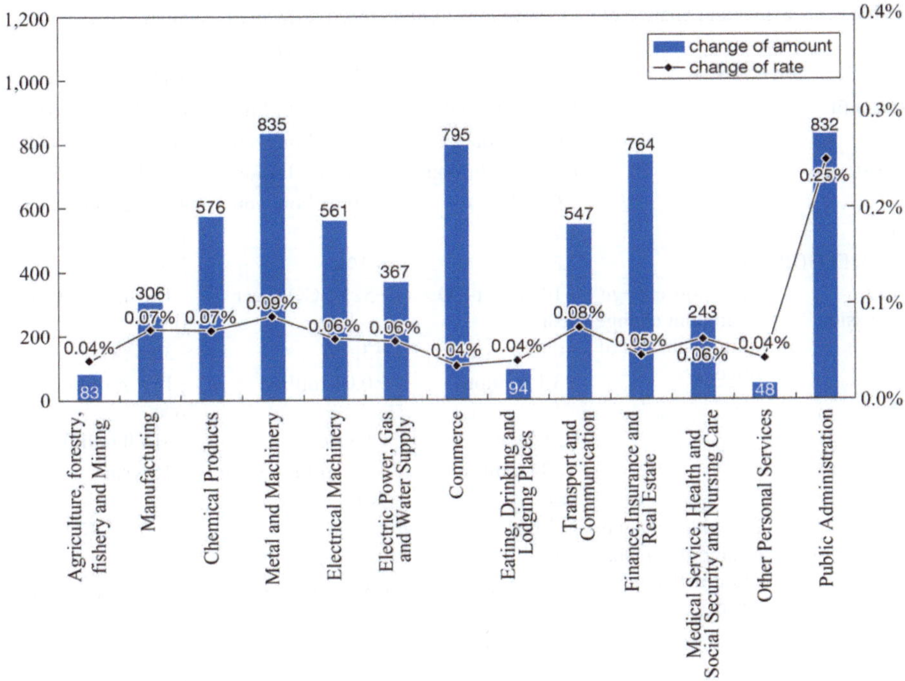

Fig. 4.13 Change in industrial outputs

Table 4.16 Summary of T-SCGE analysis (trip and CO_2 emissions)

		Without	With	Change (%)
Number of Trips (thousand)		459,229	475,407	3.5
Passenger-kilometre (million)		41,499	43,590	5.0
CO_2 emissions by transport mode (thousand ton)	Car	5,075	3,312	−34.7
	Train	83	83	0.0
	Air	422	90	−78.6
	HSR	0	400	–
	Total	5,579,430	3,884,875	−30.4
CO_2 emissions per passenger-kilometre (without=100)		100	66	−34.0
CO_2 emissions from industrial sector (thousand ton)		216,047	216,199	0.1

4.9 Comparison of Analysis Results Among Three Countries

The summary of the results of the three countries is shown in Table 4.17. Although benefit of Maglev is the largest, benefit–cost ratio (B/C) of Maglev investment is the smallest due to the huge investment cost. However, all of B/C estimates of the three

Table 4.17 Effect and change of CO_2 emissions of high-speed railway investment

		Maglev	(Reference)	
			KTX	T-HSR
Benefit		US$ 120.6 billion	US$ 41.8 billion	US$ 19.1 billion
		11.2 trillion ¥	46.7 trillion won	668.9 billion NT$
Cost		US$ 97.1 billion	US$ 19.7 billion	12.9 billion USS
		9.0 trillion ¥	22.0 trillion won	450.0 billion NT$
B/C		1.2	2.1	1.5
Benefit/GDP ratio		2.3 %	6.2 %	7.0 %
CO_2 emission	Transport change of amount change of ratio change of price	−100,621 t-CO_2/year	−525 t-CO_2/year	−1,694,555 t-CO_2/year
		−0.9 %	−3.4 %	−30.4 %
		−3.1 million	−0.01 million	−17.4 million US$/year
		US$/year	US$/year	−610.0 million
		−284.0 million ¥/year	−7.2 million won/year	NT$/year
	Industrial change of amount change of ratio change of price	137,065 t-CO_2/year	66,969 t-CO_2/year	152,387 t-CO_2/year
		0.1 %	0.3 %	0.1 %
		4.2 million US$/year	0.821 million	1.57 million US$/year
		387 million ¥/year	US$/year	55 million NT$/year
			913 million won/year	

projects are over one, and therefore all projects have satisfied social efficiency criteria.

The benefit GDP ratio (B/GDP) in KTX and T-HSR are 6.2 and 7.0 % respectively, which are much higher than Maglev (2.3 %). Since the conventional HSR (Shinkansen) has already existed in Japan, marginal effects of the additional HSR investment should be small. That is why the B/GDP in KTX and T-HSR are relatively larger.

Regarding CO_2 emissions from the transport sector, all three HSR projects can reduce the emissions successfully. Both reduction amount of CO_2 and reduction rate of CO_2 in T-HSR are the highest value among the three projects. It is because that the (exogenously given assumption) shift of modal split share from air and car to HSR, 100 and 52.7 % respectively, are much higher than other countries. On the other hand, improvement of interregional transport conditions by HSR develops the overall economic activity, and then CO_2 emissions from the industrial sector will increase.

4.10 Issues for Further Analysis

4.10.1 New Theories in Spatial Economics

Since the early 1980s, new theories have emerged in spatial economics. Urban agglomeration is a master key concept at all levels of spatial economy from local economy to global economy. The New Economic Geography has become the mainstream in spatial economics, which describes the evolutionary process in a system of cities. In this, intercity transport cost is the most critical factor for spatial structure along with production technology in each city.

Some types of SCGE models have already installed key factors of the New Economic Geography into their model structure. Increasing return to scale in production technology and consumer preferences for a variety of goods are explicitly taken into account in the SCGE model, for example, the recent version of CGEurope. The political implications from such SCGE models have not been examined sufficiently. Their merits should be tested by further experiences of application studies.

4.10.2 Dynamic Socio-Economic Trends

Ageing The ageing of the population has been a great concern in industrialised countries. The Japanese economy has already gone into the stage of declining population. Other industrialised counties in East Asia will soon come to the stage of ageing population. Age composition of the population at the national level or regional level is critical not only for productivity in each industrial sector but also for people's mobility in intercity transport. The SCGE models discussed in this section have not caught such a long-term trend in the ageing of the population.

Skilled Workers Recent theories in spatial economics have focused on the difference in worker's skills. In urban agglomeration, highly skilled workers are mobile and interact with each other. An intercity transport system, particularly a passenger transport system carries highly skilled workers who have embodied knowledge and information into themselves. The highly skilled workers then interact with others at trip destinations so that they can produce new knowledge and information.

A new intercity transport system reduces the generalised price of face-to-face communication and increases positive externality among highly skilled workers. Reduction of intercity transport cost in a system of cities may first cause the agglomeration of population into particular cities. The further reduction of the cost may result in the dispersion of population to all cities. This finding suggests that so far as the highly skilled workers have positive interactions within them, a new intercity transport system may lead to more agglomeration of highly skilled workers in a metropolitan area. To make them dispersed into peripheral cities, intercity transport cost must be reduced to almost zero.

Highly skilled workers need more communication as they become more creative. The creative class tends to agglomerate into particular cities. The impact of intercity transport policy on the nationwide distribution of workers in each class is a very interesting topic. The SCGE model is improved so as to include several classes of worker's skills as input factors and to describe the mobility of the classified workers.

4.10.3 Social Norm

Interregional or Domestic Equity The efficiency of resource allocations with environmental constraints has been ranked prior to other social norms, particularly in mainstream economics or in the recent neo-conservatism. The industrialised countries are however now putting more weight on other social norms.

Equity has been the most important social norm, along with efficiency, in any economy. The spatial distribution of economic welfare is a great political concern since all policy-makers at the nationwide level must take equity into consideration. As we mentioned, results of the SCGE can provide them with an informational basis for equity issues.

There are many studies on the rules for fair allocation of benefits and costs or transferable resources. In particular, the theoretic approach to fair allocation has turned out a variety of rules. However, there is no rule that is widely accepted in the real world. The spatial distribution of benefit or cost output by SCGE for economic impact assessment of an intercity transport policy can be used as a pay-off matrix in theoretic analysis. Political implications at the stage where fair allocation rules are tested with the SCGE pay-off matrix are very interesting not only for policymakers but also for economic modelling experts.

International or Global Equity The policy must be assessed regarding equity not only in the context of domestic policies but also in the international context. The equity norm in the international context is more complicated. Social cost of environmental emissions must be shared by countries in all economic and financial states.

An intercity transport policy in a single country is just a domestic policy. However, some environmental constraints are imposed on the policy in an international context. The sharing of environmental costs by the intercity transport policy should appear in political agendas at the global level.

World-scale economic models have already been developed and applied to policy impact analysis at the global level. The GTAP model has been applied to the economic assessment of tariff polices at WTO. The RICE model developed by Nordhaus has been applied to assessing the dynamic impacts of Global Environmental Policies including carbon tax, energy price regulation, nuclear power policy and so on. They are also the foundation of microeconomic theories that are computable enough for analysts to simulate the actual policies as well as the SCGE model that we have shown in this chapter. We would then be able to analyse the impacts of an intercity transport policy by integrating the SCGE models at different spatial coverage.

Supplementary Note: Model Assumptions The model in this chapter has the following major assumptions:

1. A nation consists of regions labelled by $\mathrm{s} \in \mathbf{S} = \{1, \cdots, \mathrm{s}, \cdots \mathrm{S}\}$.
2. Each region consists of industries (except transport sector) labelled by $\mathrm{i} \in \mathbf{I} = \{1, \cdots, \mathrm{i}, \cdots \mathrm{I}\}$, a representative household and a transport sector.

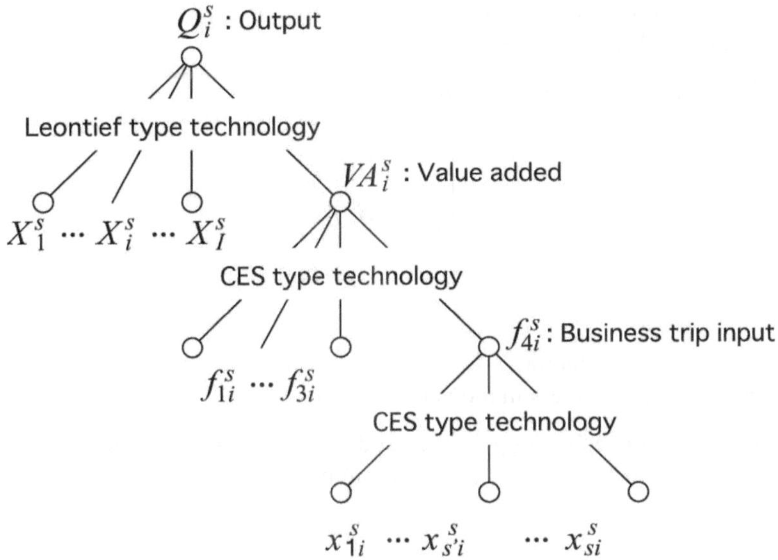

Fig. 4.14 Production function. *CES* constant elasticity of substitution

3. Each industry provides commodities using inputting factors labelled by $k \in \mathbf{K} = \{1, \cdots, k, \cdots K\}$ supplied by household and intermediate goods.
4. The household earns income by supplying factors, and consumes commodities provided by industries and private trips provided by the transport sector.
5. The transport sector provides passenger transport service to households and industries. The excess profit of the transport sector is divided and distributed to households.
6. Markets are assumed to be perfectly competitive and in long-term equilibrium.

Industries' Behaviour Industries provide commodities/services, which are regarded as production, by utilizing input factors and intermediate goods. The production technology is illustrated in Fig. 4.14.

We assume that input coefficients of intermediate goods are fixed. Hence, their production function is formulated as Leontief type technology:

$$Q_i^s = \min\left[\frac{VA_i^s}{a_{vi}^s}, \frac{X_{1i}^s}{a_{1i}^s}, \cdots \frac{X_{ii}^s}{a_{ii}^s}, \cdots, \frac{X_{Ii}^s}{a_{Ii}^s}\right], \tag{4.1}$$

Where

Q_i^s : output of industry i,
$X_{i'i}^s$: intermediate input from industry i' to industry i,
VA_i^s : value added in industry i ,

$a^s_{i'i}$: input coefficient in Leontief matrix
And a^s_{vi} :value-added rate of industry i.

VA^s_i follows CES technology such that

$$VA^s_i = \phi_{1i}\left[\sum_{k\in\mathbf{K}}\delta_{ki} f_{ki}^{s\frac{\sigma_{1i}-1}{\sigma_{1i}}}\right]^{\frac{\sigma_{1i}-1}{\sigma_{1i}}}, \tag{4.2}$$

Where

f^s_{ki} : input factor,
$\phi_{1i}, \delta_{ki}, \sigma_{1i}$,: parameters in industry i .

Taking the Leontief technology in the production function into account, we can assume that an industry behaves so as to minimise production cost as formulated below:

$$\min_{f_k}\left[\sum_{k\in K} w^s_k f^s_{ik}\right] \quad s.t. \quad VA^s_i = 1 \tag{4.3}$$

where

w^s_k : rent of input factor.

Solving the programming in (4.3), we obtain the following input factor demand function:

$$f^s_{ki} = \phi^{-1}_{1i}\delta^{\sigma_{1i}}_{ki} w_k^{s-\sigma_{1i}}\left[\sum_{k\in\mathbf{K}}\sigma^{\sigma_{1i}}_{k'i} w_{k'}^{s1-\sigma_{1i}}\right]^{\frac{\sigma_{1i}}{1-\sigma_{1i}}}. \tag{4.4}$$

We label input factor so that $k = 1$ stands for labour input, $k = 2$ for capital input, $k = 3$ for land input and $k = 4$ for business trip input. The input of business trips plays a key role in determining the productivity.

Business trip input f^s_{4i} in (4.4) is a function of trips to all regions. Destination choice for business trips is formulated as a cost-minimising problem:

$$\begin{aligned} &\min \sum_{s'\notin\mathbf{S}}\left(p^s_{s'} + t^s_{s'} w^s_1\right)x^s_{s'i} \\ &\text{s.t.} \quad f^s_{4i} = \phi_{2i}\left(\sum_{s7\in\mathbf{S}} A^s_{s'i} x_{s'i}^{s\frac{\sigma_{2i}-1}{\sigma_{2i}}}\right)^{\frac{\sigma_{2i}}{\sigma_{2i}-1}}, \end{aligned} \tag{4.5}$$

where

$p_{s'}^{s}$: transport cost s to region s',
$x_{s'i}^{s}$: business trips s to region s',
$A_{s'i}^{s}$: attractiveness level in region s',
ϕ_{2i}, σ_{2i}: parameters in industry i.

Solving the programming in (4.5), we obtain the following business trip demand function:

$$x_{s'}^{s} = \phi_{2i} A_{s'i}^{s\,\sigma_{2i}} \left(p_{s'}^{s} + t_{s'}^{s} w_{1}^{s}\right)^{-\sigma_{2i}} \left[\sum_{s' \in \mathbf{S}} A_{s'i}^{s\,\sigma_{2i}} \left(p_{s'}^{s} + t_{s'}^{s} w_{1}^{s}\right)^{-\sigma_{2i}}\right]^{\frac{\sigma_{2i}}{1-\sigma_{2i}}}. \tag{4.6}$$

The above programming also yields the rent of the business trip factor:

$$w_{4}^{s} = \left(p_{s'}^{s} + t_{s'}^{s} w_{1}^{s}\right) \frac{x_{s'i}^{s}}{f_{4i}^{s}}. \tag{4.7}$$

Since the destination choice for business trips is formulated with the CES type structure, the share of trips to each destination looks like a form of the gravity model:

$$\frac{\left(p_{s'}^{s} + t_{s'}^{s} w_{1}^{s}\right) x_{s'i}^{s}}{\sum_{s' \in \mathbf{S}} \left(p_{s'}^{s} + t_{s'}^{s} w_{1}^{s}\right) x_{s'i}^{s}} = \frac{A_{s'i}^{s\,\sigma_{2i}} \left(p_{s'}^{s} + t_{s'}^{s} w_{1}^{s}\right)^{-\sigma_{2i}}}{\sum_{s' \in \mathbf{S}} A_{s'i}^{s\,\sigma_{2i}} \left(p_{s'}^{s} + t_{s'}^{s} w_{1}^{s}\right)^{-\sigma_{2i}}}. \tag{4.8}$$

We can model the mode choice and the route choice for business trips in the same way.

The production technology in this model has the property of constant return to scale because of the structure shown in the Fig. 4.14. Hence, the price of a commodity produced by industry i equals the production cost for a unit output, or average cost:

$$P_{i}^{s} = a_{vi}^{s} \left[\sum_{k=1,2,3} w_{k}^{s} c f_{ki}^{s} + \sum_{s' \in \mathbf{S}} \left(p_{s'}^{s} + t_{s'}^{s} w_{1}^{s}\right) c x_{s'i}^{s}\right] + \sum_{i' \in \mathbf{I}} P_{i'}^{s} a_{i'i}^{s}. \tag{4.9}$$

By rewriting the right-hand side of Eq. (4.9) in matrix form, we obtain a price vector of products,

$$\begin{bmatrix} P_{1}^{s} \\ \vdots \\ P_{i}^{s} \\ \vdots \\ P_{I}^{s} \end{bmatrix}^{t} = \begin{bmatrix} a_{v1}^{s} \left\{ \sum_{k=1,2,3} w_{k}^{s} c f_{k1}^{s} + \sum_{s' \in \mathbf{S}} \left(p_{s'}^{s} + t_{s'}^{s} w_{1}^{s}\right) c x_{s'1}^{s} \right\} \\ \vdots \\ a_{vi}^{s} \left\{ \sum_{k=1,2,3} w_{k}^{s} c f_{ki}^{s} + \sum_{s' \in \mathbf{S}} \left(p_{s'}^{s} + t_{s'}^{s} w_{1}^{s}\right) c x_{s'i}^{s} \right\} \\ \vdots \\ a_{vI}^{s} \left\{ \sum_{k=1,2,3} w_{k}^{s} c f_{kI}^{s} + \sum_{s' \in \mathbf{S}} \left(p_{s'}^{s} + t_{s'}^{s} w_{1}^{s}\right) c x_{s'I}^{s} \right\} \end{bmatrix} [\mathrm{I} - \mathrm{A}]^{-1}, \tag{4.10}$$

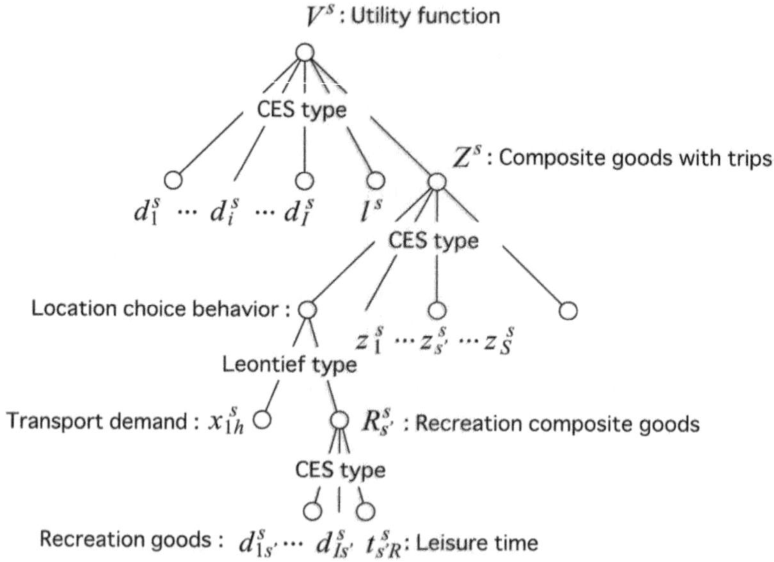

Fig. 4.15 Utility function. *CES* constant elasticity of substitution

where

I: identity matrix,
A: input coefficient matrix,
t: transposed matrix.

When the price vector of a commodity satisfies Eq. (4.10), we obtain the zero profit condition for each industry.

Household's Behaviour We assume that a representative household maximises utility. The structure of a household's consumption is illustrated in Fig. 4.15. The household consumes commodities/services, land service and self-produced composite goods with private trips in the first step. The self-production of composite goods requires not only passenger trips but also the purchase of commodities/services at the destination, such as hotel services during the journey.

The utility function is specified as a CES type. With the specification of budget and time constraints, a household's behaviour is formulated as a utility maximising problem.

$$
\begin{aligned}
V^s &= \max\left[\sum_{i\in \mathbf{I}}\left\{(\gamma_{1i}^s)^{\frac{1}{\rho_1}}(d_i^s)^{\frac{\rho_1-1}{\rho_1}}\right\}+(\gamma_{1Z}^s)^{\frac{1}{\rho_1}}\left(Z^s\left(d_{s'i}^s,x_{s'h}^s,t_{s'R}^s\right)\right)^{\frac{\rho_1-1}{\rho_1}}+(\gamma_{1l}^s)^{\frac{1}{\rho_1}}(l^s)^{\frac{\rho_1-1}{\rho_1}}\right]^{\frac{\rho_1}{\rho_1-1}}, \\
&\text{s.t.} \sum_{i\in \mathbf{I}} P_i^s d_i^s + \sum_{s'\in \mathbf{S}}\sum_{i\in \mathbf{I}} P_i^{s'} d_{is'}^s + \sum_{s'\in \mathbf{S}} p_{s'}^s x_{s'h}^s + w_3^s l^s = \sum_{k=1,2,3} w_k^s F_k^s + \pi_0^s \\
&T^s = F_1^s + \sum_{s'\in \mathbf{S}} t_{s'}^s x_{s'h}^s + \sum_{s'\in \mathbf{S}} t_{s'R}^s
\end{aligned}
\tag{4.11}
$$

Where

V^s : indirect utility function,
d_i^s : consumption level of commodity i,
Z^s : consumption level of self-produced composite goods with private trips,
l^s : consumption level of land service,
$\gamma_{1i}^s, \gamma_{1Z}^s, \gamma_{1l}^s, \rho_1$: parameters,
F_k^s : capital stock of input factor k,
π_0^s : excess profit/loss from transport sector,
T^s : total available time.

The budget and the time constraints can be combined into the following equation:

$$\sum_{i\in\mathbf{I}} P_i^s d_i^s + \sum_{s'\in\mathbf{S}}\sum_{i\in\mathbf{I}} P_i^{s'} d_{is'}^s + \sum_{s'\in\mathbf{S}}\left(p_{s'}^s + w_1^s\left(t_{s'}^s + t_{s'R}^s\right)\right)x_{s'h}^s + w_3^s l^s = w_1^s T^s + w_2^s F_2^s + w_3^s F_3^s + \pi_0^s. \tag{4.12}$$

The solutions of the programming in (4.11) give the demand functions:

$$\begin{aligned} d_i^s &= \frac{\gamma_{1i}^s\left(w_1^s T^s + w_2^s F_2^s + w_3^s F_3^s + \pi_0^s\right)}{P_i^{s\rho_1}\left[\sum_{i'\in\mathbf{I}} \gamma_{1i'}^s P_{i'}^{s1-\rho_1} + \gamma_{1Z}^s P_Z^{s1-\rho_1} + \gamma_{1l}^s w_3^{s1-\rho_1}\right]}, \\ Z^s &= \frac{\gamma_{1Z}^s\left(w_1^s T^s + w_2^s F_2^s + w_3^s F_3^s + \pi_0^s\right)}{P_Z^{s\rho_1}\left[\sum_{i'\in\mathbf{I}} \gamma_{1i'}^s P_{i'}^{s1-\rho_1} + \gamma_{1Z}^s P_Z^{s1-\rho_1} + \gamma_{1l}^s w_3^{s1-\rho_1}\right]}, \\ l^s &= \frac{\gamma_{1l}^s\left(w_1^s T^s + w_2^s F_2^s + w_3^s F_3^s + \pi_0^s\right)}{w_3^{s\rho_1}\left[\sum_{i'\in\mathbf{I}} \gamma_{1i'}^s P_{i'}^{s1-\rho_1} + \gamma_{1Z}^s P_Z^{s1-\rho_1} + \gamma_{1l}^s w_3^{s1-\rho_1}\right]}, \end{aligned} \tag{4.13}$$

where

P_Z^s : price of self-produced composite goods with private trip to region s.

The household demands for the private trip. We assume that the household chooses the destinations so as to maximise the sub-utility specified as a CES type function under the budget constraint. The maximisation of the sub-utility is formulated as:

$$\begin{aligned} Z^s &= \max_{Z_{s'}^s}\left\{\sum_{s'\in\mathbf{S}} (\gamma_{2s'}^s)^{\frac{1}{\rho_2}} (Z_{s'}^s)^{\frac{\rho_2-1}{\rho_2}}\right\}^{\frac{\rho_2}{\rho_2-1}}, \\ \text{s.t.}\quad P_Z^s Z^s &= \sum_{s'\in\mathbf{S}} P_{s'z}^s Z_{s'}^s \end{aligned} \tag{4.14}$$

where

$\gamma^s_{2s'}$: share parameter in destination choice,
$P^s_{s'Z}$: price of self-produced goods with the private trip to destination s'.

The programming in (4.14) yields the demand function for self-produced composite goods with a trip to each region:

$$Z^s_{s'} = \frac{\gamma^s_{2s'}\left(P^s_Z Z^s\right)}{P^{s\rho_2}_{s'z} \sum_{s' \in \mathbf{S}} \gamma^s_{2s'} P^{s^{1-\rho_2}}_{s'z}}. \tag{4.15}$$

We also obtain the marginal utility of self-produced composite goods with trips as the Lagrangian multiplier accompanied with the solution, or the shadow price of composite goods with trips:

$$P^s_Z = \left[\sum_{s' \in \mathbf{S}} \gamma^s_{2s'} P^{s(1-\rho_2)}_{s'z}\right]^{\frac{1}{1-\rho_2}}. \tag{4.16}$$

The household minimises the cost of self-production under the constraint of self-production technology specified as Leontief type. The behaviour is formulated as:

$$\begin{aligned} C^s_{s'} &= \min \quad \left[\left(p^s_{s'} + w^s_1 t^s_{s'}\right) x^s_{s'h} + P^s_{s'R} R^s_{s'}\right] \\ \text{s.t.} \quad z^s_{s'} &= \min \quad \left\{\frac{x^s_{s'h}}{b_1}, \frac{R^s_{s'}}{b_2}\right\}, \end{aligned} \tag{4.17}$$

Where

b_1, b_2 : parameters,
$P^s_{s'R}$: price of commodity/service to be input to the self-production.

The programming in (4.17) yields two demand functions, for passenger transport and for input to self-production

$$\begin{aligned} x^s_{s'h} &= b_1 z^s_{s'}, \\ R^s_{s'} &= b_2 z^s_{s'}. \end{aligned} \tag{4.18}$$

We obtain the marginal utility of the price of composite goods with trips to each region, as in the previous step:

$$P^s_{s'z} = b_1 p^s_{s'} + b_2 P^s_{s'R}. \tag{4.19}$$

Finally, the household determines the input of recreation goods and leisure time. We suppose that household behaviour can be formulated in the same way as the household production function in Becker (1965). The idea of household production function is that the household produces the self-produced goods by inputting its own resources. Hence, the household behaviour is expressed as minimisation of cost with constraint of production technology specified as CES type. The behaviour is formulated as:

$$\begin{aligned} &\min_{d^s_{IS'i}, t^s_{s'R}} \left(\sum_{i \in \mathbf{I}} P_i^{s'} d^s_{is'} + w_1^s t^s_{s'R} \right) \\ &\text{s.t.} \quad R^s_{s'} = \phi_h \left[\sum_{i \in \mathbf{I}} \delta_{ih} d^{s\ \frac{\sigma_h - 1}{\sigma_h}}_{is'} + \sigma_{0h} t^{s\ \frac{\sigma_h - 1}{\sigma_h}}_{s'R} \right]^{\frac{\sigma_h}{\sigma_h - 1}}, \end{aligned} \tag{4.20}$$

Where

$\phi_h, \delta_{ih}, \delta_{0h}, \sigma_h$,: parameters.

The programming in (4.20) yields the demand functions for recreation goods and leisure time:

$$\begin{aligned} d^s_{s'i} &= R^s_{s'} \cdot \phi_h^{-1} \delta_{ih}^{\sigma_H} P_i^{s' - \sigma_H} \left[\sum_{i \in \mathbf{I}} \delta_{ih}^{\ \sigma h} P_i^{s' 1 - \sigma_h} + \delta_{0h}^{\ \sigma_h} w_1^{s 1 - \sigma_h} \right]^{\frac{\sigma_h}{1 - \sigma_h}}, \\ t^s_{s'R} &= R^s_{s'} \cdot \phi_h^{-1} \delta_{0h}^{\ \sigma h} w_1^{s - \sigma_h} \left[\sum_{i \in \mathbf{I}} \delta_{ih}^{\ \sigma_h} P_i^{s' 1 - \sigma_h} + \delta_{0h}^{\ \sigma_h} w_1^{s 1 - \sigma_h} \right]^{\frac{\sigma_h}{1 - \sigma_h}}. \end{aligned} \tag{4.21}$$

Then, we obtain the price of the self-produced goods, which equals the production cost for a unit output, or the average cost,

$$P^s_{s'R} = \sum_{i \in \mathbf{I}} \left(P_i^{s'} \frac{d^s_{s'i}}{R^s_{s'}} \right) + w_1^s \frac{t^s_{s'R}}{R^s_{s'}}. \tag{4.22}$$

Transport Sector's Behaviour The transport sector produces transport services for other industries and households. We assume that the production function is formulated with the Leontief type technology,

$$Q^s_0 = \min \left[\frac{f^s_{10}}{c^s_1}, \frac{f^s_{20}}{c^s_2}, \frac{f^s_{30}}{c^s_3} \right], \tag{4.23}$$

Where

C^s_1, C^s_2, C^s_3: coefficients of input factor.

Solving the programming in (4.23), we obtain the input factor demand function as:

$$f_{k0}^{s} = c_k^s Q_0^s \text{ where, } k = 1, 2.3. \tag{4.24}$$

The price of transport service is to be set as part of a policy scenario as well as transport investment. Hence, the excess profit from the transport sector is given by the following and divided among households:

$$\pi_0^s = \sum_{s' \in \mathbf{S}} p_{s'}^s \left(\sum_{i \in \mathbf{I}} x_{s'i}^s + x_{s'h}^s \right) - \sum_{k \in 1,2,3} w_k^s f_{k0}^s, \tag{4.25}$$

Where, $Q_0^s = \sum_{i \in \mathbf{I}} x_{s'i}^s + x_{s'h}^s$

Equilibrium Conditions and Benefit Definition In this model, the industry's production function has the property of constant return to scale because of CES type technology. Each industry produces commodities/services as consumers demand, and supply-demand balance for the industry's product is maintained within each sector. The system of equations is:

$$\begin{aligned}
\sum_{i \in \mathbf{I}} w_1^s \left(a_{vi}^s Q_i^s \left(\mathbf{D}^s\right) cf_{1i}^s \right) + w_1^s \left(c_1^s Q_0^s \left(\mathbf{D}^s\right) cf_{10}^s \right) &= w_1^s \left[T^s - \sum_{s' \in \mathbf{S}} t_{s'}^s x_{s'h}^s \left(\mathbf{P}^s\right) - \sum_{s' \in \mathbf{S}} t_{s'R}^s \left(\mathbf{P}^s\right) \right], \\
\sum_{i \in \mathbf{I}} w_2^s \left(a_{vi}^s Q_i^s \left(D^s\right) cf_{2i}^s \right) + w_2^s \left(c_2^s Q_0^s \left(D^s\right) cf_{20}^s \right) &= w_2^s F_2^s, \\
\sum_{i \in \mathbf{I}} w_3^s \left(a_{vi}^s Q_i^s \left(D^s\right) cf_{3i}^s \right) + w_3^s \left(c_3^s Q_0^s \left(D^s\right) cf_{30}^s \right) &= w_3^s \left[F_3^s - l^s \right],
\end{aligned} \tag{4.26}$$

where

D^s : vector of consumption in region s,
P^s : vector of price in region s.

The final demand includes both industry's production and household's production, as in the following:

$$\begin{bmatrix} Q_1^s \\ \vdots \\ Q_i^s \\ \vdots \\ Q_I^s \end{bmatrix}^t = \left[\mathbf{I} - \mathbf{A}\right]^{-1} \begin{bmatrix} d_1^s + \sum_{s' \in \mathbf{S}} d_{1s}^{s'} \\ \vdots \\ d_i^s + \sum_{s' \in \mathbf{S}} d_{is}^{s'} \\ \vdots \\ d_I^s + \sum_{s' \in \mathbf{S}} d_{Is}^{s'} \end{bmatrix}. \tag{4.27}$$

In practice, we only have to solve the supply-demand balance equation with respect to the price of input factors. In this model, the benefit of passenger transport investment is defined as EV. Since the household's utility level has been obtained in the previous section, EV satisfies the following equation:

$$V^{s}(\mathbf{P}^{s^{WO}}, \mathbf{P}^{s'^{WO}}, \mathbf{p}_{s}{}^{WO}, \pi_{0}^{s^{WO}}, w_{1}^{s^{WO}} + EV^{s}) = V^{s^{W}}, \tag{4.28}$$

Where

WO, W: expressed without and with investment, respectively.

Chapter 5
Social Efficiency Measurement and Comparison of Airlines and Railway Companies in Japan's Domestic Intercity Travel Market

Tae Hoon Oum, Somchai Pathomsiri and Yuichiro Yoshida

Recent public awareness of the environmental impact of CO_2 emissions provides a new perspective on the discussion of modal choice between aviation and railways. This argument is becoming especially intense in Japan and Western European countries. However, this question does not have a clear-cut answer, and its answer is rather ambiguous when one compares the life cycle CO_2 emissions of these two transport modes; while the railway causes lower environmental damage relative to air transport and other modes of transport in terms of marginal CO_2 emissions for its transport service operations, railway transport incurs a substantial amount of CO_2 emission in the stage of infrastructure construction. It is thus important not only to take into account the CO_2 emission caused by transport service operations but also to include the CO_2 emission from infrastructure construction in social efficiency analysis and benchmarking. In other words, social efficiency measurement that takes into account life cycle CO_2 emissions should form a fundamental basis for discussing the mode choice argument. Another important issue when considering socially efficient choices of passenger transport modes is the value of time spent by travellers using alternative modes since time is an input for passengers to use transport services.

T. H. Oum (✉)
Sauder School of Business, University of British Columbia, Vancouver, Canada
e-mail: tae.oum@sauder.ubc.ca

S. Pathomsiri
Department of Civil and Environmental Engineering, Faculty of Engineering,
Mahidol University, Nakhon Pathom, Thailand
e-mail: somchai.pat@mahidol.ac.th

Y. Yoshida
Graduate School for International Development and Cooperation,
Hiroshima University, Hiroshima, Japan
e-mail: yuichiro@hiroshima-u.ac.jp

Y. Hayashi et al. (eds.), *Intercity Transport and Climate Change*,
Transportation Research, Economics and Policy 15, DOI 10.1007/978-3-319-06523-6_5

5.1 Introduction and Background

Recent public awareness of the environmental impact of CO_2 emissions provides a new perspective on the discussion of modal choice between aviation and railways. This argument is becoming especially intense in Japan and Western European countries. However, this question does not have a clear-cut answer, and its answer is rather ambiguous when one compares the life cycle CO_2 emissions of these two transport modes; while the railway causes lower environmental damage relative to air transport and other modes of transport in terms of marginal CO_2 emissions for its transport service operations, railway transport incurs a substantial amount of CO_2 emission in the stage of infrastructure construction. It is thus important not only to take into account the CO_2 emission caused by transport service operations but also to include the CO_2 emission from infrastructure construction in social efficiency analysis and benchmarking. In other words, social efficiency measurement that takes into account life cycle CO_2 emissions should form a fundamental basis for discussing the mode choice argument. Another important issue when considering socially efficient choices of passenger transport modes is the value of time spent by travellers using alternative modes since time is an input for passengers to use transport services.

This chapter conducts a 'social' efficiency measurement of two major transport modes, aviation and railways, in Japan's domestic intercity passenger travel market, by taking into account (a) the life cycle CO_2emissions caused by operation, construction and operation of infrastructure, and construction of aircraft/rolling stock, and (b) the time spent by passengers to use these two modes. We attempt to identify a common production frontier between these two types of transport service providers who use a set of inputs such as labour, other variable inputs and capital to produce both desirable and undesirable outputs, namely passenger transport service outputs and CO_2 emissions.

There are three major passenger railway companies in Honshu, the main island of Japan, namely Central Japan Railway Company (JR Central), West Japan Railway Company (JR West)and East Japan Railway Company (JR East). When the Japan National Rail authority was privatised some 20 years ago, it was divided into six regional companies. Among these, JR Central operates areas around Nagoya and Shizuoka, as well as the Tokaido Shinkansen, a high-speed rail (HSR) system connecting the densest intercity transport link in Japan between Tokyo and Osaka. JR West is in charge of areas around and to the west of the greater Osaka area, as well as the Sanyo Shinkansen, a HSR extending from Osaka to Hakata in the Kyushu region. JR East covers greater Tokyo as its commercial area, as well as a network of several HSR systems that expand towards the north and east from Tokyo.

On the aviation side, although our data set includes three major airlines in Japan, namely Japan Airlines (JAL), All Nippon Airways (ANA) and Japan Air System (JAS), we decided to aggregate the JAL and JAS data into a single company since JAS was merged into new JAL in 2002. Most of the airports in Japan are owned and operated by either national or regional governments except for the terminal buildings, which are private.

The analysis techniques applied in this study include various forms of the classical data envelopment analysis (DEA) models; the recently proposed model from production economics—directional output distance function; average social cost (ASC) approach; social total factor productivity (TFP) index and the simple, but perhaps easiest-to-understand, partial productivity ratios. The main reason we chose to use several methods is that none of the currently known methods allow us to compare fairly between companies in the railway industry and companies in the airline industry. The ultimate goal is to provide a fair and unbiased overall assessment of the sample firms' social efficiency performance taking into account both CO_2 emissions on the output side and passenger time inputs, in addition to the outputs and inputs normally considered for efficiency comparisons among private sector firms.

The rest of the chapter describes the details of our study, which is organised into sections as follows. In Sect. 5.2, we briefly review certain literature on the assessment of the productivity of transport systems. In particular, the review is focused on when the undesirable outputs were taken into consideration, since this is the case for our research. We then technically explain in Sect. 5.3 the five approaches used for measurement: DEA, the directional output distance function, ASC, TFP index and partial productivity ratio. The data section follows subsequently. In Sect. 5.4, we discuss the individual inputs, desirable outputs and undesirable outputs of the three railway companies and two airlines. We also give details on how to obtain a consistent aggregate output that takes into account the difference in services between railway and airline companies. The results and discussion on each analysis technique are provided in Sect. 5.5. Last, in Sect. 5.6, we summarise substantial results on social efficiency measurement and point out further research needs.

5.2 Literature Review

Literature on efficiency measurement of transport carriers has been accumulating ever since economists were interested in measuring (cost) inefficiency caused by economic regulations (mostly on entry and pricing) and predicting the effects of liberalization/deregulation. Gillen et al. (1990) and Oniki et al. (1994) are earlier examples of such regulatory efficiency studies.

Efficiency measurement through benchmarking gains relevance when markets fail to discipline the firms, for example markets with natural monopoly, government regulation and/or externalities. The transport industry is a typical example. Oum et al. (1992), Oum and Yu (1994, 1995, 1998), and Oum et al. (1999) were among the first attempts to measure and compare productivity and efficiency of railways and/or airlines. The Air Transport Research Society (ATRS 2002) initiated in 2001 its annual measurement and benchmarking of productive efficiency of the world's major airports. Yoshida (2004) and Yoshida and Fujimoto (2004) measured productive efficiency of Japanese airports using cross-sectional data. They found that recently-constructed regional airports are relatively inefficient. In all of these studies, however, externalities such as CO_2 emission, air pollution, congestion, noise, or risk of accidents have not been incorporated. Nor did they include transport users' time input.

Recent research developments expand the efficiency measurement literature in the direction of social efficiency benchmarking by incorporating undesirable outputs into efficiency measurements, though not necessarily in the field of transport industry.[1] Only recently, have there been social efficiency benchmarking studies in the field of transport using DEA that treated negative externalities as undesirable outputs.[2] McMullen and Noh (2007) evaluated the environmental efficiency of 43 US bus transit agencies in 2000. They employed DEA with directional distance function to account for vehicular emissions (i.e. hydrocarbons, CO, and NO_x) as undesirable outputs in the analysis. Yu et al. (2008) analysed productivity growth of Taiwan's four airports from 1995 to 1999 with the Malmquist-Luenberger productivity index.[3] In this study, aircraft noise as a negative externality of airport service is considered as an undesirable output. Pathomsiri (2007) and Pathomsiri et al. (2008) took number of flight delays and time delay as two undesirable outputs from airport operations, and measured social efficiency of 56 US airports between 2000 and 2003. It was shown that ignoring undesirable outputs from the efficiency measurement may not be fair for those system managers who work to keep the environmental impact as low as possible.

The literature is, however, limited to the social efficiency benchmarking of one transport mode, and no attempt has been made so far in multi-modal social efficiency benchmarking. Inter-modal comparison of social efficiency requires measuring the amounts of inputs including the time users spend using transport modes and outputs including undesirable outputs such as greenhouse gas emission in common measurement units among these heterogeneous transport modes. Given this research background, this paper makes probably the first attempt to do an inter-modal comparison of social efficiency by comparing two major public transport modes, namely aviation and railways in Japan, in order to discuss socially efficient modal choices in various market situations.

---

[1] Park and Weber (2006) applied the directional distance function approach to estimate the efficiency and productivity change of Korean banks between 1992 and 2002. In this study, loan losses are treated as an undesirable output of a bank's operation. Korhonen and Luptacik (2004) used two DEA models that incorporate undesirable output into analysis and evaluated the ecological efficiencies of 24 electric power-generating plants in a European country. In this study, the power generating process is defined to produce emission of dust, NO*x* and SO2 as by-products, which are treated as undesirable outputs.

[2] DEA was pioneered by Farrel (1957), and then a number of applications to efficiency measurement have been accumulated since the seminal paper by Charnes et al. (1978).

[3] The Malmquist index decomposes the productivity growth into technological advance in the industry and efficiency gain of each DMU. The Luenberger productivity index is a directional productivity measurement, which is neither input- nor output-oriented. See Caves et al. (1982), Färe and Grosskopf (1992), and Chap. 9 of Färe et al. (1994) for the Malmquist index, and Luenberger (1992) for the Luenberger productivity index.

5.3 Methodology

In this section we describe each of the four complementary methods via which we accomplish the social efficiency measurement of, and comparison across, two major companies (ANA, JAL+JAS) in the domestic airline sector and three major companies (JR East, JR Central and JR West) in the railway sector in Japan.

We first describe the DEA approach, as it has become a basic technique for assessing technical efficiency. The DEA concept is relatively simple. For example, in the case of a simple production process of one input and one output, if we respectively plot the input and output of all transport systems on the horizontal and vertical axes, the DEA tries to determine the outmost horizon that envelopes the cloud of all data points. Such an enveloped horizon is also known as an efficient frontier. Any company (or Decision Making Unit, DMU in DEA terminology) lying on this frontier is then regarded as efficient. On the contrary, other companies that lie under this frontier are called inefficient. However, if some outputs are socially undesirable, the DEA may not be well defined. Therefore, the directional output distance function is introduced in Sect. 5.3.2 to cope with DEA's weakness. In Sect. 5.3.3 and 5.3.4, we describe another two alternative approaches for assessing social efficiency: partial productivity factor and TFP factor.

5.3.1 Efficiency Assessment: Data Envelopment Analysis Approach

Perhaps the most popular method for assessing efficiency is DEA. Since its introduction by Charnes et al. (1978), there have been thousands of articles across all disciplines published in a variety of journals and books.

Two orientations of DEA may be applied: input and output. Input-oriented DEA focuses on the determination of an efficient company among its peers, and it uses the least economic inputs, conditioned on the same output levels. This can be accomplished by repetitively solving the following linear programs for all companies under consideration:

$$\theta^* = \min \theta,$$

subject to

$$\begin{aligned} &\sum_{j=1}^{n} \lambda_j x_{ij} \le \theta x_{io} && i = 1, 2, \ldots, m; \\ &\sum_{j=1}^{n} \lambda_j y_{rj} \ge y_{ro} && r = 1, 2, \ldots, s; \\ &\lambda_j \ge 0 && j = 1, 2, \ldots, n; \end{aligned} \tag{5.1}$$

where company o represents one of the n companies under assessment, and each company $j = 1,\ldots,n$ uses m inputs $x_{ij}(i = 1,2,\ldots,m)$ to produce s outputs $y_{rj}(r = 1,2,\ldots,s)$ where x_{ij} and y_{rj} are the ith input and r^{th} output for company j, respectively. λ_j is a non-negative scalar that allows for a linear combination of companies to create a benchmark for the company under consideration.

The linear programme in (5.1) initially assumes that each company is not efficient and needs to reduce its input by the factor of θ^* in order to become efficient. However, if solving (5.1) yields a θ^* value of exactly 1 (or should reduce its inputs by a factor of 1), the solution then contrasts with the assumption. As a result, this means that the company need not reduce any of its inputs because it is already on the efficiency frontier. And one is effectively the upper bound of θ^* value. The value of θ^* can be as low as about zero, meaning that its inputs should be decreased. Note that the linear programme in (5.1) assumes a constant returns-to-scale (CRS) production process. Afterwards, we will represent this model as input-oriented constant returns-to-scale DEA, or I-CRS for short.

Conversely, the output-oriented DEA looks for the most efficient company among many comparable peers, which produces the largest amount of outputs. It seeks to find the best producer, given the same amount of inputs. Assuming that the production process is CRS, the linear programme in (5.2) can be used to solve whether a company is efficient. An efficient company must possess a ϕ^* value of exactly 1, meaning that the company is performing so well that it does not need to expand any of its current outputs. For any companies lying below an efficient frontier, they would need to increase their outputs by a factor of ϕ^* in order to reach the efficient frontier:

$$\phi^* = \max \phi,$$

subject to

$$\begin{aligned} &\sum_{j=1}^{n} \lambda_j x_{ij} \leq \theta x_{io} && i = 1,2,\ldots,m; \\ &\sum_{j=1}^{n} \lambda_j y_{rj} \geq \phi y_{ro} && r = 1,2,\ldots,s; \\ &\lambda_j \geq 0 && j = 1,2,\ldots,n. \end{aligned} \tag{5.2}$$

The linear programme in (5.2) is usually called output-oriented constant returns-to-scale DEA, or O-CRS for short. It should be noted that either solving linear program, (5.1) or (5.2), must yield the same classification results. Only the direction of projection to the frontier makes them seem different. If we allow for scale difference by imposing a variable returns-to-scale assumption on (5.1), the linear programme then becomes:

$$\theta^* = \min \theta,$$

subject to

$$\begin{aligned}
&\sum_{j=1}^{n} \lambda_j x_{ij} \leq \theta x_{io} && i = 1, 2, \ldots, m; \\
&\sum_{j=1}^{n} \lambda_j y_{rj} \geq y_{ro} && r = 1, 2, \ldots, s; \\
&\sum_{j=1}^{n} \lambda_j = 1 && j = 1, 2, \ldots, n. \\
&\lambda_j \geq 0
\end{aligned} \tag{5.3}$$

In DEA terminology, this is called input-oriented variable returns-to-scale DEA, or I-VRS for short hereinafter. Similarly, if a variable returns-to-scale assumption is introduced into (5.2), we obtain output-oriented variable returns-to-scale DEA, or O-VRS in the following linear program:

$$\phi^* = max\phi$$

subject to

$$\begin{aligned}
&\sum_{j=1}^{n} \lambda_j x_{ij} \leq \theta x_{io} && i = 1, 2, \ldots, m; \\
&\sum_{j=1}^{n} \lambda_j y_{rj} \geq \phi y_{ro} && r = 1, 2, \ldots, s; \\
&\sum_{j=1}^{n} \lambda_j = 1 && j = 1, 2, \ldots, n. \\
&\lambda_j \geq 0
\end{aligned} \tag{5.4}$$

Note again that to distinguish whether a company is efficient, we may solve either input- or output-oriented DEA as long as the scale assumption is the same because they always give identical classification results. DEA with variable returns-to-scale assumption always identifies the same efficient companies as DEA with CRS does, perhaps plus some additional efficient companies. Nowadays, it is simple to solve any DEA model for any kind of practical problem size. There are several choices of software. In this research, we showed DEA results for the four above-mentioned DEA: I-CRS, O-CRS, I-VRS and O-VRS.

5.3.2 *Social Efficiency Assessment: Directional Output Distance Function*

In a particular circumstance when the production system produces not only desirable outputs but also release some 'undesirable' by-products, the above ordinary DEA may not be applicable due to its use of an optimisation mechanism in determining the most efficient company (Färe and Grosskopf 2003, 2004a; Seiford and Zhu 2002, 2005; Hailu and Veeman 2001; Hailu 2003). For example, the linear programme for O-CRS in (5.2) seeks to maximise (or increase as much as possible) all outputs, whether desirable or undesirable, simultaneously to check whether a company is on the efficient frontier. This defies common sense. A rational decision maker should wish to increase only desirable outputs while preventing undesirable outputs.

This is the case for our analysis in which the three railways and the two airlines always produce both desirable and undesirable outputs. The desirable may include number of passengers and the amount of freight transported, whereas the obvious undesirable by-product is CO_2 emission. Directly applying O-CRS in this case will yield nonsensical results such that an efficient company is the one which has the highest passenger patronage and emitted the most CO_2. We therefore need to take into account that an efficient company should be the one that maximises the desirable and minimises the undesirable simultaneously. The applicable analysis technique must be able to consider the joint production of both desirable and undesirable outputs.

In order to measure the social efficiency, we adopt the social efficiency benchmarking methodology proposed by Chung et al. (1997) and Färe and Grosskopf (2004b), the so-called directional output distance function. During the past few years, this method has been applied to assess the social efficiency of several transport systems, including Weber and Weber (2004: for the US trucking industry), Yu (2004: for Taiwanese airports), McMullen and Noh (2007: for US transit agencies), Pathomsiri (2007: for US airports) and Pathomsiri et al. (2008: for 56 US airports).

In the directional output distance function approach, the CO_2 emissions are treated as an undesirable output.[4] The inefficiency is computed as the minimum distance to the production possibility frontier (PPF) from the output combination in the direction in which the desirables are increasing and the undesirables are decreasing. This is due to the fact that production technology exhibits a trade-off between the amount of desirables and reduction of undesirables. That is, a company can increase the desirables and undesirables at the same time to a certain extent, while keeping the amount of inputs constant. This further implies that the output-oriented efficiency is not well defined, and thus calls for a directional distance function approach.

[4] Another alternative is to treat an economic bad as a technology shifter. See Atkinson and Dorfman (2005), for example.

We briefly review the methodology of Chung et al. (1997) and Färe and Grosskopf (2004b) here. Let x, y and b denote vectors of inputs, desirable outputs and undesirable outputs, respectively, in R_N^+, R_M^+, and R_J^+. Then the production possibility set is described as

$$P = \{(x, y, b) : x \text{ can produce } y \text{ and } b\}. \tag{5.5}$$

This set is assumed to be convex and compact. P also satisfies the following assumptions:

- $x = 0$ implies $y = 0$ and $b = 0$;
  the production of desirable outputs always accompanies the production of a positive amount of undesirable outputs as by-products, meaning that $b = 0$ implies $y = 0$.
- Inputs and desirable outputs are strongly disposable. So if $(x, y, b) \in P$, then for $y' \leq y$, $(x, y', b) \in P$ and for $x' \geq x$, $(x', y, b) \in P$.
- Undesirable outputs are weakly disposable; a reduction of undesirable outputs requires reallocation of inputs from the production of desirable outputs, which leads to proportional reduction in the production of desirable outputs holding the amount of inputs constant. So if $(x, y, b) \in P$, then $(x, \theta y, \theta b) \in P$, where $0 \leq \theta \leq 1$.

The social efficiency of each company is measured as the distance from each observation to the frontier in the direction specified by the directional vector $g = (-g_x, g_y, -g_b)$ for some $g_x, g_y, g_b \geq 0$. This is a direction in which inputs are decreasing, desirable outputs are increasing, and undesirable outputs are decreasing. The directional distance function, say $D(x, y, b)$, can be considered as a more general expression of the Shephard distance function with the above argument, and is defined as

$$D(x, y, b; -g_x, g_y, -g_b) = \max\left\{\beta : (x - \beta g_x, y + \beta g_y, b - \beta g_b) \in P\right\}. \tag{5.6}$$

Output-oriented efficiency based on the directional distance function is measured by setting g_x element in the directional vector as zero. When $g_x = 0$ and the other elements of the directional vector g_y and g_b are just y and b respectively, then inefficiency of the company is computed as the distance to the frontier in the direction of $g = (0, y, -b)$, divided by the length of the directional vector g itself.

In order to assess the output distance function of the kth company, we may adopt a non-parametric approach and use the DEA concept to estimate the value of β_k or social efficiency score. Specifically, we solve the following linear programme in (5.7) to obtain the optimal value of β_k:

$$\begin{aligned}
&\max \beta_k \\
&s.t. \\
&\sum_{i\in K} \lambda_i y_{im} \geq (1+\beta_k) y_{mk}, && m = 1,\cdots,M, \\
&\sum_{i\in K} \lambda_i b_{ij} = (1-\beta_k) b_{jk}, && j = 1,\cdots,j, \\
&\sum_{i\in K} \lambda_i x_{in} \leq x_{nk}, && n = 1,\cdots,N, \\
&\sum_{i\in K} \lambda_i \geq 0, && i = 1,\cdots,K.
\end{aligned} \tag{5.7}$$

The major advantage of the DEA approach is that it does not require any a priori assumptions about the functional form of production technology; instead, it non-parametrically approximates the technology by a piecewise-linear PPF.[5] The solution to the above problem takes the minimum value of zero when the company is efficient. A higher value of β_k indicates a lower level of efficiency. That is, β_k is regarded as an inefficiency score. To assess the social efficiency of the three railway companies and two airlines in Japan, we then must compute β_k by solving the above linear programme 45 times (5 companies × 9 years) for individual companies in each year. Since the directional output distance function is still not available as a software package, we coded it especially for this research.

5.3.3 *Social Efficiency Assessment: Partial Productivity Factor*

The above two approaches provide the overall measurement of productivity considering multiple inputs and outputs. On one hand, it is a comprehensive measure for overall assessment of the performance of the firm. On the other hand, its complexity sometimes is not easy for the management to understand. So, we may alternatively provide also the simple but easy-to-understand partial productivity factor in the form of ratios. For example, in addition to showing the overall efficiency score, we may also compute the ratio between certain output and input such as passengers per labour input, passengers per capital input and passengers per variable cost. For social efficiency measurement, it is possible to look at the ratio between the amounts of CO_2 emission per unit of output. However, we must keep in mind that the partial productivity factor can only measure efficiency of the company based on a single perspective. It may not help us to rank overall performance of firms unless a certain firm dominates other firms in all dimensions.

[5] It is not that we assume the actual production frontier is piecewise linear, but rather that DEA is best understood as an approximation of an unknown frontier. In fact, there is a trade-off between the use of DEA and the parametric method; the parametric method with a priori assumption of the functional form will result in a significant bias if the specified functional form is far from the true functional form, while DEA does not suffer from this problem.

5.3.4 Social Efficiency: Total Factor Productivity Method

As a non-parametric approach, the Index Number Method directly defines the productivity as output index over input index. The method is easy to use for single output and single input firms. However, airlines and railway companies utilise multiple inputs such as labour, capital and other resources to produce various services. Since middle of the last century, a number of productivity index numbers have been developed for multiple output and input firms. This study chooses the multilateral index number method proposed by Caves et al. (1982), as it is the dominant method with strong underlying economic theoretical support.

A typical multilateral TFP model, which involves all factors of outputs and inputs, is defined as:

$$\ln \mathrm{TFP}_k - \ln \mathrm{TFP}_j = (\ln Y_k - \ln Y_j) - (\ln X_k - \ln X_j) = \sum_i \frac{R_{ik} + \bar{R}_i}{2} \ln \frac{Y_{ik}}{\tilde{Y}_i} - \sum_i \frac{R_{ij} + \bar{R}_i}{2} \ln \frac{Y_{ij}}{\tilde{Y}_i} - \sum_i \frac{W_{ik} + \bar{W}_i}{2} \ln \frac{X_{ik}}{\tilde{X}_i} + \sum_i \frac{W_{ij} + \bar{W}_i}{2} \ln \frac{X_{ij}}{\tilde{X}_i}, \tag{5.8}$$

where TFP_k is the productivity of kth firm; Y_{ik} and X_{ik} represent the ith output and input of kth firm, respectively; R_{ik} and W_{ik} are the weights for the ith output and input of kth firm, respectively; a bar over weights presents sample arithmetic mean, while a tilde demonstrates geometric mean.

As implied from Eq. (5.8), the TFP index is formed by a series of binary comparisons between each observations and the sample mean. The productivity of a firm is defined as the ratio of aggregate outputs over aggregate inputs. In order to utilise the model, output and input quantities must be obtained, as well as the aggregation weight for each output and input. Ideally, revenue and cost elasticities should be used as weights for output and input, respectively. However, as those numbers are usually not obtainable for most industries, including airlines and railways, Caves et al. (1982) suggest using revenue and cost shares as approximations. This adjustment comes with further assumptions on CRS across all outputs and revenue maximisation (or cost minimisation) behaviour of sample firms. In practice, these assumptions are reasonable for relatively large firms under commercial management, which includes most of our sample companies in Japan. Further, the TFP model is capable of applying both time series and cross-sectional data, and it allows for comparing the productivity of the same firm in different periods, or different firms across time.

As for the social efficiency comparisons, we stressed the importance of including externality as a negative output. Although we are aware of the criticism that one should not to be tempted to use externality costs as an input for production, we are left with no choice but use this approach, given that we found none of the currently available DEA-based methodologies allow us to make proper comparison of social efficiency between airlines and railway companies. So, we decided to use CO_2 emissions as an input when computing social TFP.

5.4 Data Construction

Our data set includes three JR railway companies, JR Central, JR West and JR East[6], and initially three full service airlines, namely JAL, ANA and JAS. However, since JAL and JAS were merged to form the new JAL in 2003, we combined the JAL and JAS data for the entire sample period as if JAL+JAS were a single company since 1999. We collected the annual data for these five firms for the 1999–2007 period. Therefore, our sample size is 45 observations: five firms spanning 9 years (45 DMUs in DEA terminology). For each observation, we have one desirable output (passenger-kilometre, PKM), and one undesirable output (the amount of CO_2 emissions measured in metric tonnes) and four inputs (labour, capital, total travel time and other variable input cost deflated by a price index). The ways we constructed these outputs and inputs are described below.

5.4.1 Desirable Output: Equivalent Passenger-Kilometres

The airline companies (JAL+JAS and ANA) produce both passenger and freight services while the three JR companies (JR Central, JR West and JR East) produce passenger services only. Therefore, there is a need to aggregate the airlines' domestic passenger services output (PKMs) and air cargo tonne-kilometres (TKMs) into a single measure of domestic services output: 'PKM equivalent output'. Furthermore, the three JR companies produce significantly different proportions of PKMs in HSR (Shinkansen) markets and in conventional rail markets (see Fig. 5.1 for details). However, in our social efficiency study, we also measure passenger travel time as an input. Since passengers pay higher charges for HSR services in order to reduce their travel time (and railways incur higher unit costs in order to provide HSR services), there is no need to take into account the higher quality HSR PKM output as long as the passenger travel time costs are accounted for as an input. Therefore, in order to avoid double counting the HSR benefits, we generate each railway firm's output simply by adding PKM outputs of HSR and of conventional railways.

For aggregating airline passenger outputs (PKMs) and cargo TKMs, we use the average revenue per PKM and the average revenue per air cargo TKM as the weight of aggregation.[7] We believe that this is a better way of aggregating passenger and freight services than the so-called WLU (Work Load Unit) for which 10 PKMs are treated the same as 1 TKM of cargo. It is evident that our method of aggregating PKMs (PKMs) and freight TKM into an aggregate output (hence-forth referred to as 'equivalent PKMs') is better than WLU because the average revenue for 1 TKM of

---

[6] In 1987, the Japanese national railway was privatized and divided into six companies, each operating a nonoverlapping railway network in Japan. These six companies are, from north to south, JR Hokkaido, JR East, JR Central, JR West, JR Shikoku, and JR Kyushu. Among these six, JR Central, JR East, and JR West are operating on the main island of Japan, and are included in our data set.

[7] The data on PKMs and the cargo TKMs are obtained from annual financial reports for JAL, ANA, JR (Central, West, East) and the ICAO database for JAS. Figure 5.2 shows the equivalent PKM data for all observations.

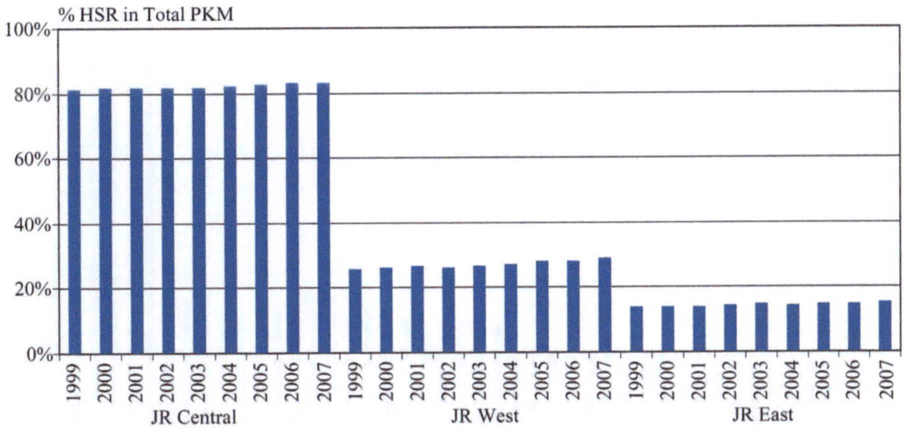

Fig. 5.1 Proportion of high-speed railways (HSR) in *Japan Railway (JR)* companies

air cargo output is only slightly more than the average revenue for one PKM of passenger output. Therefore, use of WLU as airline outputs would result in a seriously upward bias for domestic airlines' outputs vis-à-vis the three railway companies' outputs.

Figure 5.2 shows the size of equivalent PKM outputs for our sample observations. This graph shows that JR East is by far the largest firm in terms of PKMs produced, followed by JR West, JR Central, ANA and JAL+JAS. The three JR companies experienced slow output growth over our sample period. ANA's output stayed at essentially the same level without any growth while JAL+JAS began to reduce their combined output from 2002.

5.4.2 Input Data Construction

As for the three JR companies, non-negligible parts of labour and capital inputs are shared between HSR, conventional rail and commuting rail lines near major metropolitan regions like Tokyo–Yokohama and Kyoto–Osaka–Kobe areas. The three airlines refer to the domestic air transport services provided by JAL, JAS and ANA including their subsidiary airlines. To be consistent with our output aggregation between JAL and JAS, we add also all inputs for JAL and JAS for the entire 1999–2007 period, although the actual JAL+JAS merger occurred in 2003. Other new airlines are small enough to be excluded from our analysis as these new airlines cover less than 10% of the domestic air transport market. Airlines do not possess airport facilities as their assets, and instead airports are developed and operated by central or regional governments under an airport improvement special account in the Ministry of Land, Infrastructure, Transport and Tourism (MLIT). As airlines share the airport facilities, airport facilities are considered in our analysis as part of airline capital input. Also, international and domestic air transport services pro-

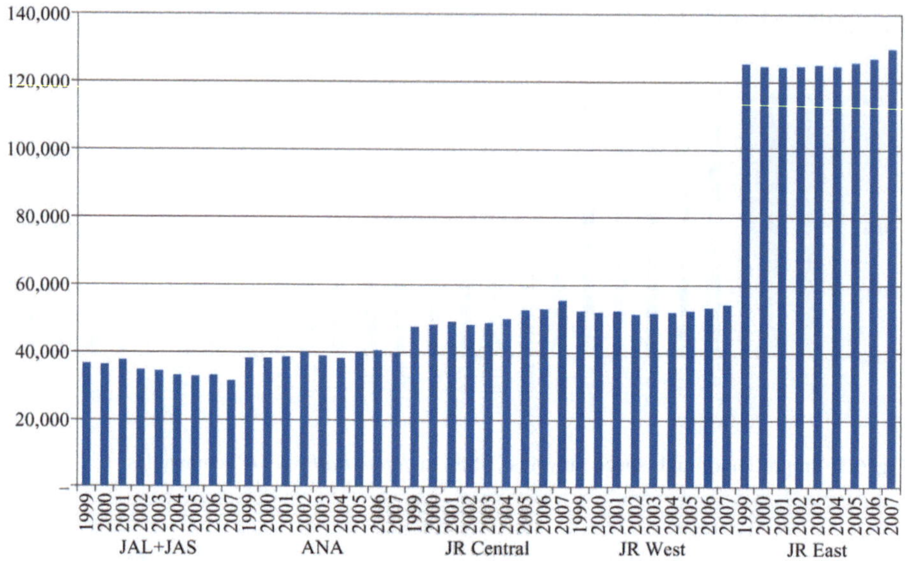

Fig. 5.2 Equivalent passenger-kilometres (million) of sample observations, 1999–2007

vided by these airlines share common factors of input. In our analysis, the input for domestic air transport is separated from the international air services. The airport inputs (labour input, other variable cost input, capital input) are allocated to airlines in proportion to their respective equivalent PKMs.

5.4.2.1 Labour Input

Labour input is represented by the number of full-time equivalent employees found in the annual financial reports. For the three airlines, this includes summary of the numbers of employees in the 'air transport business' section. Since the employees in this section serve both domestic and international operations, the number obtained is multiplied by the share of domestic passenger traffic. The numbers of employees for JR companies are found in the 'transport business' section of their annual financial reports. The data are summarised in Fig. 5.3. This figure shows that JR East and JR West reduced their labour input significantly during our study period while JR+JAS have reduced their combined labour input since 2003, the year of their merger. ANA increased its labour input in 2006 and 2007.

5.4.2.2 Annualised Capital Input

Capital input is represented by the summary of depreciation cost and the opportunity (interest) cost of the capital stock, based on the total book value of the fixed asset

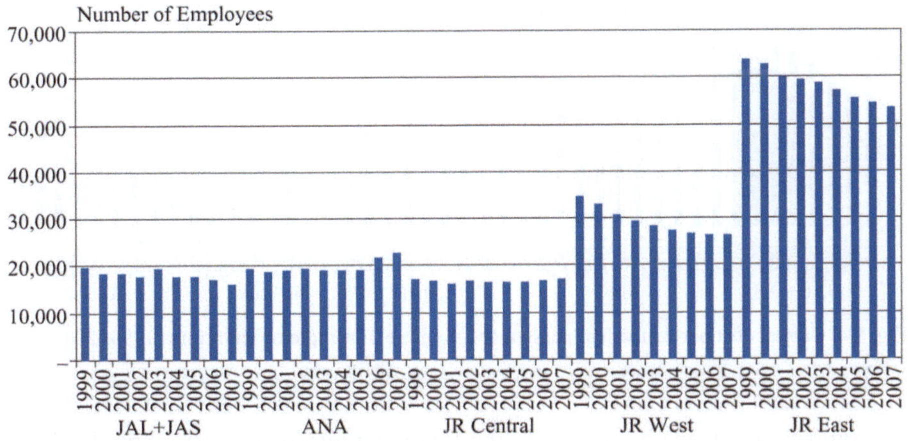

Fig. 5.3 Labour input of sample observations, 1999–2007

accounts in the balance sheet, assuming an annual rental rate of 4 %, and deflated by the domestic corporate goods price index. As for the airlines, the estimated value of the airport infrastructure capital input is included in the airline capital input by allocating the annualised total airport capital input costs to each airline based on its share of traffic. The airport infrastructure cost is calculated based on the investment data series (from 1970) available in the Special Account for Airport Investment, with the depreciation cost computed assuming economic life of 60 years on capital assets. Figure 5.4 shows the average capital input (price deflated) per equivalent PKM for all observations (DMUs).

The 'capital' is computed by total capital of the company. However, the capital may be used not only for transport production but also for other business of railway companies and airlines. For example, JR East is very active to use the space of stations for various kind of business, and they are bearing about one third of the total revenue of JR East. The problem is that it is unknown how much capital is used for transport production, and thus, we could not exclude the capital input costs used only for non-transport production. As a result, our empirical results may have been tilted somewhat against the firms with a higher proportion of non-transport business (such as JR East) than other firms.

5.4.2.3 Passengers' Total Travel Time

Airline passengers usually spend a considerable amount of time for access to and egress from the airports, at check-in counters and for security checks at the airport; these times must be added to the in-vehicle transit time in order to get passengers' total travel time. Similarly, rail passengers' access/egress/waiting time at rail stations must be added to their in-train transit times in order to obtain the passengers total time spent for travelling via railways. Tokyo University's ITPU (International

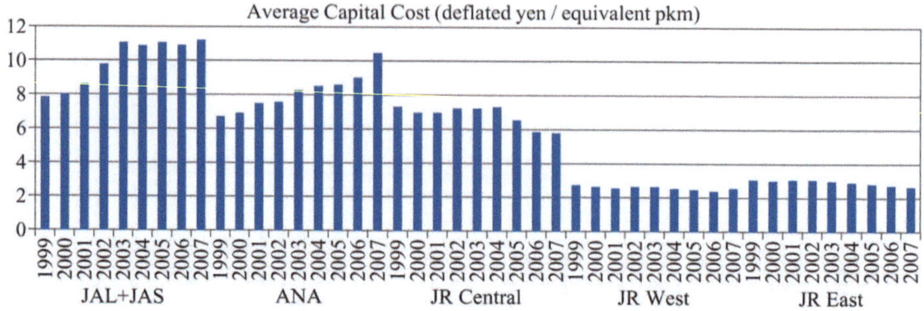

Fig. 5.4 Capital cost per equivalent passenger-kilometre of sample observations, 1999–2007

Transport Policy Unit) has provided the following information based on the 2005 Japan Transportation Survey data:

Air travellers' average time spent on access/egress/at airport: 131.52 min
Train travellers' average time spent on access/egress/at station: 48.96 min
Average between air travellers and train travellers: 82.56 min

It is expected that an average air traveller in Japan spends about 83 min more access/egress/airport time than an average train travellers' access/egress/station time. The total travel time is then computed as the summation of the products of total travel time including in-vehicle transit time and the number of passengers on all routes, shown in the KokuYuso Tokei Chosa (Annual Report of Statistics on Air Transport) published by MLIT. For JR companies, the total travel time is the summation of the division of PKM by the average speed, (which is 200 km/h for HSR and 60 km/h for local trains) and the above average time spent on access/egress and at station. The average total travel time per passenger of all DMUs is shown in Fig. 5.5.

5.4.2.4 Other Variable Inputs

Operating cost, excluding labour cost, depreciation and lease fee, is used to capture all the other types of inputs for operation. One major component is the inputs, specifically, energy and electricity. The cost is calculated based on financial reports, deflated by the domestic corporate good price index. Figure 5.6 shows the average price-deflated variable input per equivalent PKM.

5.4.3 *CO_2 Emission Amount as the Undesirable Output*

Based on the concept of input-output-table life cycle CO_2 assessment (IO-LCA), the amount of CO_2 emissions in tonnes is calculated as the summation of the estimated emissions amount from the activities in three different stages: operation, infrastruc-

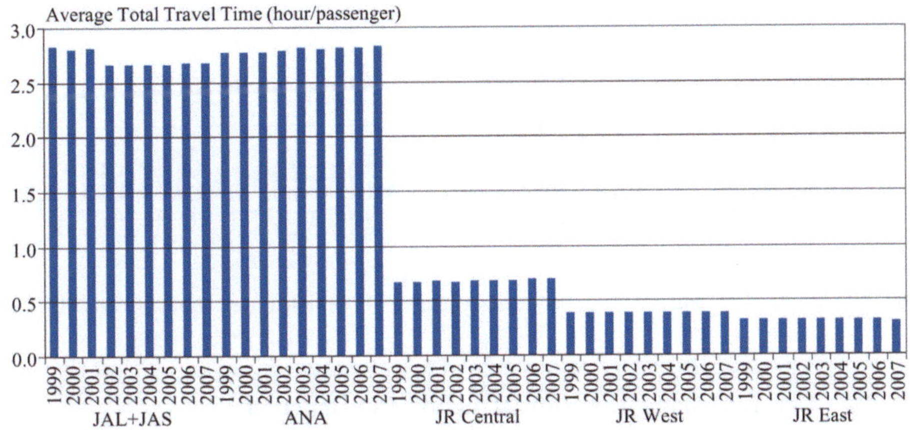

Fig. 5.5 Average total travel time per passenger of sample observations, 1999–2007

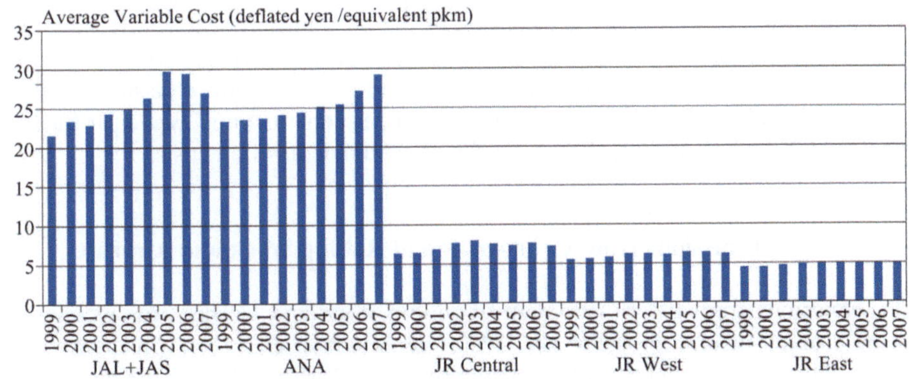

Fig. 5.6 Average variable input (price-deflated) per equivalent passenger-kilometre for sample observations, 1999–2007

ture construction and construction of rail rolling stock and aircraft. Figure 5.7 shows the average CO_2 emissions per equivalent PKM amount data for all observations.

5.4.3.1 CO_2 Emissions from Operation

CO_2 emissions from operation of airlines and railways are calculated as the product of PKM and the unit emission amount estimated by MLIT, available in 'Quantitative Look at Air Transport' (Suji De MiruKoKu)', etc. These values are for air transport 113 g CO_2/PKM and for railways 18 g CO_2/PKM.

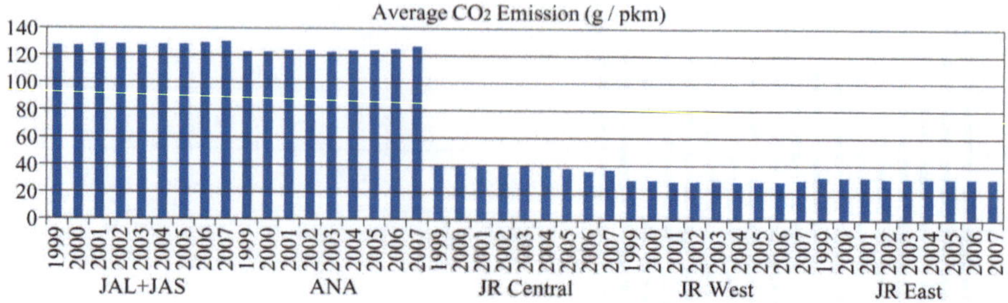

Fig. 5.7 Average CO_2 emissions per passenger-kilometre for sample observations, 1999–2007

5.4.3.2 CO_2 Emissions from Infrastructure Construction

CO_2 emissions from the construction of infrastructure are calculated as the product of unit emission per construction cost (tonnes CO_2/million Japanese yen, JPY) computed from the Input-Output table published by the Center for Global Environmental Research, and the depreciation cost (million JPY) of the corresponding fixed asset categories as well as leased assets. For the air transport sector, this value is 3.612 t CO_2/million JPY for construction of buildings and other facilities of airlines and investment for airport infrastructure. For railways, the unit emission is also 3.612 t CO_2/million JPY for the buildings and other constructions, while for trackage, it is 4.791 t CO_2/million JPY, excluding the land acquisition costs.

5.4.3.3 CO_2 Emissions from Construction of Rolling Stock and Aircraft

CO_2 emissions from the construction of rolling stock and aircraft are calculated as the product of unit emission per JPY of construction cost found in the Input-Output table by the Center for Global Environmental Research, and the depreciation cost of the corresponding fixed asset categories as well as leased assets available in annual financial reports of respective DMUs. The unit emission values are 1.549 t CO_2/million JPY for aircraft construction, and 4.746 t CO_2/million JPY for construction of train cars.

5.4.4 Summary of the Data

Table 5.1 summarises the descriptive statistics of our data set during the period 1999–2007. At a glance, we may see that the operational scales of the six companies are quite different in several aspects. For example, JR East employed 58,377 employees, much more labour input than JR West (29,240) and JR Central (16,694). The two airlines, the merged JAL+JAS and ANA, seem to have similar operational characteristics, but rather differ from the three JR companies.

Table 5.1 Descriptive statistics of the data set, 1999–2007

Company	Measure	Min	Max	Range	Sum	Median	Average	Std	COV (%)
JAL+JAS	Labour (persons)	16,181	19,768	3587	163,388	17,975	18,154	1098	6.05
	Variable cost (million JPY)	802,150	990,738	188,587	8,004,382	869,700	889,376	61,079	6.87
	Capital input (million JPY)	291,098	385,428	94,330	3,110,397	360,908	345,600	32,747	9.48
	Total user time (× 1000 h)	112,188	126,339	14,151	1,084,806	122,851	120,534	4464	3.70
	CO_2 (tonnes)	4,201,956	4,886,118	684,163	40,590,220	4,483,061	4,510,024	231,027	5.12
	Equivalent PKMs (× 1,000,000)	*32,084*	*37,900*	*5816*	*314,468*	*34,945*	*34,941*	*1984*	*5.68*
ANA	Labour (persons)	18,800	22,817	4017	178,396	19,121	19,822	1432	7.22
	Variable cost (million JPY)	904,452	1,176,816	272,364	8,983,129	975,305	998,126	92,607	9.28
	Capital input (million JPY)	263,085	422,434	159,349	2,932,320	326,229	325,813	49,737	15.27
	Total user time (× 1000 h)	124,850	131,691	6841	1,150,140	127,186	127,793	2370	1.85
	CO_2 (tonnes)	4,781,180	5,153,009	371,830	44,502,274	4,869,785	4,944,697	1481558	3.00
	Equivalent PKMs (× 1,000,000)	*38,742*	*40,940*	*2198*	*356,673*	*39,193*	*39,630*	*876*	*2.21*
JR Central	Labour (persons)	16,228	17,244	1016	150,248	16,696	16,694	355	2.13
	Variable cost (million JPY)	305,048	411,707	106,659	3,317,050	382,737	368,561	38,085	10.33
	Capital input (million JPY)	317,355	372,889	55,534	3,110,508	348,188	345,612	16,752	4.85
	Total user time (× 1000 h)	338,815	380,280	41,465	3,173,164	346,137	352,574	14,384	4.08
	CO_2 (tonnes)	1,883,661	2,048,525	164,864	17,525,571	1,911,445	1,947,286	64,119	3.29
	Equivalent PKMs (× 1,000,000)	*47,900*	*55,800*	*7900*	*456,600*	*49,600*	*50,733*	*2712*	*5.35*
JR West	Labour (persons)	26,390	34,532	8142	263,159	28,359	29,240	2999	10.28
	Variable cost (million JPY)	290,949	351,914	60,965	2,920,779	328,101	324,531	21,261	6.55
	Capital input (million JPY)	131,714	145,934	14,220	1,251,175	139,290	139,019	4587	3.30
	Total user time (× 1000 h)	691,398	711,837	20,439	6,314,788	703,507	701,643	6588	0.94
	CO_2 (tonnes)	1,472,760	1,587,140	114,380	13,578,920	1,497,586	1,508,769	34,549	2.29
	Equivalent PKMs (× 1,000,000)	*51,674*	*54,584*	*2910*	*475,231*	*52,587*	*52,803*	*855*	*1.62*
JR East	Labour (persons)	53,511	63,597	10,086	525,389	58,900	58,377	3,511	6.01
	Variable cost (million JPY)	569,649	658,036	88,388	5,543,512	622,557	615,946	31,592	5.13
	Capital input (million JPY)	351,168	389,991	38,823	3,371,658	384,481	374,629	14,865	3.97
	Total user time (× 1000 h)	1,870,632	1,942,432	71,800	16,993,186	1,881,151	1,888,132	22,741	1.20
	CO_2 (tonnes)	3,812,584	3,940,500	127,915	34,833,778	3,366,778	3,870,420	45,783	−1.18

Table 5.1 (continued)

Company	Measure	Min	Max	Range	Sum	Median	Average	Std	COV (%)
	Equivalent PKMs (× 1,000,000)	*124,916*	*130,557*	*5641*	*1,136,704*	*125,751*	*126,300*	*1794*	*1.42*
Overall	Labour (persons)	16,181	63,597	47,416	1,280,580	19,453	28,457	15,906	55.80
	Variable cost (million JPY)	290,949	1,176,816	885,866	28,768,852	622,557	639,308	277,964	43.48
	Capital input (million JPY)	131,714	422,434	290,720	13,776,059	345,976	306,135	90,157	29.45
	Total user time (×1000 h)	112,188	1,942,432	1,830,244	28,716,085	346,137	638,135	667,325	104.57
	CO_2 (tonnes)	1,472,760	5,153,009	3,680,250	151,030,762	3,866,778	3,356,239	1,400,622	41.73
	Equivalent PKMs (×1,000,000)	*32,084*	*130,557*	*98,473*	*2,739,676*	*49,600*	*60,882*	*33,804*	*55.52*

Std standard deviation, *JAL* Japan Airlines, *JAS* Japan Air System, *ANA* All Nippon Airways, *PKMs* passenger-kilometres, *JR* Japan Railway, *JPY* Japanese yen, *COV* covariance

The data description in this section indicates the following patterns quite clearly:

- Equivalent PKMs (our desirable output measure) show the following order of output size: JR East (largest), JR West, JR Central, ANA and JAL+JAS (smallest).
- Labour input size (number of employees): JR East (largest), JR West, ANA, JAL+JAS, JR Central (smallest).
- Annualised average capital input cost per PKM: JAL+JAS (highest), ANA, JR Central, JR East, JR West (lowest).
- Average non-labour variable input cost per PKM: JAL+JAS = (almost the same as ANA; highest), JR Central, JR West, JR East (lowest).
- Average CO_2 emissions (grams per PKM): JAL+JAS (almost the same as ANA: highest), JR Central, JR West, JR East (lowest).

The above patterns of the output and input data for the five companies already show that neither ANA nor JAL+JAS can be better in social efficiency than any of the three JR companies because JR companies dominate the two airlines in all of the above five output and inputs criteria. This means we can be sure that neither ANA nor JAL+JAS can have higher social efficiency than JR companies. We know this even before we actually compute any formal efficiency indices.

5.5 Results and Discussion

In this section, we report the empirically measured efficiency indicators using various methods discussed in Sect. 5.3.

5.5.1 Results Obtained from Alternative Forms and the Distance Function Approach

First of all, we estimated inefficiency scores of the five companies without undesirable outputs of CO_2 as if we were not concerned about it. All four traditional DEA models (I-CRS, O-CRS, I-VRS and I-CRS) were solved for each company. The resulting inefficiency scores are shown in the first five columns of Table 5.2. The value of 1.0 indicates that the company appeared on the efficient frontier. In the table, we highlighted all efficient companies in bold typeface. It can be observed that ANA appeared to be efficient more often than the others. The efficient score 1.0 appears least often for JAL+JAS. Furthermore, absolute differences between efficiency scores from 1.0 are much smaller for ANA, JR Central and JR West than for JAL+JAS or JR East. Roughly speaking, when CO_2 emissions were ignored, the ranking of efficiency might be in the ascending order of ANA, JR Central and then JR West. Meanwhile JR East and JAL+JAS were at the bottom.

Next, we estimated the social efficiency scores where CO_2 emissions were taken into consideration simultaneously. The directional output distance function was also

Table 5.2 Efficiency measures by data envelopment analyses (DEAs) and directional distance function with CO_2

DEA without CO_2					Directional distance function with CO_2
Company	I-CRS	O-CRS	I-VRS	O-VRS	O-CRS
JAL 1999	1.0000	1.0000	1.0000	1.0000	0.0000
JAL 2000	0.9814	1.0189	0.9985	1.0039	0.0176
JAL 2001	1.0000	1.0000	1.0000	1.0000	0.0000
JAL 2002	0.9471	1.0559	0.9861	1.0238	0.0385
JAL 2003	0.9442	1.0590	0.9832	1.0283	0.0367
JAL 2004	0.9402	1.0636	0.9892	1.0167	0.0388
JAL 2005	0.9388	1.0652	0.9921	1.0123	0.0397
JAL 2006	0.9394	1.0646	0.9895	1.0163	0.0393
JAL 2007	0.9379	1.0662	1.0000	1.0000	0.0421
ANA 1999	1.0000	1.0000	1.0000	1.0000	0.0000
ANA 2000	1.0000	1.0000	1.0000	1.0000	0.0000
ANA 2001	0.9980	1.0020	0.9981	1.0018	0.0018
ANA 2002	1.0000	1.0000	1.0000	1.0000	0.0000
ANA 2003	1.0000	1.0000	1.0000	1.0000	0.0000
ANA 2004	0.9976	1.0024	1.0000	1.0000	0.0018
ANA 2005	1.0000	1.0000	1.0000	1.0000	0.0000
ANA 2006	0.9995	1.0005	1.0000	1.0000	0.0000
ANA 2007	1.0000	1.0000	1.0000	1.0000	0.0000
JR Central 1999	1.0000	1.0000	1.0000	1.0000	0.0000
JR Central 2000	1.0000	1.0000	1.0000	1.0000	0.0000
JR Central 2001	1.0000	1.0000	1.0000	1.0000	0.0000
JR Central 2002	0.9681	1.0329	1.0000	1.0000	0.0299
JR Central 2003	0.9663	1.0349	0.9973	1.0047	0.0293
JR Central 2004	0.9780	1.0225	0.9949	1.0124	0.0183
JR Central 2005	0.9911	1.0090	1.0000	1.0000	0.0000
JR Central 2006	1.0000	1.0000	1.0000	1.0000	0.0000
JR Central 2007	1.0000	1.0000	1.0000	1.0000	0.0000
JR West 1999	1.0000	1.0000	1.0000	1.0000	0.0000
JR West 2000	1.0000	1.0000	1.0000	1.0000	0.0000
JR West 2001	1.0000	1.0000	1.0000	1.0000	0.0000
JR West 2002	0.9919	1.0081	1.0000	1.0000	0.0043
JR West 2003	0.9916	1.0084	0.9972	1.0056	0.0047
JR West 2004	1.0000	1.0000	1.0000	1.0000	0.0000
JR West 2005	0.9997	1.0003	1.0000	1.0000	0.0002
JR West 2006	1.0000	1.0000	1.0000	1.0000	0.0000
JR West 2007	1.0000	1.0000	1.0000	1.0000	0,0000
JR East 1999	1.0000	1.0000	1.0000	1.0000	0.0000
JR East 2000	1.0000	1.0000	1.0000	1.0000	0.0000
JR East 2001	0.9947	1.0053	0.9949	1.0050	0.0038
JR East 2002	0.9913	1.0087	0.9930	1.0066	0.0057
JR East 2003	0.9876	1.0126	0.9903	1.0091	0.0090
JR East 2004	0.9917	1.0084	0.9924	1.0072	0.0060
JR East 2005	0.9946	1.0054	0.9955	1.0044	0.0034
JR East 2006	1.0000	1.0000	1.0000	1.0000	0.0000
JR East 2007	1.0000	1.0000	1.0000	1.0000	0.0000

Table 5.2 (continued)

DEA without CO_2					Directional distance function with CO_2
Company	I-CRS	O-CRS	I-VRS	O-VRS	O-CRS
Min	0.9379	1.0000	0.9832	1.0000	0.0000
Max	1.0000	1.0662	1.0000	1.0283	0.0421
Range	0.0621	0.0662	0.0168	0.0283	0.0421
Sum	44.4708	45.5549	44.8923	45.1580	0.3709
Median	1.0000	1.0000	1.0000	1.0000	0.0000
Average	0.9882	1.0123	0.9976	1.0035	0.0082
Std	0.0203	0.0215	0.0043	0.0067	0.0141
COV (%)	2.05	2.13	0.43	0.66	171.48
Number of efficient companies	23	23	30	30	25

I-CRS input-oriented constant returns-to-scale DEA, *O-CRS* output-oriented constant returns-to-scale DEA, *I-VRS* input-oriented variable returns-to-scale DEA, *O-VRS* output-oriented variable returns-to-scale DEA, *JAL* Japan Airlines, *ANA* All Nippon Airways, *JR* Japan Railway, *Std* standard deviation, *COV* covariance

solved 45 times for all companies in each year. The resulting social inefficiency scores are shown in the last column of Table 5.2. Note that the efficient company must have a score of zero in this case. It seems that even with the inclusion of CO_2 in the analysis, there is very little change in the empirical results as compared to the case of using the traditional DEA models. The best is ANA, followed by JR West and then JR Central. However, it becomes clearer that JAL+JAS alone are at the bottom. While JR East is much better than JAL+JAS, it still has more cases of inefficiency than JR West or JR Central. Both DEA and directional output distance function reported largely consistent results (classifications). In general, the companies, including JAL+JAS, were far less than 10% away from the efficient frontier.

Table 5.3 explores the detailed results from the O-CRS model. In particular, the table reports the benchmark for inefficient companies. Recall that the scalar λ allows for the linear combination of peer companies to identify the ideal or benchmark (frontier) company. It shows that for JAL in the year 2000 to be efficient, the benchmark company should look like a combination of 0.219 of JAL 1999, 0.140 of JAL 2001 and 0.622 of ANA 2000. Similarly, the projected ANA 2001 onto the efficient frontier is a combination of 0.047 of ANA 1999, 0.791 ANA 2000 and 0.171 ANA 2003.

If we look at the results for railway companies, we can see that the efficient benchmark for JR Central 2002 should be 0.011 of JAL 1999 combined with 0.891 of JR Central 2007. The results suggest that JR Central 2002 should be more like JR Central 2007, rather than JAL 1999. When we explore the whole table, it is interesting to see that the results suggest stronger benchmarking within the same business, that is, JAL with ANA, whereas JR Central, JR West and JR East should be benchmarked against each other. Seemingly, the DEA mechanism did not really compare the performance between railway companies and airlines.

Table 5.3 Peer companies based on output-oriented constant returns-to-scale DEA results

Company	CRS efficiency	Benchmarks								
JAL + JAS 1999	*1.0000*	1.000	1.000	JAL+JAS 1999						
JAL + JAS 2000	1.0189	0.980	0.219	JAL+JAS 1999	0.140	JAL+JAS 2001	0.622	ANA 2000		
JAL + JAS 2001	*1.0000*	1.000	1.000	JAL+JAS 2001						
JAL + JAS 2002	1.0559	0.953	0.005	JAL+JAS 1999	0.086	JAL+JAS 2001	0.848	ANA 2000	0.014	JR Central 2007
JAL 2003	1.0590	0.940	0.341	ANA 1999	0.580	ANA 2003	0.019	JR Central 2007		
JAL 2004	1.0636	0.907	0.888	ANA 2003	0.018	JR Central 2007				
JAL 2005	1.0652	0.897	0.882	ANA 2003	0.015	JR Central 2007				
JAL 2006	1.0646	0.905	0.891	ANA 2003	0.014	JR Central 2007				
JAL 2007	1.0662	0.850	0.065	ANA 2003	0.772	ANA 2005	0.013	JR Central 2007		
ANA 1999	*1.0000*	1.000	1.000	ANA 1999						
ANA 2000	*1.0000*	1.000	1.000	ANA 2000						
ANA 2001	1.0020	1.008	0.047	ANA 1999	0.791	ANA 2000	0.171	ANA 2003		
ANA 2002	*1.0000*	1.000	1.000	ANA 2002						
ANA 2003	*1.0000*	1.000	1.000	ANA 2003						
ANA 2004	1.0024	0.986	0.713	ANA 2003	0.273	ANA 2005				
ANA 2005	*1.0000*	1.000	1.000	ANA 2005						
ANA 2006	1.0005	1.021	0.820	ANA 2005	0.201	ANA 2007				
ANA 2007	*1.0000*	1.000	1.000	ANA 2007						
JR Central 1999	*1.0000*	1.000	1.000	JR Central 1999						
JR Central 2000	*1.0000*	1.000	1.000	JR Central 2000						
JR Central 2001	*1.0000*	1.000	1.000	JR Central 2001						
JR Central 2002	1.0329	0.901	0.011	JAL+JAS 1999	0.891	JR Central 2007				
JR Central 2003	1.0349	0.918	0.017	ANA 2003	0.900	JR Central 2007				
JR Central 2004	1.0225	0.930	0.013	JAL+JAS 1999	0.917	JR Central 2007				
JR Central 2005	1.0090	0.957	0.957	JR Central 2007						
JR Central 2006	*1.0000*	1.000	1.000	JR Central 2006						
JR Central 2007	*1.0000*	1.000	1.000	JR Central 2007						
JR West 1999	*1.0000*	1.000	1.000	JR West 1999						

Table 5.3 (continued)

Company	CRS efficiency	Benchmarks								
JR West 2000	*1.0000*	1.000	1.000	JR West 2000						
JR West 2001	*1.0000*	1.000	1.000	JR West 2001						
JR West 2002	1.0081	0.980	0.502	JR West 2001	0.478	JR West 2006				
JR West 2003	1.0084	0.988	0.001	JR Central 2007	0.451	JR West 2001	0.535	JR West 2006		
JR West 2004	1.0000	1.000	1.000	JR West 2004						
JR West 2005	1.0003	0.987	0.002	JR Central 2007	0.130	JR West 2001	0.855	JR West 2006		
JR West 2006	*1.0000*	1.000	1.000	JR West 2006						
JR West 2007	*1.0000*	1.000	1.000	JR West 2007						
JR East 1999	*1.0000*	1.000	1.000	JR East 1999						
JR East 2000	*1.0000*	1.000	1.000	JR East 2000						
JR East 2001	1.0053	1.014	0.013	JR Central 2007	0.025	JR West 2000	0.647	JR East 1999	0.329	JR East 2006
JR East 2002	1.0087	1.088	0.009	JR Central 2007	0.149	JR West 2000	0.425	JR East 1999	0.505	JR East 2006
JR East 2003	1.0126	1.110	0.030	JR Central 2007	0.155	JR West 2000	0.326	JR East 1999	0.599	JR East 2006
JR East 2004	1.0084	1.044	0.019	JR Central 2007	0.068	JR West 2000	0.292	JR East 1999	0.664	JR East 2006
JR East 2005	1.0054	1.046	0.004	JR Central 2007	0.083	JR West 2000	0.078	JR East 1999	0.880	JR East 2006
JR East 2006	*1.0000*	1.000	1.000	JR East 2006						
JR East 2007	*1.0000*	1.000	1.000	JR East 2007						

Numbers in bold typeface indicate that the companies are relatively efficient
CRS constant returns to scale, *JAL* Japan Airlines, *JAS* Japan Air System, *JR* Japan Railway, *ANA* All Nippon Airways

Table 5.4 Classification of companies by three different clustering methods

Clustering method	Companies in each cluster		
	Cluster 1	Cluster 2	Cluster 3
Two-step cluster	JAL 1999-2007 ANA 1999-2007	JR East 1999-2007	JR West 1999-2007 JR Central 1999-2007
K-means cluster	The same	The same	The same
Hierarchical cluster	The same	The same	The same
Number of observations in the cluster	18	9	18

JAL Japan Airlines, *ANA* All Nippon Airways, *JR* Japan Railway

Furthermore, for our data set, DEA appears to determine basically the corner points of the efficient frontier within time-series data of the same firm. The efficient observation (DMU) may be just the best as compared to itself in other years. We tested this hypothesis by running three clustering analyses: two-step clustering, k-means clustering and hierarchical clustering. The clustering analysis tries to classify firms into a certain number of dissimilar groups, based on specified variables. Here we used the four inputs and two outputs as variables. The results are shown in Table 5.4, which confirms that our data set can be clustered into three distinct groups. The first group consists of all airlines for all years, that is, JAL 1999–2007 and ANA 1999-2007. This is not very surprising. In fact, we observed earlier from the descriptive statistics that these two airlines have similar operational characteristics in both inputs and outputs.

The second group consists of only JR East 1999–2007. This is why the DEA results suggest that the inefficient JR East should imitate the efficient JR East operated in other years. Lastly, JR West 1999–2007 and JR Central 1999–2007 belong to the same group. Figure 5.8 plotted the dendrogram based on hierarchical cluster analysis to show the level of similarity of all data points. It is evident that airlines were benchmarked among themselves exclusively. DEA tended to treat railway companies and airlines as two different things and never compared their performance.

The lesson learned here is important for assessing efficiency of too few companies with time-series data by the DEA approach. Even though our sample size passed the stringent DEA rule of thumb (Cooper et al. 2000, p. 103) of $n \geq \max\{m\ x\ s, 3(m+s)\}$, where m and s are the numbers of inputs and outputs, respectively, it does not necessarily guarantee the effectiveness of applying DEA for classification due to too many data points; in fact, they look very similar as if they were the same data points. As a result, such time-series data effectively reduce the sample size inadvertently and hence limit the discriminatory power of DEA.

The dendrogram based on hierarchical cluster analysis presented in Fig. 5.8 shows a clear picture where each of the three JR companies is compared to different years of itself, and each of ANA and JAL is compared to different years of itself and/or each other. This clearly shows difficulty in applying DEA-based methods for comparing efficiencies across different industries and between firms with somewhat different attributes even within a single industry (the railway industry in our case).

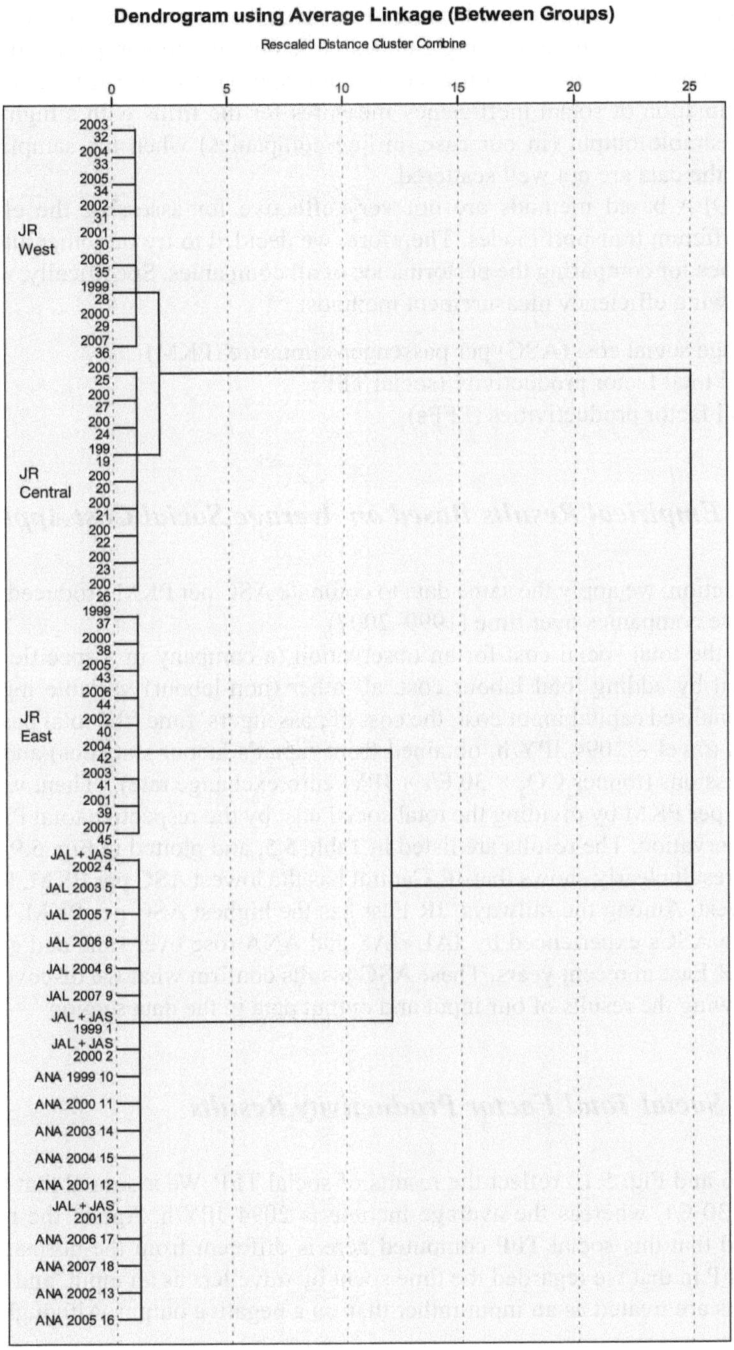

Fig. 5.8 Dendrogram based on hierarchical cluster analysis

In addition, the supplementary note to this chapter argues that, for the case of a single desirable output and a single undesirable output, the non-parametric methods including the directional distance function approach can lead to a significant underestimation of social inefficiency measures for the firms with a high level of the undesirable output (in our case, airline companies) when the sample size is small or the data are not well scattered.

The DEA-based methods are not very effective for assessing the efficiency across different transport modes. Therefore, we decided to try out other alternative approaches for comparing the performance of all companies. Specifically, we apply the following efficiency measurement methods:

- Average social cost (ASC) per passenger-kilometre (PKM)
- Social total factor productivity (social TFP)
- Partial factor productivities (PFPs)

5.5.2 *Empirical Results Based on Average Social Cost Approach*

In this section, we apply the same data to compute ASC per PKM produced by each of the five companies over time (1999–2007).

First, the total social cost for an observation (a company in a specific year) is computed by adding total labour cost, all other (non-labour) variable input cost, total annualised capital input cost, the cost of passengers' time (the total passengers spent on travel × 2094 JPY/h, obtained from Japan's labour statistics) and cost of CO_2 emissions (tonnes CO_2 × 30 €/t × JPY–euro exchange rate).[8] Then, we obtain the ASC per PKM by dividing the total social cost by the respective total PKMs for each observation. The results are listed in Table 5.5, and plotted in Fig. 5.9.

This result clearly shows that JR Central has the lowest ASC per PKM, followed by JR West. Among the railways, JR East has the highest ASC per PKM. Interestingly, the ASCs experienced by JAL+JAS and ANA rose over time and exceeded that of JR East in recent years. These ASC results confirm what we discovered just by reviewing the results of our input and output data in the data section.

5.5.3 *Social Total Factor Productivity Results*

Table 5.6 and Fig. 5.10 reflect the results of social TFP. We assumed that the CO_2 price is 30 €/t, whereas the average income is 2094 JPY/h. Again, the reader is reminded that this social TFP computed here is different from the normal private firms' TFP in that we regarded the time spent by travellers as an input, and the CO_2 emissions are treated as an input rather than as a negative output. Although we are

[8] The highest value for a CO2 certificate in the ETS market so far was about 30 €/t. Since this study is being done for examining future policy issues, we believe the use of 30 € is justified.

Table 5.5 Average social cost (ASC) per equivalent passenger-kilometre

	JAL+JAS	ANA	JR Central	JR West	JR East
1999	42.3	42.2	31.6	41.2	42.9
2000	44.4	42.7	31.3	41.1	42.8
2001	43.5	42.8	31.0	40.7	42.7
2002	46.1	42.6	31.9	40.9	42.5
2003	48.1	43.3	32.0	40.7	42.4
2004	49.7	44.6	31.7	40.4	42.4
2005	53.7	45.4	30.7	40.4	42.3
2006	53.4	48.6	30.7	40.4	42.2
2007	51.7	53.6	30.3	40.2	42.2
Average	*48.1*	*45.1*	*31.3*	*40.7*	*42.5*

JAL Japan Airlines, *JAS* Japan Air System, *ANA* All Nippon Airways, *JR* Japan Railway

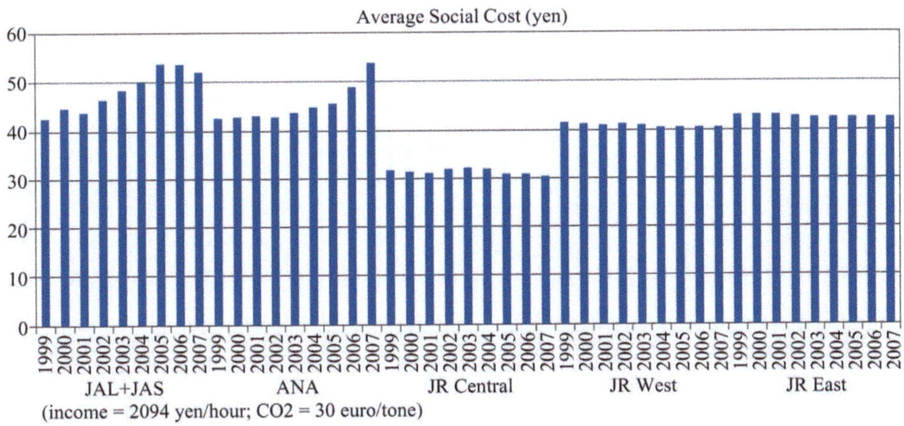

Fig. 5.9 Average social cost per equivalent passenger-kilometre

well aware of the criticism against treating CO_2 emissions as an input in the literature, this is one of the several imperfect measures we attempted to use to compare the social efficiencies of railway companies versus airlines as well as within each group of airlines and railway companies.

Again, our social TFP results show that JR Central is by far the most socially efficient firm among the five firms compared in this chapter. JR West and JR East have similar social TFP levels following JR Central. JAL+JAS and ANA have the lowest social TFP performances, and their social TFP level has been decreasing over our study period, 1999–2007.

Table 5.6 Total factor productivity (TFP) index

	JAL+JAS	ANA	JR Central	JR West	JR East
1999	1.000	0.999	1.303	1.006	0.971
2000	0.955	0.993	1.314	1.012	0.973
2001	0.960	0.974	1.312	1.017	0.972
2002	0.893	0.965	1.264	1.008	0.971
2003	0.852	0.942	1.256	1.013	0.972
2004	0.835	0.925	1.273	1.025	0.976
2005	0.780	0.918	1.326	1.030	0.981
2006	0.790	0.869	1.340	1.033	0.987
2007	0.824	0.806	1.371	1.041	0.990
Average	*0.877*	*0.932*	*1.307*	*1.020*	*0.977*

JAL Japan Airlines, *JAS* Japan Air System, *ANA* All Nippon Airways, *JR* Japan Railway

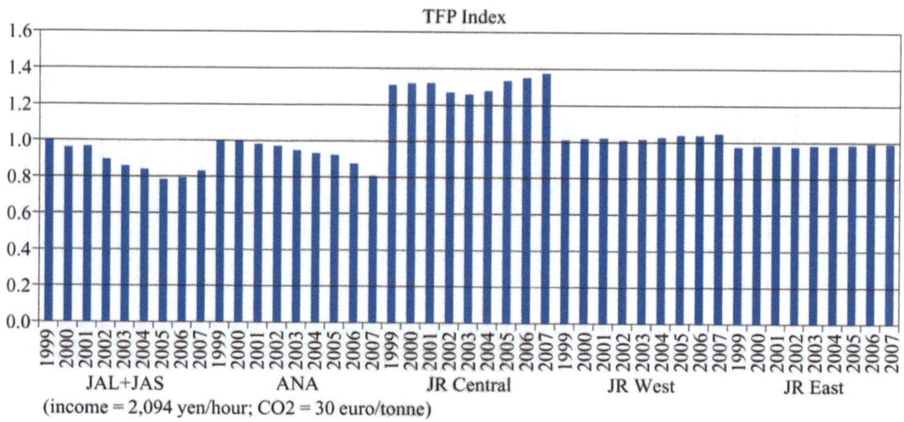

Fig. 5.10 Total factor productivity (TFP) index

5.5.4 *Partial Productivity Results*

None of the above methodologies are perfect for generating a combined efficiency ranking of airlines and railways. Therefore, it may be also useful to look at various partial productivity indicators.

Table 5.7 reports the five partial productivity results, namely, output/labour, output/variable cost, output/capital input, output per hour of travel time, and CO_2/output. As the name implies, partial productivity gives only partial measures ignoring all other aspects. A firm may look better in one ratio, but not necessarily in the others. For example, JR Central (JR East) is the best in output per employee (output per unit of non-labour variable input) but not other indicators, while ANA is best on output per passenger time unit, but not for other input productivities. Overall, for the partial productivity measures other than labour productivity and outputs

Table 5.7 Partial productivity ratios

Company	Output/labour	Output/other variable input	Output/capital input	Output/total user time PKMs per hour of travel time	CO_2/output
JAL 1999-2007	1.92	0.04	0.10	290	129.11
ANA 1999-2007	2.00	0.04	0.12	*310*	124.75
JR Central 1999-2007	*3.04*	0.14	0.15	144	38.44
JR West 1999-2007	1.82	0.16	*0.38*	75	*28.57*
JR East 1999-2007	2.17	*0.21*	0.34	67	30.65
Overall	*2.19*	*0.12*	*0.22*	*177*	*70.30*

JAL Japan Airlines, *ANA* All Nippon Airways, *JR* Japan Railway, *PKMs* passenger-kilometres

per hour of traveller's time, railway companies are more efficient than airlines: that is, non-labour variable input and capital input productivities, and CO_2 emission per PKM.

5.6 Summary of Findings and Further Research Needs

5.6.1 Summary of Findings

This chapter compares the social economic efficiency performance of two airlines (ANA and JAL+JAS) and three railways (JR East, JR Central and JR West) in Japan's domestic intercity passenger transport markets, using yearly panel data of these firms for 1999–2007. We incorporate the CO_2 emissions as an undesirable output and passenger travel time as an input, in addition to the conventional outputs and inputs normally considered in efficiency measurement and benchmarking. Furthermore, in order to put air transport on an equal footing with railways regarding social cost, we also measured and allocated all air transport infrastructure costs to airlines since railways pay for their own infrastructure.

In Sect. 5.3 and 5.5, we attempted to demonstrate methodologically and empirically why the DEA-based efficiency measurement methods (including the distance function approach widely being adopted for efficiency measurement and comparisons when one or more undesirable outputs are involved) are not adequate for comparing efficiencies between airlines and railways that need to be included in our sample of peer firms. The supplementary note to this chapter also points out that the directional distance function method is likely to give underestimation of the inefficiency for the firms with relatively large undesirable output (e.g. airlines in our case).

The other methods of measurement used in this chapter are:

- Comparison of the social average costs (SAC)
- Social TFP
- Six partial productivity measures

Our empirical results can be summarised as follows:

All DEA-based methods, including the distance function approach indicate that:

- ANA is socially more efficient than JAL+JAS.
- JR West is socially slightly more efficient than JR Central or JR East; JR Central is slightly more efficient than JR East.
- But these DEA-based methodologies do not allow us to compare between airline companies and railway companies; if we insist on using these DEA-based methods, then ANA is socially slightly more efficient than any of the JR companies.

Our careful examination of the extensive data and partial productivity measure indicates that for comparing social efficiency between airlines and railways, our approach of using SAC and social TFP methods is better than the DEA-based approach because DEA is not designed to compare and/or rank performances across heterogeneous companies such as firms in different industries.

The SAC estimate approach gives the following very different results from those of the DEA and its variant methods:

- JR Central has by far the lowest ASC per PKM, followed by JR West. Among the railways, JR East has the highest ASC per PKM.
- ASC experienced by JAL+JAS and ANA rose over time and exceeded those of railway companies. ANA had a lower ASC per PKM than JAL+ANA until 2006.

The social TFP approach gives the following results which are quite consistent with the above SAC approach:

- JR Central achieved by far the highest social TFP among the five firms compared in this study. JR West has higher social TFP than JR East.
- ANA's social TFP level was lower than the one of the JR companies, while JAL+JAS achieved the lowest social TFP levels among the five companies. Consistently with the SAC results, the social TFP levels of both ANA and JAL-JAS declined during our study period, 1999–2007.

The results of our five partial productivity measures (labour productivity, non-labour variable input productivity, capital input productivity, CO_2 emission per unit of output and output per hour of traveller's time) give the following results:

- JR Central enjoyed by far the highest labour productivity among the five companies compared. JR Central has by far the lowest capital input productivity of all three railway companies because Shinkansen services account for the majority of their outputs, and Shinkansen is more capital intensive than conventional railways. JR Central enjoys the highest output per hour of passenger travel time among the three JR companies (also because of the high Shinkansen share of their business).

- JR East has the highest productivity of non-labour variable inputs while JR West enjoyed the highest capital input productivity. Railway companies' capital input productivity is lower than those of airlines because railways are a far more capital intensive industry than airlines, even after we included the cost of airport infrastructure in airlines' capital input.
- Airlines enjoy higher output to 1 h of passenger travel time than railways because air travel is a faster travel mode than railway, and airline travellers in average travel longer distances than rail travellers.

The results described above are consistent with the pattern we observed in reviewing our input and output data in our data section.

All of the above empirical results together give a combination of evidence which led us to conclude that JR Central is the most socially efficient firm among the five peers compared. JR West is probably the second most socially efficient company. ANA is clearly more socially efficient than JAL+JAS. Our current evidence shows that JR East appears to be the least socially efficient among the three railways. However, this last observation on JR East needs further investigation because our study did not consider the effects of non-transport business that JR companies conduct on their costs and productivities because we were not able to separate JR companies inputs into transport business and non-transport business (it is our understanding that JR East does proportionally more non-transport commercial business than other JR companies.)

5.6.2 Further Research Needs

In order to incorporate the undesirable output (CO_2 emission) in efficiency measurement, we first employed the directional distance function approach in the DEA framework. We also computed various forms of DEA efficiency indices, while ignoring the undesirable output (CO_2 emissions). However, by examining our empirical results from all of these DEA-based methods and our data on outputs and inputs it became apparent that none of these DEA-based methods are capable of correctly comparing the efficiencies of airlines and railways, although these methods are capable of comparing efficiencies of fairly homogeneous firms within an industry. In fact, DEA-based methods are not designed to rank efficiencies of firms in different industries.

Since there is not a single perfect method for comparing efficiency between a firm in the airline industry and a firm in the railway industry, we chose to use several alternative methods as complementary tools for comparing social efficiency across the firms in our sample. The complementary methods we used include the SAC method, the social TFP method and various partial productivity measures.

Although we are confident that our results and rankings of social efficiency of these five firms (ANA, JAL+JAS, JR East, JR Central and JR West) are reliable because all of these methods produced largely consistent rankings, there is a need to improve methodologies for measuring and comparing social efficiencies of firms or

DMUs, especially when these observations are not from a homogeneous industry. Development of such a methodology would also help in measuring and comparing social efficiency between companies in Japan and those in other countries such as railway companies in Europe.

Another issue deserving future attention is to include the social efficiency of expressway modes (car drivers and buses) in the intercity market. Additional evaluation of expressway modes by social efficiency is necessary in Japan, since the former government drastically reduced tolls on expressways, and the current government is willing to mitigate.

The third, but not the least important issue is the extension of this research to include other types of externality costs such as air pollution, noise, congestion, risk of accidents, obstruction of views and landscapes.

Chapter 6
Policy Proposal for Intercity Passenger Transport

Shigeru Morichi and Surya Raj Acharya

The transport sector contributes about 20 % of total greenhouse gas emissions. Despite various policy initiatives, transport sector emissions have an increasing trend, primarily due to an increasing travel demand globally. However, the travel demand and emission pattern vary widely across countries in the world. Even among OECD countries, significant differences can be observed in the overall travel demand, choice of travel modes, and resulting loads of CO_2 emission. As shown in Fig. 2.1 (Chap. 2), distance travelled per capita per year is around 30,000 km in the USA, whereas the figure for France is around 15,000 km, Germany is around 13,000 km, and Japan is only around 10,000 km. Moreover, the share of railways in total passenger travel (passenger-km) is much higher in Japan, accounting for 30 % of total. In comparison, the railway share in EU countries is around 10 %, while in the USA railways account for a negligibly low share of passenger transport. In the USA, passenger transport is dominated by private automobiles and air transport. As a result of such different patterns, the intensity of transport sector CO_2 emissions across these countries varies significantly. Figure 6.1 shows CO_2 emission by sector. In the transport sector, the USA records the highest emissions, followed by France, the UK, and Japan. Germany's transport sector has the lowest emissions.

S. Morichi (✉)
Policy Research Center, National Graduate Institute for Policy Studies (GRIPS), Minato-ku, Japan
e-mail: smorichi.pl@grips.ac.jp

S. R. Acharya
Institution for Transport Policy Studies, Minato-ku, Japan
e-mail: suryabani@gmail.com

Y. Hayashi et al. (eds.), *Intercity Transport and Climate Change*,
Transportation Research, Economics and Policy 15, DOI 10.1007/978-3-319-06523-6_6

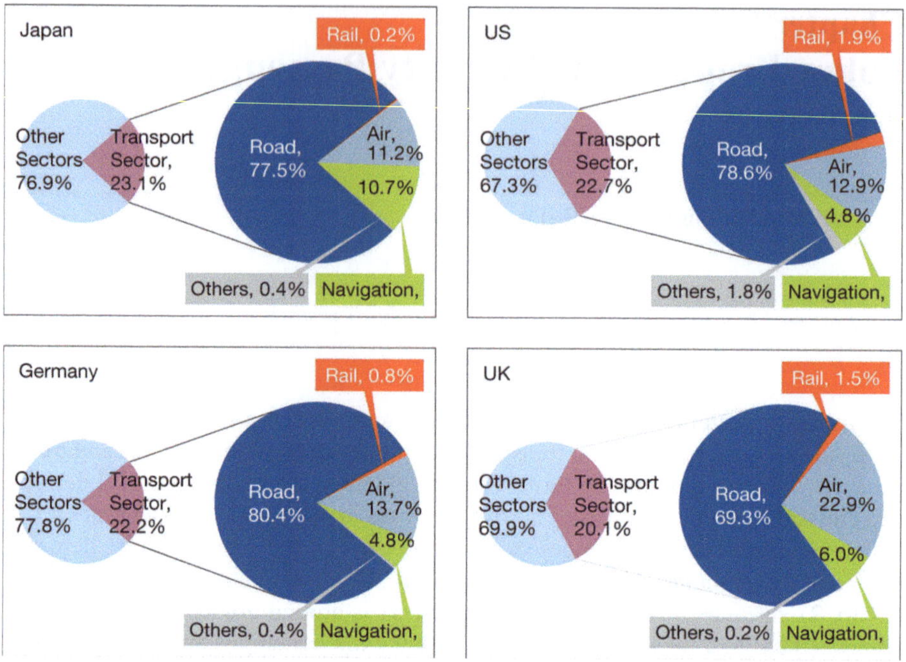

Fig. 6.1 CO_2 emission by sector. (Source: International Transport Forum (ITF): Key Transport and Greenhouse Gas Indicators by Country, available at http://www.internationaltransportforum.org/statistics/CO2/index.html)

6.1 Intercity Passenger Transport and Greenhouse Gas Emission

The transport sector contributes about 20 % of total greenhouse gas (GHG) emissions. Despite various policy initiatives, transport sector emissions have an increasing trend, primarily due to an increasing travel demand globally. However, the travel demand and emission pattern vary widely across countries in the world. Even among Organisation for Economic Cooperation and Development (OECD) countries, significant differences can be observed in the overall travel demand, choice of travel modes, and resulting loads of CO_2 emission. As shown in Fig. 2.1 (Chap. 2), distance travelled per capita per year is around 30,000 km in the USA, whereas the figure for France is around 15,000 km, Germany is around 13,000 km, and Japan is only around 10,000 km. Moreover, the share of railways in total passenger travel (passenger-km) is much higher in Japan, accounting for 30 % of total. In comparison, the railway share in EU countries is around 10 %, while in the USA railways account for a negligibly low share of passenger transport. In the USA, passenger transport is dominated by private automobiles and air transport. As a result of such different patterns, the intensity of transport sector CO_2 emissions across these countries varies significantly. Figure 6.1 shows CO_2 emission by sector. In the transport sector, the USA records the highest emissions, followed by France, the UK, and Japan. Germany's transport sector has the lowest emissions.

In the wake of increasing concern about global environmental impacts of GHG emissions, various policy efforts have been made to reduce transport sector emissions. However, such efforts are largely focused on the urban transport sector. This is because the emissions from urban transport are larger than that from intercity transport, with significant impacts on the living environment of urban areas. Therefore, it received policy responses from decision makers long before the issue of global environment came into picture. Reduction in demand for passenger and freight transport, control on the use of automobiles, promotion of public transport modes, and improvement in fuel efficiency of motorised vehicles are some of the common measures, which have been in practice for quite a few years in the urban transport sector. Recently, non-motorised modes, such as bicycles and walking have been promoted to minimise use of motorised vehicles for both private and public transport in order to maximize emission reduction. Details of urban transport and environmental issues and related policy measures have been presented in Nakamura, et al. (2004), which reports the output of international collaborative research jointly undertaken by the Institute for Transport Policy Studies (ITPS) and the World Conference on Transport Research Society (WCTRS).

In response to the increasing trend of GHG emissions from the transport sector, the intercity transport sector has recently been targeted to reduce CO_2 emissions. The policy debate for intercity transport is primarily focused on reducing emissions from each mode with various kinds of regulatory and fiscal measures to restrict transport demand or promote the use of efficient modes. Also promoted is the multimodality approach in freight transport to facilitate rationality in mode choice and improve efficiency of the overall system. Despite such initiatives, the overall policy approach in intercity transport is not as comprehensive as it is in the urban transport sector. For example, in intercity passenger transport, policy debate is limited to selected issues, such as an emission trading scheme for air transport, expressway toll-related issues in EU countries and recently in Japan, and the proposal of high-speed rail development by US President Obama. In response to the Kyoto Protocol, Japan has introduced a package of transport policies, which includes policy measures to control GHG emissions from the intercity passenger transport sector.

However, there is still much confusion to be clarified in the environmental aspects of intercity transport policies. For example, EU countries are giving a higher priority to high-speed rail development than to expressways, primarily on environmental grounds, with formulation of higher-level strategic plans (EC 2001), whereas in Japan, the expansion of the high-speed rail network is being discussed with completely different rationales. Proponents of high-speed rail highlight its role in stimulating economic growth in local areas. Sceptics point out that poor profitability and financial constraints do not justify high-speed rail investment. In such a situation, the environmental dimension takes a back seat, only being occasionally invoked by proponents to garner broader political support for the high-speed rail agenda.

Likewise, despite the strong theoretical appeal of expressway tolls to internalise external effects of road transport, EU countries face political opposition in adopting this measure. On the other hand, Japan is able to introduce and maintain a higher level of expressway tolls, which is arguably one of the factors contributing to the higher modal share of railways in Japan—a subject of envy for other counties. However, major political parties of Japan appear to be less aware of the effective role of expressway tolls in restraining the use of cars. Rather, with a simplistic view that the toll rate in Japan is higher in international comparisons, the Democratic Party of Japan (DPJ), as an opposition party, spearheaded a debate for reduction of toll rates including toll-free expressways. The debate, however, is completely ignoring many relevant issues like how the toll-reduction policy would impact the share of public transport, and how it would impact competition between the trucking industry and freight railways, both of which can have a big implication on CO_2 emissions. Similarly, the policy debate about merits or demerits of the currently imposed provisional fuel tax is basically focused on its possible impacts in terms of reducing government tax revenue or relieving the burden on car users. Environmental repercussions of abolishing the provisional fuel tax seem to be of secondary importance. On the other hand, the Japanese government adopted a policy of reducing 25 % of GHG emissions, which shows a clear inconsistency in policy goals, and adopted instruments for addressing the problem of GHG emissions from the intercity transport sector.

In the aftermath of the Fukushima nuclear accident, nuclear power plants have been closed down across the country. After the change of the ruling party, the future energy policy has remained vague, mainly because of newly emerging discussion about variety of perspectives on the share of energy sources including renewable one.

6.2 Overview of the Previous Chapters

After a broad overview of the world's GHG emission by region, Chap. 1 discusses common policy measures for GHG reduction in the transport sector, and summarises international efforts.

Chapter 2 examines the present situation of intercity transport in relation to the climate change agenda through international comparative review and analysis. The review reveals that potential impacts on global warming are explicitly considered in Europe during the planning process in the intercity transport sector. Particularly, in the UK, higher attention is given to achieving effective integration between regional planning and policies for multimodal transport. Moreover, EU directives require taking CO_2 emissions into full account during the process of Strategic Environmental Assessment (SEA). In contrast, the planning system in Japan is by mode, and, as a result, is highly fragmented. Development of an environmental assessment system, which takes into account the CO_2 emission, also lags far behind. There is also a need to initiate discussion on a planning system that can deal with planning for an intercity transport network covering multiple regions.

Chapter 3 briefly reviews the concept of externality as it relates to the transport industry, and discusses approaches for estimating the external cost of GHG emissions from intercity transport, including common mitigation measures. Section 3.2 analyses the external cost characteristics and makes actual estimations. This section shows quantitatively that externalities of railways are remarkably lower than those of road and air transport. The estimation also shows the significance of external costs of intercity transport in terms of climate change effects.

Section 3.3 estimated and compared lifecycle CO_2 emission for the high-speed rail (Shinkansen) and air modes considering travel between different city pairs. Compared to the air mode, CO_2 emissions of high-speed rail are higher during the construction phase, and the scale of passenger demand for the Shinkansen should be high enough to environmentally outperform the air mode. Most of the currently planned new Shinkansen lines in Japan can meet such conditions. A case study of Japan was also conducted in Sect. 3.4 to analyse the impact of a carbon tax on regional travel demand. The analysis shows that introduction of a carbon tax produces the expected modal shift to alternative transport modes in metropolitan areas. However, the analysis finds negative economic impacts for Hokkaido and Okinawa mainly due to a decrease in long-distance trips. This may require some provision for redistributing tax revenues as a compensatory measure.

Chapter 4 analyses a wider range of social and economic impacts caused by policy efforts aimed at reducing GHG emissions of intercity transport, and puts forward a discussion from multiple perspectives. The methodological approach utilises a spatial general equilibrium model to grasp the economic effects of high-speed rail development at the regional and national levels. As expected, the analysis shows substantial economic benefits from the first high-speed rail route along high-density corridors in countries like Taiwan and Korea. Even in the case of expansion of the high-speed rail network in Japan with new Shinkansen lines, the analysis indicated sufficient economic benefits.

Chapter 5 compares social efficiency of three Japan Railway (JR) companies and Japanese airlines at the microeconomic level on the basis of each company's inputs and outputs as well as CO_2 emissions by the data from 1999 to 2007. Using a number of approaches for efficiency measurement, the analysis consistently shows that efficiency of air transport is lower compared to intercity rail transport with a lower efficiency rate for Japan Airlines (JAL), while JR Central records the highest efficiency in this term.

6.3 Policy Proposals for Intercity Transport to Address Climate Change Issues

Based on the discussion and analysis in previous chapters, various policy proposals are presented in the following paragraphs mainly focusing on the strategic need of intercity transport sector in Japan.

Proposal 1: In-depth policy debate on climate-change related intercity transport issues should be promoted As discussed in Sect. 6.1, in the transport sector, climate change issues have been responded to mostly through urban transport and intercity freight transport policies, while intercity passenger transport policies have yet to take the issues into full account. At the international level, measures for reducing GHG from intercity transport are required for the rapidly growing air transport. Major policy instruments in intercity transport include tolling expressways, emission trading in air transport, and development of high-speed rail. In Japan, many passengers choose railways for intercity travel, and the rail share is much higher than in countries in Europe and North America. As a result, higher energy efficiency and lower CO_2 burden has been achieved in intercity passenger transport. In the freight sector, even though rail share is very low, the share of ship transport is high. As for CO_2 emissions, both passenger and freight transport thus show some encouraging trends. The emission efficiency of Japan's logistics and distribution industry has been significantly improved in order to meet the emission target of the Kyoto Protocol.

However, as mentioned in Sect. 6.1, when it comes to the climate change agenda, intercity passenger transport policies in Japan face significant confusion due to lack of consistency in policy approach. Chapter 2 further reinforced this argument by showing that in comparison with the EU, Japan lags far behind in reforming the process of spatial planning and transport planning for reducing GHG emissions.

In urban transport, various policy strategies evolved over time in the face of the difficult challenge of securing necessary infrastructure capacity to accommodate rapidly increasing motorised traffic during the wave of motorisation. Major strategies in response to such challenges included transport system management (TSM) in the 1970s, transport demand management (TDM) in the 1980s, and transit-oriented development (TOD) in the 1990s. The environmental objectives always remained as a core element, directly or indirectly, in all these policy strategies. On the other hand, policy measures in intercity transport are limited to a narrow scope of operational efficiency, for example, improvement in energy efficiency, technological improvement in railway vehicles and aircraft, and their control systems, and improvement in efficiency of logistics systems. Most of these measures are implemented by service operating companies on their own mainly in response to some regulation or market incentives. These observations lead to a logical conclusion that when compared with the urban transport sector, the policy approach in intercity passenger transport is largely firm-driven and fragmented, lacking much-needed comprehensiveness in scope and approach, which is indispensible to effectively address climate change issues. The case of Japan's domestic airlines service illustrates one of such phenomena more strikingly.

In fact, the domestic airline industry in Japan differs significantly in several areas when compared internationally. Firstly, the proportion of large aircraft in the total fleet is much higher compared with Europe and America (Fig. 6.2). Secondly, the frequency of flights, as a result of large capacity aircraft, is much lower. Thirdly, service withdrawal, particularly from routes serving local areas, continued as a result of deregulation, which is counterintuitive since deregulation is normally expected to increase new entries and network expansion. Finally, operational efficiency of Japanese airlines is significantly lower by international comparison. These factors can somehow be seen as interrelated. Figure 6.3 illustrates the relationship among

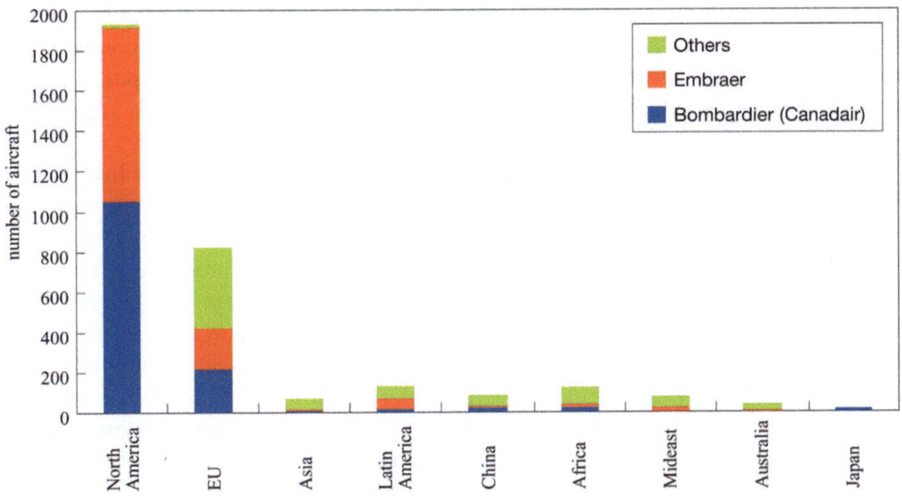

Fig. 6.2. Number of regional jet aircraft compared among countries (2007). (Source: Japan Aircraft Development Corporation)

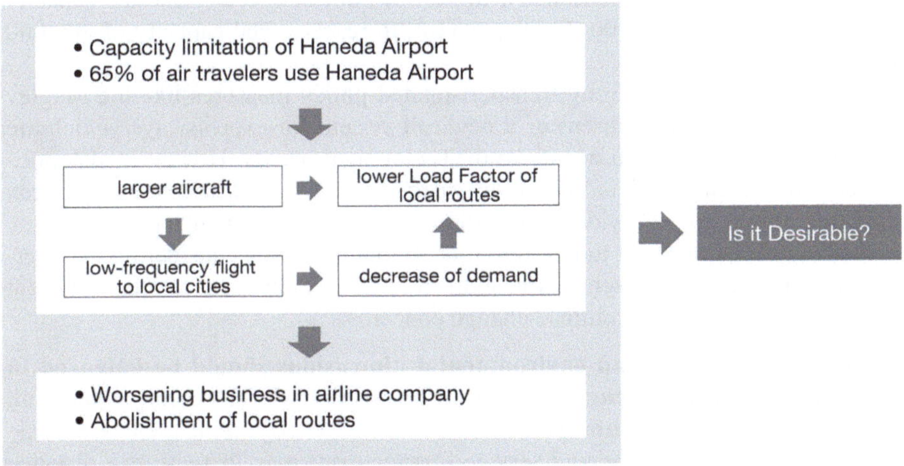

Fig. 6.3 Problems of aviation service and global climate

the above factors. There is severe capacity constraint in Tokyo International Airport (Haneda Airport). Limited availability of airport slots forces carriers to operate service with large aircraft in order to maximise passenger load per flight. As Haneda Airport serves 65 % of total domestic air passengers (Fig. 6.3), airlines have to maintain a disproportionally large fleet of large aircraft. Airlines then have to use large aircraft also for local routes with moderate demand, which results in a lower load factor along with lower frequency service. Lower frequency service is less attractive and reduces demand, which compels the airline to cut the frequency further. This sets a powerful vicious cycle in motion, and eventually compels the airline to withdraw service from such routes. The domestic air transport market, as a result, shrinks, which is very much against the worldwide trend of significant improvement in air transport services.

The USA enacted the Airline Deregulation Act in 1978, which resulted in significant improvement in the efficiency of the airline industry. The benefit was also passed on to users in terms of lower fares and improved services. The successful experience in the USA, in fact, triggered a wave of deregulation internationally. In the international aviation market, the open sky policy became a worldwide trend. The strong competition in the aviation market along with lowering of costs and entry of low-cost carriers (LCCs) changed the aviation market completely. The innovative approach of LCCs, which include use of smaller aircraft, and many cost minimisation measures, made it possible to offer higher frequency services with lower fares. Currently, LCCs account for almost half of air travellers in the EU and the USA.

In the context of such encouraging worldwide trends in the aviation industry, it was decided in 2007 to introduce open sky system to airports located out of the Tokyo metropolitan areas. In 2010, the capacity of Haneda Airport was increased by 100,000 flights per year, and slots for international flights also increased to serve 60,000 flights per year. Additionally, it has been decided to increase the capacity of Narita Airport by 100,000 flights per year, to reach a total capacity of 300,000 flights per year.

On the other hand, intercity transport-related policy measures like the Maglev Line (Linear Chuo-Shinkansen) or a new toll system for expressways, which are currently under debate, can transform the sector significantly. However, such policies are conceived individually without adequate analysis of interaction between different policies, system-wide optimisation or global environmental consideration.

It is therefore important to have an in-depth policy debate considering the overall system of intercity passenger transport in order to effectively address different strategic aspects including climate change concerns.

Proposal 2: Economic and environmental dimensions should be balanced in intercity transport policies Transport infrastructure development along with improvement in different transport services is a basic requirement for economic growth. Network expansion and service improvement may increase trip distance and total transport demand which may cause higher transport emission. In case of urban transport, long-term trends however suggest that the number of trips per person remain almost stable over time (Fig. 6.4). Urban transport basically serves daily activities, such as commuting, schooling, and shopping, and therefore, the number of trips and trip distance do not increase significantly due to economic growth and transport investment in the case of a developed country. Additionally, the time pressure on daily life may also have an influence. Improvement in urban transport systems can therefore contribute to both economic growth and emission reduction through its effects on reducing traffic congestion or improving public transport efficiency. In contrast, intercity transport services play a critical role in driving the

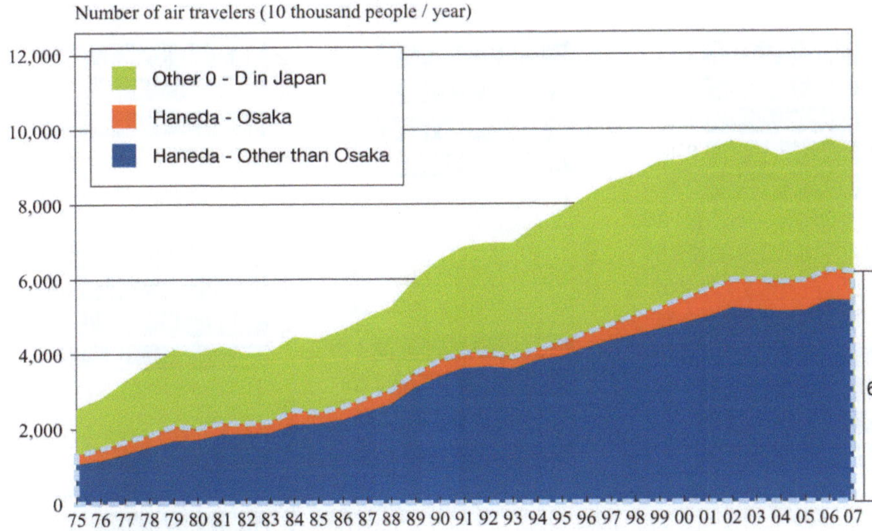

Fig. 6.4 Trend of domestic airlines passenger volume. (Source: MLIT, 2008)

process of economic growth. Intercity transport development and economic growth, in combination, generate huge transport demand—a dynamic process much stronger than in urban transport. Moreover, even if high-speed railway development is promoted to reduce transport emission through modal shift from automobiles and air, some increase in emissions is unavoidable due to the increase in transport demand as a result of improvements in transport services (Figs. 6.5 and 6.6).

Economic development is indeed a high-priority issue for the developing countries. Moreover, economic growth in developing countries greatly influences economic growth in the developed countries. In addition, the globalisation of the economy inevitably causes an increase in transport demand. The so-called North–South problem therefore should be addressed in the context of not only global environment but also balanced economic growth, reduction of economic disparities, and harmony with the environment.

Globalisation of the economy has, in fact, contributed positively to addressing the North–South problem in terms of reducing gaps between nations. However, globalisation is likely to increase regional disparity within a country. The National Spatial Planning Act (NSPA), which was enacted in 2009, has realised such mechanisms, and accordingly put much priority on revitalisation of local regions, identifying key strategic issues as: 1) strengthening international competitiveness of Japan's regional economic blocks to effectively face challenges from globalisation of the economy, 2) maintenance and improvement of social services to respond to the needs of the aging and declining population in Japan.

It has been realised that without changing the regional structure, these strategic issues cannot be effectively addressed. As for the first issue, political and

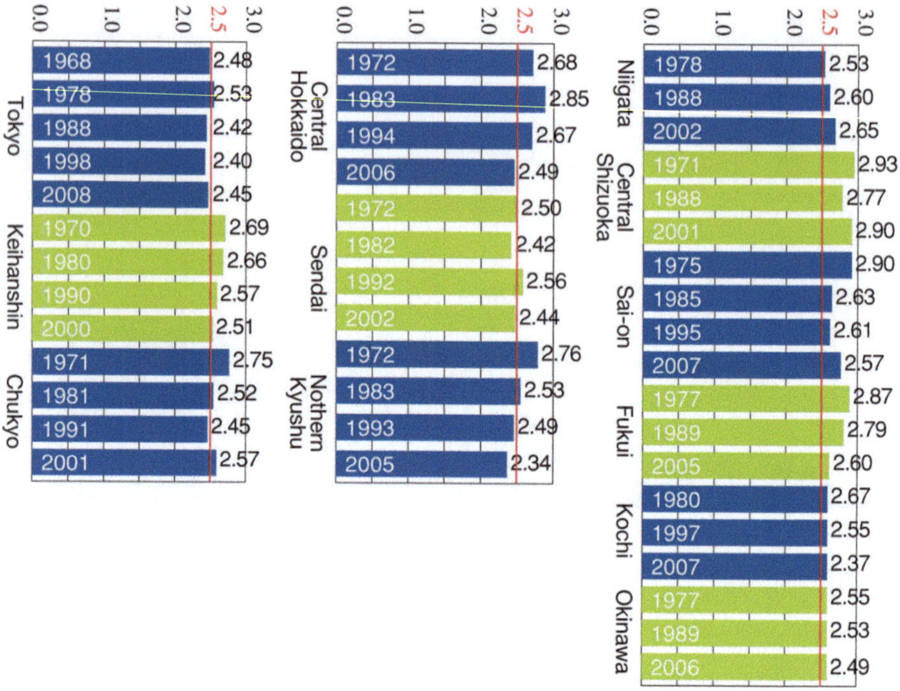

Fig. 6.5 Trend of trips per passenger in urban transport. (Source: Person Trip Survey)

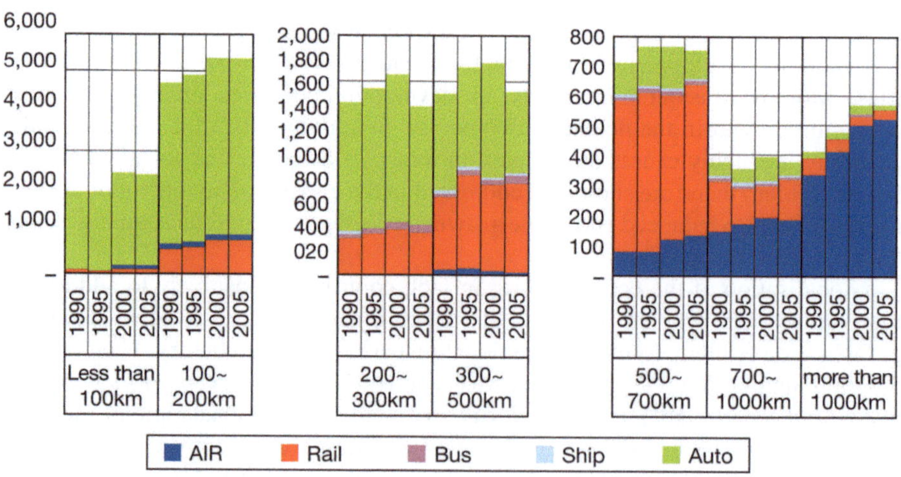

Fig. 6.6 Trend of passenger volume of each mode. (Source: Kansen Ryokaku Junryudo Chosa 2005 (MLITT))

administrative restructuring is necessary to reorganise current prefectural units into wider regional economic blocks to fully utilise the potential of each regional block for international competition. The reorganisation of prefectural units is currently under debate within the so-called Doushuusei, which is an administrative reform proposal, involving integration of prefectures into seven or nine states. As for the second issue, it is expected that social services through public or private entities (e.g., hospitals, schools, businesses) will not be able to maintain the existing service level due to declining population in local towns and villages. Thus, it is proposed to share, merge, and consolidate service facilities in a coordinated way so as to efficiently serve a larger population with a higher level of services. This proposal is consistent with administrative restructuring, which seeks merging of municipalities. Improvements of transport services are necessary in order to facilitate reorganisation of both service facilities and administrative units. Intercity transport policies, in particular, greatly influence the progress of restructuring through their multiplier effects on mobility patterns and spatial development. Since the residential population is decreasing in local areas, it is important for economic vitality to increase the non-resident population, such as business travellers or tourists. Thus, the improvement of intercity transport services is one of the important agenda items for every region from both economic and quality-of-life (social) perspectives.

The deregulation wave, which came as a strategy for broader economic reform, has significant influence on both transport and environmental policies. Various new businesses have been developed due to deregulation. Increases in LCCs and expressway bus services are typical examples. Although opening up new local air routes or increasing service frequency can drastically improve the level of transport service, they sometimes lead to an increase in environmental burden. Generally, the increasing regional disparity causes a decline of population and economic stagnation in local areas, and as a result the quality of public transport service is continually degraded.

Deregulation is also important for economic revitalisation in local areas, but in some cases local economies suffer from deregulation. Especially, the deregulation in transport may cause a decline in quality of public transport or withdrawal of service from unprofitable routes. Usually, the operation of unprofitable routes is supported through cross-subsidy under the franchise regulation, which is no longer possible under declining demand and deregulation. Direct public subsidy is also difficult due to increasing budgetary constraints. That is why EU countries, along with deregulation, usually adopt a policy measure called public service obligation (PSO), which ensures maintenance of basic services in local areas. As for the air transport business in the USA, a similar provision called Essential Air Service (EAS) has been made. These policies systematically address the issues of maintaining basic services, controlling monopolistic practices, ensuring efficient subsidy through bidding, and allocating financial burden fairly between the local and national governments. In Japan, local governments have made provisions to secure minimum load factor for airlines. In case of local rail and bus transport, provisions have been made to involve local stakeholders for public consultation, and receive assistance from the national government.

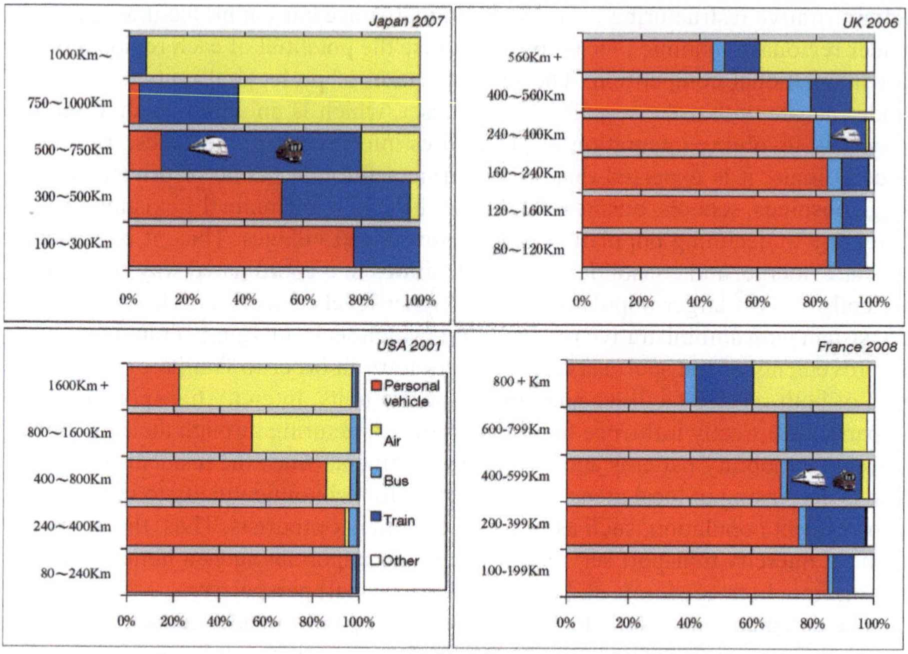

Fig. 6.7 Modal share comparison by trip distance. (Source: MLIT (2009), Department for Transport (2006), BTS (2007), Commissariat general au développement durable (2010)

Policy measures are aiming at improving the intercity transport system, stimulating economic growth in local areas, and reducing regional disparity, but at the same time may increase transport emissions. Keeping this fact in view, it is therefore important to seek optimal decision making, which can strike a judicious balance among economic, social, and environmental aspects.

Proposal 3: High-speed rail services should be improved in a sustained way Intercity transport systems need to serve trips of greatly varied travel distances, which typically range from 100 km to over 1500 km depending upon the country's size and geography. Conceptually, each intercity transport mode can efficiently serve a particular market niche, which usually depends on travel distance, demand density, and trip purpose. However, in practice, countries show very diverse patterns in terms of the share of different modes. As shown in Fig. 6.7, in the USA, the share of railways in intercity passenger transport is extremely low, while private cars account for over 80 % even for the 400–800-km segment of travel distance. In the UK, for 400–500-km travel distance, the modal share of cars is about 70 %, but that of railways is just 10 %. France records a higher share of railway than the USA and the UK for all segments of travel distance, yet car is dominant. However, the pattern in Japan is drastically different, as the railway share is relatively much higher. For a distance range of 500–700 km, railway takes a 70 % share, while air and private cars take 20 % and 10 % shares, respectively.

Table 6.1 Results of estimation for logit model

Year	Parameters		Time/Fare
	Time	Fare	
1965 data	−0.0042	−0.0007	6.00
1975 data	−0.0077	−0.0006	12.83
1985 data	−0.0082	−0.0004	20.50
1998 data	−0.0077	−0.0003	25.67
Pooling data	−0.0056	−0.0004	14.00

In Japan, as already described in Sect. 2.3.1.3 (Chap. 2), the railway share is largely dependent on the availability (or lack of it) of high-speed railway service and the nature of trip purposes (see Figs. 2.23 and 2.24).

Another notable phenomenon is the tendency of a shrinking mode share of railways over time (Fig. 2.22). Growing competition in intercity passenger transport due to deregulation is one of the causes, but the key factor is the rising value of time, and as a result, increasing preference for the high-speed mode, in which the air mode outperforms railways. Table 6.1 shows results of estimation for a logit model using Japanese data of different years, in which dependent variables are mode shares of railway and air, and the independent variables are trip time and fare. The result shows that the value of time increases significantly over time. When a high-speed rail line is opened, a substantial degree of modal shift from air to rail can be observed. However, if the high-speed rail service is not continuously improved, the railway may gradually lose market share.

$$Share_{railway} = \frac{e^{U_{railway}}}{e^{U_{railway}} + e^{U_{air}}}$$

$$U_{railwaay} = \alpha.Time + \beta.Fare + \varepsilon$$

From the emission reduction aspect, the policy for higher usage of railway transport is justified mainly on the basis of a lower amount of emissions per passenger-km for railway as compared with other modes. Figure 6.8 shows that railways in Japan are the most efficient in terms of emission per passenger-km, but the US passenger rail is inferior to buses and van pools. Moreover, as pointed out in Chap. 3, if life-cycle emission, which also includes GHGs generated from construction work, is taken into consideration, the effectiveness of modal shift policies for emission reduction depends on whether or not the high mode share of railway can be sustainably maintained.

In fact, the relative competitiveness of high-speed rail and conventional rail against air or car mode varies considerably. Moreover, the competitive strength of even high-speed rail is likely to diminish gradually, since the dynamics of intermodal competition usually work to the advantage of the air or auto modes. Thus, in order to sustain competitiveness of railways against other intercity transport modes, first, development of a high-speed rail network is indispensible. Then, it is important to maintain superiority of railways in the intercity passenger market by continually improving service quality through innovative measures, such as speeding up of

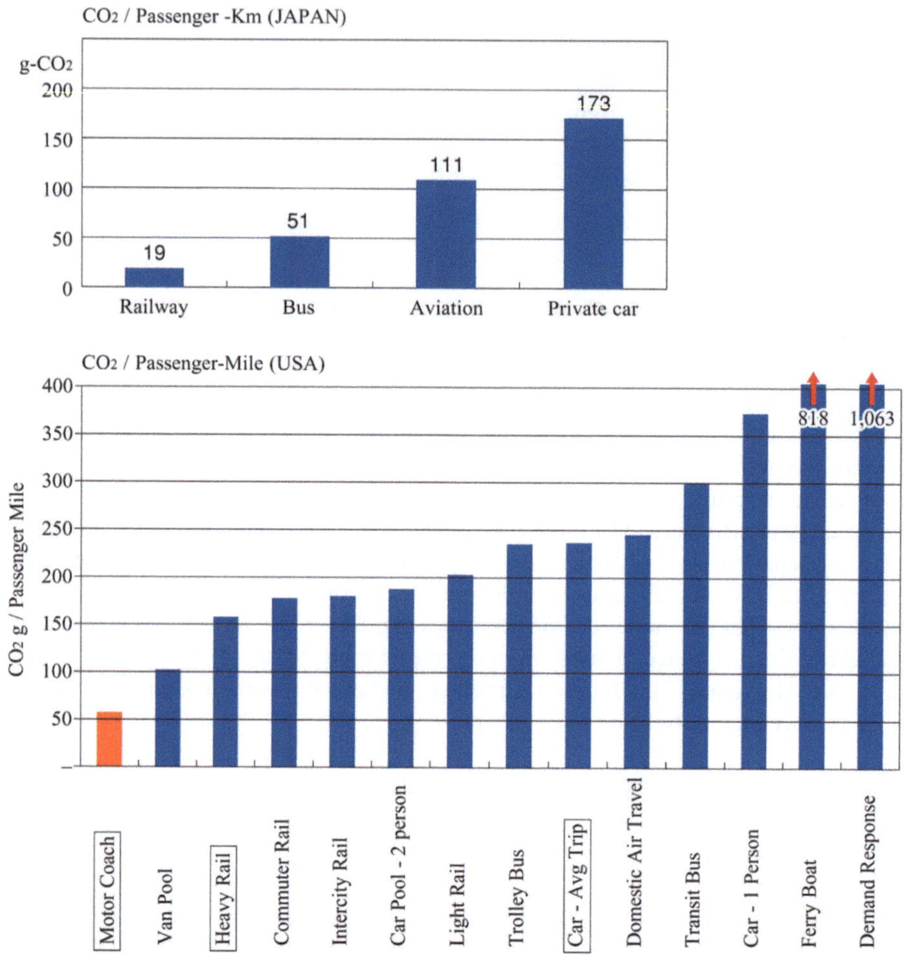

Fig. 6.8 Difference in the CO_2 load of transport modes (Source: American Bus Association (2007). Comparison of Energy Use & CO_2 Emissions From Different Transportation Modes, available at http://www.buses.org/files/ComparativeEnergy.pdf

services, provision of convenient transfer facilities, introducing comfortable train cars, and providing convenient facilities at the stations.

As mentioned above under Proposal 2, it is increasingly difficult to maintain and continue conventional railway services in areas with declining populations. Unlike in Europe and the USA, where substantial rail freight revenue makes it possible to maintain the network, Japan's conventional railway is facing a range of problems in maintaining the current level of service. First, the declining population in local areas is causing a decrease in railway demand. Second, expansion of the expressway system is likely to further reduce the demand. Third, the discounted tolls or toll-free expressways certainly make automobiles more attractive. Fourth, expansion of expressway bus services and introduction of low-cost tour buses as a result of deregulation are also shifting demand. All these factors individually or in combination will adversely impact the usage, management, and operation of railways significantly.

A high share of the railway mode in intercity passenger transport is one of the important factors behind the relatively superior performance of Japan in transport emission. The issue here for Japan is how to realise the continuation and improvement of this performance, while the experience of Japan in terms of successful or failed policy measures may be of use for the USA, the EU, and developing countries.

Proposal 4: Profitability of transport enterprises should be ensured Discussion on the profitability of transport enterprises has intensified in relation with policy debate on major agenda items like privatisation of national rail, private finance initiative (PFI), public–private partnership (PPP), and so forth. This has created a great deal of awareness, particularly realisation that profitability is a sign of increased efficiency. In fact, most of the initiatives for environmental protection at the firm level are taken by highly profitable enterprises. Transport enterprises with weak profitability are more conservative in spending or investing on environmental measures, as this may pose substantial risk, which may further deteriorate their competitiveness, and can even lead to bankruptcies. On the other hand, highly profitable companies can further improve their profitability as the environmental measures can generate new business opportunities or strengthen their market competitiveness.

Since the profitable enterprises can allocate substantial resources for capital investment and research and development, they can further strengthen environmental protection activities.

As for Japanese railway companies, the highly profitable enterprises, such as East Japan Railway Company (JR East), Central Japan Railway Company (JR Central), Tokyu Corporation, and Odakyu Electric Railway Co. Ltd., are proactively undertaking a range of initiatives directly or indirectly related to environmental protection, such as introduction of energy-efficient rolling stock and control systems, and other research activities.

In most countries, urban and intercity rail systems receive public subsidies, and operate with lower profitability. Under such conditions, it is very difficult for the transport enterprises themselves to undertake reform measures and improve efficiency. Under the prevailing notion that efficiency can be improved through market competition and that low profitability makes it impossible to take drastic reform measures, at least the Japanese railway enterprises survived the competition and improved efficiency and profit level—a phenomenon well observed after the quarter century of privatisation of Japan National Railway.

Proposal 5: Planning and project systems should be properly set up in an integrated way Objectives of planning systems for transport and other social overhead capital (infrastructure) include making plans across time to meet future needs, assessing environmental conditions of various activities and carrying them out, and ensuring consistency between different components of the infrastructure system. Likewise, urban planning systems first articulate a vision of a desirable city form and structure, and then accordingly guide land use development to ensure consistency over the time. Since the planning involves the interests of many stakeholders, the planning system should systematise a procedure for consensus building. On the other hand, a project system should involve the process of implementing infrastructure, and urban and regional development projects, including decision-making rules for burden sharing and regulation.

The planning system, in fact, involves wider areas, broader fields, and upper-level planning with a long-term perspective, which gradually moves down to tackle more specific elements and short-term planning following a step-by-step procedure.

Institutional restructuring has been called for, since the existing planning and project systems cannot address a range of emerging situations, such as environmental restrictions, consensus building among the public and stakeholders, a globally competitive market, financial constraints, role sharing between public and private entities, and decentralization. Institutional designs related to provisions like Public Involvement (PI) and PPP are the typical patterns of restructuring.

As described in Chap. 2, the review of planning systems such as regional planning and infrastructure development planning in the USA and Europe reveals that the provisions for addressing the global environmental concerns have been carefully built into the upper-level plan. They also have an institutional framework to make planning decisions in a step-by-step manner so as to avoid repeated cycles of negotiation to build consensus among the stakeholders, including residents. For instance, the needs assessment for transport facilities and division of roles and responsibilities are first decided in the upper-level plan, and specific details for each transport facility are then worked out in the lower-level plans following the provisions made in the upper-level plans. Through such frameworks, consistency can be maintained among the different plans in the lower-level planning system, and consensus among key stakeholders can be achieved without the costly process of repeated consultation and negotiation.

Recently, the planning systems in Japan are being extensively reviewed. However, provisions concerning the global environment that are under debate are inadequate, and, as mentioned before, there is much confusion to be clarified, particularly in various political arguments.

Proposal 6: Possibility to support intercity rail systems in developing countries should be considered As discussed under Proposal 3, the mode share structure of intercity passenger transport in Japan is significantly different from that in Europe and America. As for developing countries, intercity passenger transport is heavily dependent on railways and buses mainly because of the delay in motorisation. But, mode share structure is likely to change in the face of rapid growth in the vehicle ownership rate for motorbikes and cars, development of expressway networks, and the rising value of time. The intercity transport system in developing countries, in fact, is at a crossroad and has to make a choice of the path ahead, which may lead to a mode share pattern closer either to that in Japan or that in Europe and America (Fig. 6.9). For urban mass transit systems, suitability of alternative technologies such as bus rapid transit (BRT), light rail transit (LRT), elevated railways and subways, and appropriate timing of development have been discussed vigorously.

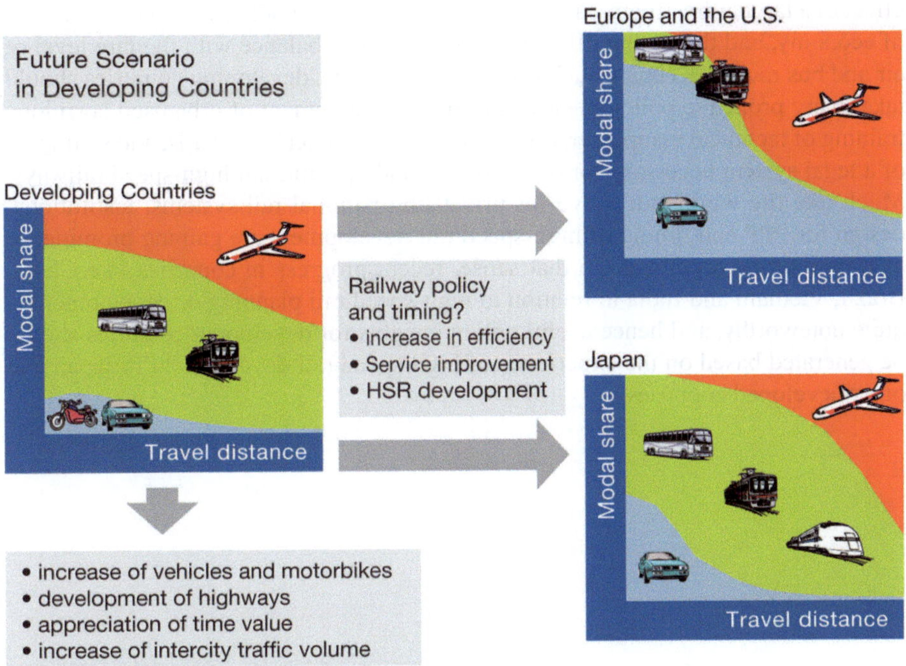

Fig. 6.9 Future scenario in developing countries. (Source: Acharya and Morichi 2013)

Especially for megacities in Asia, recently there is a growing realisation that it is difficult to efficiently manage an urban transport system without developing urban railways as the backbone of the system hierarchy, since the scale and growth rate of the urban population is much higher than in cities in Europe and America. As for intercity railway development, since the scale of investment is very large, decision making is more complicated than for urban railways. However, since intercity railways generally come under the jurisdiction of the national government, there are both advantages and disadvantages. An advantage is that the national government can mobilise better skilled human resources, while the disadvantage is lower management efficiency since resistance to reform at the level of the national government is much stronger as commonly observed in national rail systems.

The superiority of high-speed rail over air transport largely depends on both accessibility from the centre of a city and the concentration of population along the high-speed rail corridor. An airline service route serves only one pair of origin–destination, but in the case of railway service, a single route can serve multiple origin–destination pairs, and as a result, railway has the advantage of providing high-frequency services and better accessibility. In most Asian countries, population density and distribution patterns are more favourable for railways than in EU countries. It is therefore possible for Asian countries to adopt policies in order to move towards the Japanese pattern. However, it is critically important to consider the effect of a large infrastructure investment in a high-speed railway for the given scale of economy, and the timing of investment to achieve balance with the fare level of air and bus modes. Keeping such timing stage in view, developing countries should take many proactive policy measures such as development of urbanised corridors, training of technical manpower, establishment of technical standards, and setting up of a legal system necessary for development and operation of high-speed railways. Much like the worldwide wave for privatising national rail systems, institutional design for PPP and timing of high-speed rail development are gaining prominence as important policy issues. In that sense, recent progress in countries like China, Brazil, Vietnam and India in relation to high-speed rail planning or development is quite noteworthy, and hence useful policy insights for developing countries should be generated based on the experience of high-speed rail development in Japan and other developed countries.

References

Acharya, S. R., & Morichi, S. (2013). Introducing High Speed Rail (HSR) system in developing Asia: Issues and prospects. Proceedings of the Eastern Asia Society for Transportation Studies, 9.

Air Transport Research Society. (2002). Airport benchmarking report: Global standards for airport excellence 2002. Part I (Summary Report: 80 pages), Part II (Full Results and Analysis: 235 pages), Part III (Airport Profiles, Methodology and Data Description: 95 pages). http://www.atrsworld.org. Accessed 1 Feb 2011.

Atkinson, A. E., & Dorfman, J. H. (2005). Bayesian measurement of productivity and efficiency in the presence of undesirable outputs: Crediting electric utilities for reducing air pollution. *Journal of Econometrics, 126,* 445–468.

Baumol, W. J., & Oates, W. E. (1971a). *Economics, environmental policy and the quality of life.* New York University Press.

Baumol, W. J., & Oates, W. E. (1971b). The use of standards and prices for protection of the environment. *Swedish Journal of Economics, 73,* 42–54.

Baumol, W. J., & Oates, W. E. (1988). The theory of environmental policy.

Becker, G. S. (1965). A theory of the allocation of time. *The Economic Journal, 75*(299), 493–517.

Better place. (Online). http://japan.betterplace.com/. Accessed 25 Jan 2011.

Bosch&Partner, et al. (2010). Erarbeitung eines Konzepts zur "Integration einer Strategischen Umweltprüfung in die Bundesverkehrswegeplanung". Development of a Concept for "Integration of SEA into the Federal Transport Infra-structure Plannning. Project for the German Ministry of Transport. Hannover.

Bröcker, J. (1998). How would an EU-membership of the Visegrád- countries affect Europe's economic geography? *The Annals of Regional Science, 32*(1), 91–114.

Bröcker, J., Meyer, R., Schneekloth, N., Schürmann, C., Spiekermann, K., & Wegener, M. (2004). Modeling the Socio-economic and spatial impacts of EU Transport Policy, IASON, Deliverable 6.

BTS. (2007). US The 2001 national household travel survey, preliminary long distance file.

Cabinet Office. (2010). Economic and Social Research Institute. National accounts of Japan. (Online). http://www.esri.cao.go.jp/jp/sna/menu.html. Accessed 15 Nov 2010.

Cabinet Office. (n.d.a). National Accounts of Japan. http://www.esri.cao.go.jp/jp/sna/data/data_list/sokuhou/files/2014/qe142_2/gdemenuja.html. Accessed 16 June 2014.

Cabinet Office. (n.d.b). Economic and Social Research Institute. Annual report on prefectural accounts. (Online). http://www.esri.cao.go.jp/jp/sna/sonota/kenmin/kenmin_top.html. Accessed 29 Feb 2012.

Caves, D. W., Christensen, L. R., & Diewert, W. E. (1982). The economic theory of index numbers and the measurement of input, output, and productivity. *Econometrica, 50,* 1393–1414.

CE, et al. (2008). Handbook on estimation of the external costs of the transport sector. Study for the EU Commission, Delft.

Y. Hayashi et al. (eds.), *Intercity Transport and Climate Change,*
Transportation Research, Economics and Policy 15, DOI 10.1007/978-3-319-06523-6

Central Japan Railway Company. (2007). Environmental report 2007.
Central Japan Railway Company. (2009). Environmental report 2009.
Central Japan Railway Company. (2010). Realization of superconducting Maglev Chuo Shinkansen. (Online). http://jr-central.co.jp/news/release/nws000516.html. Accessed 25 Jan 2011.
Chang, J. S., & Lee, J.-H. (2008). Accessibility analysis of Korean high-speed rail: A case study of the Seoul metropolitan area. *Transport Reviews, 28*(1), 87–103.
Charnes, A., Cooper, W. W., & Rhodes, E. (1978). Measuring the efficiency of decision making units. *European Journal of Operational Research, 2,* 429–444.
Chicago Climate Exchange. (Online). http://www.chicagoclimatex.com/. Accessed 31 Jan 2011.
Chung, Y. H., Fare, R., & Grosskopf, S. (1997). Productivity and undesirable outputs: A directional distance function approach. *Journal of Environmental Management, 51*(3), 229–240.
Coase, R. H. (1960). The problem of social cost. *Journal of Law and Economics, 3,* 1–44.
Commissariat general au developpement durable. (2010). La mobilité des Français: panorama issu de l'enquête nationale transports et déplacements 2008.
Commission Mobilité 21. (2013). Pour un schéma national de mobilité durable-Rapport au ministre chargé des transports, de la mer et de la pêche-.
Commission of the European Communities. (1992). The future development of the common transport policy—A global approach to the construction of a community framework for sustainable mobility. Brussels: CEC, COM (92) 494.
Commission of the European Communities. (2001). White paper European transport policy for 2010: Time to decide.
Commission of the European Communities. (2003). Extended impact assessment of the proposal amending the amended proposal for a decision amending decision. *The trans-European transport network*, No. 1692/96/EC.
Committee on Climate Change. (2010). Meeting carbon budgets-ensuring a low-carbon recovery 2nd progress report to parliament.
Communities and Local Government. (2010). Policy statement on regional strategies.
Cooper, W. W., Seiford, L. M., & Tone, K. (2000). *Data envelopment analysis: A comprehensive text with models, applications, references and DEA-solver software*. New York: Kluwer.
Dasgupta, P. S., & Heal, G. M. (1980). *Economic theory and exhaustible resources*. Cambridge University Press.
DATELINE. (2003). Social data, et al.: Design and application of a travel survey for European long distance trips based on an international network of expertise: Deliverable 5. Final report on data collection. http://cgi.fg.uni-mb.si/eö, os/docs/DATALINE_del5.pdf.
Department for Transport. (2000). Transport ten year plan 2000.
Department for Transport. (2004). The future of rail. *White Paper CM 6233*.
Department for Transport. (2006a). National travel survey.
Department for Transport. (2006b). UK national travel survey, long distance journey 2006.
Department for Transport. (2007a). Delivering a sustainable railway: Summary of key research and analysis.
Department for Transport. (2007b). Towards a sustainable transport system.
Department for Transport. (2008a). Carbon pathways analysis: Informing development of a carbon reduction strategy for the transport sector.
Department for Transport. (2008b). Delivering a sustainable transport system.
Department for Transport. (2010). High speed rail.
DESUCA. (2009). (Online). http://www.desuca.co.jp/. Accessed 26 Jan 2011.
Dewar, J. A., & Wachs, M. (2009). *Transportation planning, climate change, and decision making under uncertainty*. Paper presented to the TRB Conference, Washington.
East Japan Railway Company. (2009). Sustainability Report 2009.
Edenhofer, D. (2007). *Costs and strategies or climate stabilization*. Presentation to the Symposium on climate change and transport strategy, Tokyo.
Ernst & Young and Urban Land Institute: Infrastructure. (2010). *Investment imperative*. London: Ernst & Young and Urban Land Institute: Infrastructure.
ETIS plus. (2012). Database Manual. Annex Report D6. Brussels.
EU Commission. (2012). *EU Transport: Facts and figures*. Brussels.

European Commission. (2001). White paper—European transport policy for 2010: Time to decide.
European Commission. (2006). *Draft directive for integrating aviation into the emission trading scheme*. Brussels.
European Commission. (2008). *Handbook on internalisation of external costs*. Brussels.
European Commission. (2010). *High-speed Europe, a sustainable link between citizens*. Luxembourg: Publications Office of the European Union.
European Commission. (2011). White paper-roadmap to a single European transport area-towards a competitive and resource efficient transport system, COM 144.
European Commission. (2013). The core network corridors-Trans European transport network 2013.
European Commission Projects: UNITE, GRACE, CAPRI, PETS, HEATCO, Extern.E, NEEDS, OSD, NewExt. Website of the Commission.
European Environmental Agency. (2008). Climate for a transport change. Copenhagen.
EUROPEAN PARLIAMENT. (2013). REPORT on the proposal for a regulation of the European Parliament and of the Council on Union guidelines for the development of the trans-European transport network.
European Parliament, Council of the European Union. (1996). Decision No 1692/96/EC of the European Parliament and of the council of 23 July 1996 on community guidelines for the development of the trans-European transport network.
EWS. (2005). *Empfehlungen für Wirtschaftlichkeitsrechnungen im Straßenwesen*. Köln: Forschungsgesellschaft für Straßen- und Verkehrswesen.
Färe, R., & Grosskopf, S. (1992). Malmquist productivity indexes and Fisher ideal indexes. *The Economic Journal, 102*(410), 158–160.
Färe, R., & Grosskopf, S. (2003). Nonparametric productivity analysis with undesirable outputs: Comment. *American Journal of Agricultural Economics, 85*(4), 1070–1074.
Färe, R., & Grosskopf, S. (2004a). Modeling undesirable factors in efficiency evaluation: Comment. *European Journal of Operational Research, 157,* 242–245.
Färe, R., & Grosskopf, S. (2004b). *New directions: Efficiency and productivity*. New York: Kluwer.
Färe, R., Grosskopf, S., & Lovell, C. A. K. (1994). *Production frontiers*. Cambridge: Cambridge University Press.
Farrel, M. J. (1957). The measurement of productive efficiency. *Journal of the Royal Statistical Society, Series a General, 120*(3), 253–290.
Friedrich, R., & Bickel, P. (2001). *Environmental external costs of transport*. Stuttgart: Springer.
Fujisaki, N. (2007). US aviation, demand, capacity and climate change-can it all be fixed in time? International Symposium Proceedings of "Climate Change and Transport Strategy", p. 19.
Fujisaki, K. (2013). Effect of fluctuation of gasoline prices on travel behavior: An empirical analysis using regional data in Japan and its policy implications. *Transport Policy Studies' Review, 16*(1), 4–16.
Fujisaki, K., Morichi, S., & Itoh, M. (2011). Effect of fluctuation of gasoline prices on transport behavior—An empirical analysis using transport statistics in Japan. *Journal of the Eastern Asia Society for Transportation Studies, 9,* 354–369.
Futamura, M. (2008). Global warming issue international trends: Application of economic tools, Institute of transportation economics. *Transportation and Economy, 68*(1), 38–45.
Gillen, D. W., Oum, T. H., & Tretheway, M. W. (1990). Airline cost structure and policy implications: A multiproduct approach for Canadian airlines. *Journal of Transport Economics and Policy, 24*(1), 9–34.
Greenhouse Gas Inventory Office of Japan, National Institute for Environmental Studies. (2013). The GHG's emissions data of Japan (1990–2011), sec. 4. (Online). *http://www-gio.nies.go.jp/aboutghg/data/data-archives_e.html*. Accessed 31 Dec 2013.
Gunn, H. (2004). *SCGE models: Relevance and accessibility for use in the UK, with emphasis on implications for evaluation of transport investments, final report to department of transport, London*. Cambridge: RAND Europe.
Hailu, A. (2003). Non-parametric productivity analysis with undesirable outputs: Reply. *American Journal of Agricultural Economics, 85*(4), 1075–1077.

Hailu, A., & Veeman, T. S. (2001). Non-parametric productivity analysis with undesirable outputs: An application to the Canadian pulp and paper industry. *American Journal of Agricultural Economics, 83*(3), 605–616.
Hashimoto, S. (1994). *Transport policy of European community*. Tokyo: Hakutou-Shobo Publishing.
Hayashi, Y., & Nakamura, K. (2013). Climate change, mobility and cost of transport. *Transport Policy, 29,* 261–263.
High Level Group. (1999). The expert advisors to the high level group on infrastructure charging (Several Working Groups; Several Reports for the EU Commission). Brussels.
HM Government. (2009a). Low carbon transport: A greener future.
HM Government. (2009b). The UK low carbon transition plan.
HM TREASURY. (2006). Stern review on the economics of climate change.
Hohmeyer, O. (1996). *Social costs and sustainability. Economics, ethics, and environmental policy: contested choices* (pp. 181–202). Heidelberg.
Holmberg, J. (1998). Backcasting: A natural step when making sustainable development operational for companies. *Greener Management International, 23,* 30–51.
Horridge, M., & Wittwer, G. (2007). *The economic impacts of a construction project, using Sino TERM, a multi-regional CGE model of China—Centre of policy studies and impact project.* General working paper No. G–164.
Hugh, G. (2004). SCGE models: Relevance and accessibility for use in the UK, with emphasis on implications for evaluation of transport investments. RAND Europe Cambridge.
IADF. (2009). Future of aviation fuel. (Online). http://www.iadf.or.jp/8361/LIBRARY/MEDIA/H21_dokojyoho/H21-5.pdf. Accessed 25 Jan 2011.
IATA. (2009). The IATA technology roadmap report issued June 2009, 3rd edition.
ICAO. (2005). Global air traffic management operational concept.
ICCT International Council of Clean Transportation. (2013). From laboratory to road. A comparison of official and 'real-world' fuel consumption and CO2 values for cars in Europe and the United States. Washington.
IEA. (1971–2004, 2005, 2009, 2012). CO2 emissions from fuel combustion. International Energy Agency.
IEA. (2008). Energy technology perspectives 2008. International Energy Agency.
IEA. (2009). Energy technology perspectives 2009. International Energy Agency.
IEA. (2013a). World energy outlook.
IEA. (2013b). GLOBAL LAND TRANSPORT INFRASTRUCTURE REQUIREMENTS Estimating road and railway infrastructure capacity and costs to 2050. International Energy Agency, Paris.
IFEU. (2008). Transport in China. Study for the KfW, Heidelberg.
INFRAS/IWW. (1995). External effects of transport. Study for the UIC, Zürich and Karlsruhe.
INFRAS/IWW. (2004). Facts on the competition in the European transport market. Research Project for the UIC. Zürich and Karlsruhe.
Institute of Energy Economics, Japan. (2005). Research on related data such as global warming.
Intergovernmental Panel on Climate Change. (2007). *Fourth assessment report on climate change 2007*. Further: Technical Reports, Geneva.
International Union of Railways. (2013). High speed lines in the world. http://www.uic.org/IMG/pdf/20131101_high_speed_lines_in_the_world.pdf. Accessed 17 June 2014.
INVERMO. (2004). Chlond, B., Last, J., Manz, W. and D. Zumkeller: Research Project for the German Ministry of Transport. In Cooperation between University of Karlsruhe, Deutsche Lufthansa AG, Deutsche Bahn AG and NFO Infratest. Karlsruhe.
IPCC. (2007). Climate change 2007: Synthesis report. http://www.ipcc.ch/publications_and_data/publications_ipcc_fourth_assessment_report_synthesis_report.htm.
IPCC-NGGIP. (1997). Revised 1996 IPCC guidelines for national greenhouse gas inventories: Reference manual (Vol. 3). INTERGOVERNMENTAL PANEL ON CLIMATE CHANGE: National Greenhouse Gas Inventories Programme. http://www.ipcc-nggip.iges.or.jp/public/gl/invs6.html, 2008/4/27.

ITF. (2008). Key transport and greenhouse gas indicators: Information by country. Greenhouse Gas Reduction Strategies in the Transport Sector.
ITPS. (2010). Urban transport annual report 2009 ITPS, Tokyo.
Ivanova, O., Vold, A., & Jean-Hansen, V. (2002, revised in 2006). PINGO—A model for prediction of regional and interregional freight transport. Institute of Transport Economics, TOI report, Vol. 578.
IWR. (2007). Internationales Wirtschaftsforum Alternative Energien. CO2- Ausstoß weltweit, IWR On-line, Münster.
IWW, IFEU, KuP, PÖU, PTV. (1999). Entwicklung eines Verfahrens zur Aufstellung umweltorientierter Fernverkehrskonzepte als Beitrag zur Bundesverkehrswegeplanung. (Development of a Methodology for Designing Sustainable Transportation Systems. A Contribution to the Federal Transport Investment Planning Methodology), Berlin.
Jang-Ho, L., & Chang, J. S. (2006). Effects of high-speed rail service on shares of intercity passenger ridership in South Korea. *Transportation Research Record, 1943,* 31–42.
Japan Aircraft Development Corporation. (2010). Status and trends in private aircraft. Data collection for commercial aircraft in FY 2009, Chap. 2, pp. 22.
Japan Automobile Manufacturers Association. (2009). Suggestion of reduce CO2 emissions in global road transport sector.
Japan Automobile Dealers Association. (n.d.). Monthly sales of new and used normal and small passenger cars. (Online). http://www.jada.or.jp/contents/data/index.html. Accessed ca. Nov 2009. (data disappeared).
Japan Carbon Fiber Manufactures Association. (2008). Report of quantify the effect of environment improvement through the use of carbon fiber. Japan Carbon Fiber Manufactures Association LCA Model. http://www.jemai.or.jp/lcaforum/seminar/pdf/lca090128_3.pdf, p. 11. Accessed 29 Jan 2009.
Japan Light Motor Vehicle and Motorcycle Association. (n.d.). Monthly sales of new and used light cars. (Online). http://www.zenkeijikyo.or.jp/statistics/index.html. Accessed 26 Nov 2009.
Japan Maritime Center. (2010). Revolution marine environment, maritime community and the global warming issue.
Japan Railway Construction, Transport and Technology Agency (JRTT). (n.d.). http://www.jrtt.go.jp/01Organization/Csr/pdf/k2012_04.pdf. Accessed 16 June 2014.
Japanese Ministry of Economy, Trade and Industry. (n.d.). Act on the rational use of energy. METI related Laws, No. 125. http://www.meti.go.jp/english/information/data/laws.html.
Jochem, P. (2009). *Impacts of a CO2 emission trading scheme in German road transport*. Paper Presented to the TRB, 2008.
JPDO. (2008). Next generation air transportation system integrated work plan version 1.0.
Kato, H., Shibahara, N., Osada, M., & Hayashi, Y. (2005). A life cycle assessment for evaluating environmental impacts of inter-regional high-speed mass transit projects. *Journal of the Eastern Asia Society for Transportation Studies, 6,* 3211–3224.
Kawasaki, H. I. (2010). *Urban transportation friendly for people and the environment: Bogie for the SWIMO-X low-floor battery-driven light rail vehicle*. KHI Technical Report, Vol. 170.
Kim, E., & Kim, K. (2003). Impacts of the development of large cities on economic growth and income distribution in Korea: A multiregional CGE model. *Papers in Regional Science, 82,* 101–122.
Knaap, T., & Oosterhaven, J. (2000). *The welfare effects of new infrastructure: An economic geography approach to evaluating new Dutch railway links*. Paper to the North American RSAI Meeting, Chicago.
Knaap, T., & Oosterhaven, J. (2001). *On the development of RAEM: The Dutch spatial general equilibrium model and its first application to a new railway link*. Paper presented to the 41th Congress of the European Regional Science Association.
Knaap, T., & Oosterhaven, J. (2002). The welfare effects of new infrastructure: An economic geography approach to evaluating new Dutch railway links, Mimeo.

Koike, A., & Ueda, T. (2005). Spatial economic damage assessment of great earthquake, welfare-metrics by the SCGE model. Economics of Disaster Prevention, Chap. 8, (Tatano, Y., Takagi, A. (2005), Economic Analysis of Disaster, Keiso Shobo).

Korhonen, P. J., & Luptacik, M. (2004). Eco-efficiency analysis of power plants: An extension of data envelopment analysis. *European Journal of Operational Research, 154,* 437–446.

Kotz, R., Müller, P., & Rothengatter, W. (1987). *Entwicklung eines Verfahrens für dynamische Investitionsplanung und Ermittlung des bei der Fortschreibung der BVWP anzuwendenden Zinssatzes (Development of a dynamic investment planning method and determination of the interest rate to be applied to Federal Transport Plan (BVWP) revaluations).* Expert report commissioned from Ulm University by the federal Minister of Transport, Ulm.

Kuosmanen, T., & Podinovski, V. (2009). Weak disposability in nonparametric production analysis: Reply to färe and grosskopf. *American Journal of Agricultural Economics, 91*(2), 539–545.

LeBlanc, L. J., & Rothengatter, W. (1982). Equilibrium in transportation networks—The basic model and extensions for integrated transportation planning. *German Journal of Regional Science, 3,* 82–115.

Lockwood, S. (2009). *Operational responses to climate change impacts.* Paper Presented to the TRB Conference, Washington.

Luenberger, D. G. (1992). Benefit functions and duality. *Journal of Mathematical Economics, 21*(5), 461–481.

Madala, G. S. (2001). *Introduction to econometrics* (3rd ed.). West Sussex: Wiley. (Translated by Saeki, C. (2004). Analysis method of Madala's econometrics (3rd ed.). EconomistSha Publishing).

Marshall, A. (1920). *Principles of economics* (8th ed.). London.

McMullen, B. S., & Noh, D. (2007). Accounting for emissions in the measurement of transit agency efficiency: A directional distance function approach. *Transportation Research, Part D, 12,* 1–9.

MIAC. (2010). Statistics Bureau. Consumer price index, index time series for Japan (2005 = 100, general). (Online). http://www.e-stat.go.jp/SG1/estat/ListE.do?lid=000001062590. Accessed 27 March 2010.

MIAC. (n.d.). Statistics Bureau. Population estimates. (updated base population: 2005 Population Census). (Online). http://www.e-stat.go.jp/SG1/estat/NewList.do?tid=000000090001. Accessed 16 April 2010.

MID. (2008). DLR and Infas: Mobilität in Deutschland. Project for the German Ministry of Transport. Berlin.

Ministry of Economy, Trade and Industry. (2010). Carbon footprint of products. (Online). http://www.cfp-japan.jp/. Accessed 31 Jan 2011.

Ministry of Land, Infrastructure, Transport and Tourism. (2005). Net flow survey of national main line passenger, 4th.

Ministry of Land, Infrastructure, Transport and Tourism. (2007). White paper of land, infrastructure, transport and tourism 2007.

Ministry of Land, Infrastructure, Transport and Tourism. (2008a). White paper of land, infrastructure, transport and tourism 2008.

Ministry of Land, Infrastructure, Transport and Tourism. (2008b). Survey of region passenger flow.

Ministry of Land, Infrastructure, Transport and Tourism. (2009a). Modal share by travel distance (kyoritaibetsuyusou kikan buntanritsu) 2007.

Ministry of Land, Infrastructure, Transport and Tourism. (2009b). Scale of assessments for different transportation distance range.

Ministry of Land, Infrastructure, Transport and Tourism. (2009c). White paper of land, infrastructure, transport and tourism 2009. Gyosei Corporation.

Ministry of Land, Infrastructure, Transport and Tourism. (n.d.). Person trip survey.

Ministry of the Environment. (2005). Global warming reader for executives.

Ministry of the Environment. (n.d.). Kyoto protocol to the United Nations framework convention on climate change. (Online). http://www.env.go.jp/earth/ondanka/cop.html. Accessed 24 Jan 2011.

Ministry of the Environment, Global Environment, Market Mechanism. (2010). Implementation and Investigation Status of Emissions Trading in Other Country.

Miyashita, M., & Koike, A. (2007). Evaluation of linear Chuo express improvement by SCGE model. *Railway Technology Symposium, 14,* 469–472.

Miyoshi, H., & Tanishita, M. (2008). *Automotive innovation and economic welfare*. Tokyo: Hakutou-Shobo Publishing.

MLIT. (2006). http://www.mlit.go.jp/singikai/koutusin/koutu/kankyou/8/shiryou2-2.pdf. Accessed 16 June 2014.

MLIT. (2008). Intercity passenger survey of 2005.

MLIT. (2009a). Script of senior vice minister of MLIT at the 1st ministerial conference on global environment and energy in transport, January 2009. (Online). http://www.mlit.go.jp/kokusai/MEET/result_en.html. Accessed 3 May 2014.

MLIT. (2009b). Monthly gazette *Kokudokotsu*. 62(3). https://www.mlit.go.jp/page/kanbo01_hy_000296.html. Accessed 2 June 2014.

MLIT. (2010). Inter-regional travel survey in Japan. MLIT Policy Bureau. Tokyo.

MLIT. (2012a). Reference material no. 1 in the 21st joint session of the council for infrastructure, environment committee and the council for transport policy transport system subcommittee, environment committee. http://www.mlit.go.jp/policy/shingikai/sogo10_sg_000073.html. Accessed 25 May 2014.

MLIT. (2012b). Statistical information packet for transportation.

MLIT. (n.d.a). White paper on land, infrastructure, transport and tourism in Japan, 2008. (Online). http://www.mlit.go.jp/english/white-paper/mlit-index.html. Accessed 23 Sept 2011.

MLIT. (n.d.b). Policy bureau. Annual statistical report on railway transport. http://www.mlit.go.jp/k-toukei/. Accessed 25 Nov 2010.

MLIT. (n.d.c). Policy bureau. Annual statistical report on motor vehicle transport. http://www.mlit.go.jp/k-toukei/. Accessed 25 Nov 2010.

MLIT. (n.d.d). Road transport bureau. Motor vehicles owned. In print version of MLIT (n.d.c), 25–46 (13).

MLIT Civil Aviation Bureau. (2008). Trends in domestic air passenger transport.

MLIT Civil Aviation Bureau. (2010). Collaborative actions for renovation of air traffic systems (Plan), Workshop on future air traffic system.

MLIT Road Transport Bureau. (2010). Automobile fuel efficiency standards. http://www.mlit.go.jp/jidosha/sesaku/environment/ondan/ondan.htm. Accessed 26 Jan 2011.

MLIT, Shikoku Transport & Tourism Bureau. (2009). https://wwwtb.mlit.go.jp/shikoku/soshiki/koutsuu/img/actionplan_h21.pdf. Accessed 16 Jun 2014.

MOE. (2008). Kyoto protocol target achievement plan. Formulated in 2005, totally revised in 2008 (provisional translation). http://www.env.go.jp/en/earth/. Accessed 3 May 2014.

Morichi, S., & Acharya, S. R. (2011). Strategic perspectives on intercity transport development in Asian countries. *Proceedings of the Eastern Asia Society for Transportation Studies, 8*.

Musso, A., & Rothengatter, W. (2013). Internalisation of external costs of transport-A target driven approach with a focus on climate change. Transport Policy. Special issue on climate change, mobilityx and cost of transport.

Nakamura, H., Hayashi, Y., & May, A. D. (Eds.). (2004). *(World Conference on Transport Research Society and Institute for Transport Policy Studies). Urban transport and the environment -An international perspective*. Bingley: Elsevier.

Nakano, S. (2009). Prospects for California high-speed railway construction project, focusing on environmental issues' background and trends in the automotive industry. *Transportation and Economic, 69*(3), 77–85.

National Bureau of Statistics of China. (2002–2011). *China statistical year book 2001–2010*. Beijing: China Statistics Press.

Nordhaus, W. (2007). *The stern review on the economics of climate change*. New Haven: Yale University CT 06511—8268.

North, D. C. (1990). *Institutions, institutional change and economic performance*. Cambridge: Cambridge University Press.

OECD Publishing. (2007). *Cutting transport CO2 emissions: What progress?* European Conference of Ministers of Transport.

OECD/ITF. (2008). Greenhouse gas reduction strategies in the transport sector: Preliminary report. http://www.internationaltransportforum.org/pub/pdf/08ghg.pdf.

OECD/ITF. (2011). Transport greenhouse gas emissions county data 2010.

Oil Information Center. (n.d.). Institute of energy economics of Japan. Monthly gasoline (Regular) retail price. http://oil-info.ieej.or.jp/price/price.html.

Okahisa, K. (2009). 2008 climate change law in UK, basis aimed at low-carbon economy. 240 Foreign Legislation, National Diet Library: Research and Legislative Reference Bureau.

Oniki, H., Oum, T. H., & Stevenson, R. (1994). The productivity effects of the liberalization of Japanese telecommunications policy. *Journal of Productivity Analysis, 5*(1), 63–79.

OSD. (2006). Externe Kosten des Straßen- und Schienenverkehrs 20Im Auftrag des Schweizer Bundesamtes für Raumentwicklung, INFRAS, Bern.

Oum, T. H., & Yu, C. (1994). Economic efficiency of railways and implications for public policy: A comparative study of the OECD countries' railways. *Journal of Transport Economics and Policy, 28*(2), 121–138.

Oum, T. H., & Yu, C. (1995). A productivity performance of the world's major airlines. *Journal of Air Transport Management, 2*(3, 4), 181–195.

Oum, T. H., & Yu, C. (1998). *Winning airlines: Productivity and cost competitiveness of the world's major airlines* (p. 240). New York: Kluwer. http://www.wkap.nl/prod/b/0-7923-8010-X. Accessed 1 Feb 2011.

Oum, T. H., & Zhang, Y. (1995). Competition and allocative efficiency: The case of competition in the U.S. telephone industry. *Review of Economics and Statistics, 77*(1), 82–96.

Oum, T. H., Tretheway, M. W., & Waters, W. G. II (1992). Concepts, methods, and purposes of productivity measurement in transportation. *Transportation Research A, 26*(6), 493–505.

Oum, T. H., Waters, W. G. II, & Yu, C. (1999). A survey of productivity and efficiency measurement in rail transport. *Journal of Transport Economics and Policy, 33*(I), 9–42.

Park, K. H., & Weber, W. L. (2006). A note on efficiency and productivity growth in the Korean banking industry, 1992–2002. *Journal of Banking and Finance, 30,* 2371–2386.

Pathomsiri, S. (2007). *Assessment of airport productivity: Nonparametric directional output distance function approach and its robustness* (pp. 29–36). Washington, DC.: Transportation Research Board of the National Academies. (Transportation Research Record: Journal of the Transportation Research Board, No. 2007).

Pathomsirid, S., Haghani, A., Windle, R. J., & Dresner, M. (2008). Impact of undesirable outputs on the productivity of U.S. airports. *Transportation Research Part E: Logistics and Transportation Review, 44*(2), 235–259.

Pigou, A. C. (1924). *The economics of welfare*. London.

Progtrans, I. W. W. (2007). Wegekostenrechnung für die Bundesfernstraßen. Study on Behalf of the German Ministry of Transport, Basel and Karlsruhe.

Quinet, E. (2010). *The practice of cost-benefit analysis in transport: The case of France*. Discussion paper 2010-17, OECD/ITF 2010.

Railway Technical Research Institute. (2002). RTRI report.

Rhoades, D. L., & Williams, M. J. Intercity transport policy and systems in the United States: Changing the balance to address climate change.

Rhoades, D. L., Williams, M. J., & Green, D. J. (2006). Imperfect substitute: Competitive analysis failure in US intercity passenger rail. *World Review of Intermodal Transportation Research, 6,* 82–93.

Rothengatter, W. (1999). *External effects of transport*. Paper for the ECMT, Paris.

Rothengatter, W. (2000). External effects of transport. In J. B. Polak & A. Heertje (Eds.), *Analytical transport economics* (pp. 47–60). Cheltenham.

Rothengatter, W., & Szimba, E. (2009). *Transport demand, transport and climate policy in Europe—Status quo and developments*. Background paper, Prepared for ITPS and JR Central, Karlsruhe.

RUG/CBS. (1999). Regionale samenhang in Nederland. Bi-regionale input-output tabellen en aanboden, gebruiktabellen voor de 12 provincies en de twee mainport regio's. Rijksuniversiteit, Groningen/Centraal Bureau voor de Statisitiek, REG-publicatie 20, Stichting Ruimtelijke, Economie Groningen.
Sato, T. (2007). Changes in demand for passenger train study. *Transport Policy Studies' Review, 10*(2), 71–74.
Schade, W., Senger, F., Rothengatter, W., Meyer-Rühle, O., & Brouwer, I. S. (2013). Ten-T large projects-Investments and costs, European Parliament.
Schafer, A., & Victor, D. G. (2000). The future mobility of the world population. *Transportation Research, Part A, 34,* 171–205.
Schreckenberg, M., et al. (1999). *Traffic and mobility: Simulation-economics-environment.* Duisburg: Springer.
Schweizer, U. (1988). Externalities and the Coase theorem: Hypothesis or result? *Journal of Institutional and Theoretical Economics, 144,* 245–266.
Seiford, L. M., & Zhu, J. (2002). Modeling undesirable factors in efficiency evaluation. *European Journal of Operational Research, 142,* 16–20.
Seiford, L. M., & Zhu, J. (2005). A response to comments on modeling undesirable factors in efficiency evaluation. *European Journal of Operational Research, 161,* 579–581.
SESAR Consortium. (2007). The ATM target concept deliverable 3 at a glance.
Sperling, D., & Gordon, D. (2009). *Two billion cars.* New York: Oxford University Press.
State of California. (2010). California high-speed rail authority. http://www.cahighspeedrail.ca.gov/. Accessed 24 Jan 2011.
Stern, R. (2006). *The economics of climate change.* Report to the treasury, London.
Stiglitz, J. E. (Ed.). (1990). On the economic role of the state. *The Economic Role of the State.* Oxford University Press.
Strategic, R. A. (2003). West coast main line strategy.
The White House. (2009). http://www.whitehouse.gov/blog/09/04/16/A-Vision-for-High-Speed-Rail/. Accessed 18 Feb 2010.
Tojo, S. (2008). Study on nation's part in urban planning, refer to the system in France. *Shin-Toshi, 62*(5).
TOYOTA Motor Corporation. (2009). Sustainability report 2009.
Tsujimura, T., Miyauchi, T., Nagatomo, T., & Hashimoto, A. (1998). LCA case studies and environmental efficiency of Shinkansen. *Railway Technology Symposium*, 601–604.
Tyndall Centre by A. Bristow, et al. (2004). *How can we reduce carbon emissions of transport?* Technical Report 15, Norwich.
Ueda, T. (2010). *Regional and urban economics study in excel.* Tokyo: Corona Publishing.
UIC. (2010). *High speed rail: Fast track to sustainability.* Paris.
UIC. (2011). *High speed rail & sustainability.* Paris.
US Climate Change Science Program. (2006). CCSP Fact Seetl: CCSP Overview. http://www.climatescience.gov/infosheets/factsheet1/default.htm. Accessed 24 Jan 2011.
US Department of Transportation. (2001). National household travel survey.
US Department of Transportation. (2006). Findings from the national household travel survey.
US Department of Transportation. (2006–2010). Center for climate change and environmental forecasting, strategic plan.
US Department of Transportation. (2009). National transportation statistics 2009.
Vold, A., & Jean-Hansen, V. (2007). *PINGO-A model for prediction of regional and interregional freight transport in Norway.* Institute of Transport Economics, TOI report, Vol. 899.
Wardrop, J. G. (1952). Some theoretical aspects of road traffic. *Road Papers No. 36, Proceedings of the Institution. of Civil Engineers, Part II, 1*(2), 325–365.
WCTRS, & ITPC. (2003). *Urban transport and the environment.* Amsterdam.
Weber, M. M., & Weber, W. L. (2004). Productivity and efficiency in the trucking industry accounting for traffic fatalities. *International Journal of Physical Distribution & Logistics Management, 34*(1), 39–61.

Wiederkehr, P., Gilbert, R., Crist, P., & Caid, N. (2004). Environmentally sustainable transport (EST): Concept, goal, and strategy-the OECD's EST project. *Economic Journal, 4*(1), 11–25.

Yamazaki, S. (2003). *Automotive industrial policy in postwar Japan*. Kyoto: Houritsu Bunka Sha.

Yoshida, Y. (2004). Endogenous-weight TFP measurement: Methodology and its application to Japanese airport benchmarking. *Transportation Research, Part E, 40*(2), 151–182.

Yoshida, Y., & Fujimoto, H. (2004). Japanese airport benchmarking with the DEA and endogenous-weight TFP methods: Testing the criticism of overinvestment in Japanese regional airports. *Transportation Research, Part E, 40*(6), 533–546.

Yu, M. M. (2004). Measuring physical efficiency of domestic airports in Taiwan with undesirable outputs and environmental factors. *Journal of Air Transport Management, 10*(5), 295–303.

Yu, M. M., Hsu, S. H., Chang, C. C., & Lee, D. H. (2008). Productivity growth of Taiwan's major domestic airports in the presence of aircraft noise. *Transportation Research, Part E, 44*(3), 543–554.

Zierock, K. H., & Rothengatter, W. (2007). Presentation on concepts for CO2 reduction policy. *Transport and Climate Change*, (Html version), Karlsruhe.

Index

Y. Hayashi et al. (eds.), *Intercity Transport and Climate Change*, Transportation Research, Economics and Policy 15, DOI 10.1007/978-3-319-06523-6

MIX
Papier aus verantwortungsvollen Quellen
Paper from responsible sources
FSC® C105338

If you have any concerns about our products,
you can contact us on
ProductSafety@springernature.com

In case Publisher is established outside the EU,
the EU authorized representative is:
Springer Nature Customer Service Center GmbH
Europaplatz 3, 69115 Heidelberg, Germany

Printed by Libri Plureos GmbH
in Hamburg, Germany